With fond memories of
wisdom teeth, monkeys
and Beckenham pubs!

Dan Z.

June 20, 2002

Notes on the
Elements of Behavioral Science

Notes on the
Elements of Behavioral Science

Doris Zumpe
and
Richard P. Michael
Emory University School of Medicine
Atlanta, Georgia

Kluwer Academic / Plenum Publishers
New York   Boston   Dordrecht   London   Moscow

Library of Congress Cataloging-in-Publication Data

Zumpe, Doris, 1940–
    Notes on the elements of behavioral science/Doris Zumpe and Richard P. Michael.
    p. cm
    Includes bibliographical references and index.
    ISBN 0-306-46577-9
    1. Psychology. 2. Social sciences. 3. Psychology, Comparative. I. Michael, Richard P. (Richard Phillip) II. Title.
    [DNLM: 1. Behavioral Medicine. 2. Behavioral Sciences. WB 103 Z94n 2001]
BF121 .N68 2001
150—dc21
                                                                                                   2001020256

ISBN: 0-306-46577-9

©2001 Kluwer Academic / Plenum Publishers, New York
233 Spring Street, New York, New York 10013

http://www.wkap.nl/

10 9 8 7 6 5 4 3 2 1

A C.I.P. record for this book is available from the Library of Congress

All rights reserved

No part of this book may be reproduced, stored in a retrieval system, or transmitted in any form or by any means, electronic, mechanical, photocopying, microfilming, recording, or otherwise, without written permission from the Publisher.

Printed in the United States of America

In fond memory of my parents and of my first great teacher, Konrad Lorenz.

D.Z.

To my parents: they found life joyful; and to three of my fine teachers, Sir Aubry Lewis, G. W. Harris, and D. W. Winnicott.

R.P.M.

Preface

These notes are intended to help undergraduates who need to understand something of behavior both for its intrinsic interest and for their future careers in medicine, biology, psychology, anthropology, veterinary medicine, and nursing. In Emory University's Biology Department, a single-semester course called *Evolutionary Perspectives on Behavior* is given to undergraduates. It amounts to four, not eight months of study, so a great deal of compression is essential. There are several excellent textbooks available that deal with behavioral science from different perspectives, but we have found them too compendious for use in a short course when students are so heavily burdened; it is unsatisfactory to direct them to a chapter here and there in several different books or to this or that review article and original paper. In this volume, we have tried effectively and inexpensively to put in one place what we know is needed. The topics we have selected deal with their subjects in a simple, straightforward way without being too superficial. We could not cover everything and the gaps are not entirely idiosyncratic but reflect what students are given very well in other courses. Thus, there is no mention of the physiology of the axon and synapse; learning, memory, cognition, and basic genetics are hardly touched upon because students know about these matters from elsewhere. This volume particularly emphasizes certain physiological mechanisms when they are known, and it also draws attention to the application of what is known about animal behavior to the human, sometimes even to the clinical situation. The illustrative examples are taken from both the classical behavioral literature and newer work.

It has given us considerable pleasure to write this volume and we trust it will find a useful niche in the education of our students.

ACKNOWLEDGMENTS

We are grateful to Drs. Andrew Clancy and Darrell Stokes for kindly reading the manuscript and making helpful suggestions. We are also grateful to F. Cawthon and M. Maddox for continuing help and, particularly, to F. Cawthon, who is responsible for all the illustrative material. We also thank several generations of Emory students for stimulatng us with their interest and many useful questions.

Contents

1	**The Study of Behavior: History**	1
	A Brief History	1
	Classical Ethology	1
	Comparative Psychology	4
	Emergence of Modern Behavioral Research	5
	Some Theory and Terminology	6
	Fitness and Inclusive Fitness	6
	Benefit–Cost Ratios	7
	Optimality Theory	7
	Game Theory	8
	Evolutionarily Stable Strategies	9
	General Life Strategies: K and r Selection	10
	Warnings, Fallacies, and Pitfalls	11
	Correlational Studies	11
	Genes: What They Do and Do Not Do	12
	Understanding Optimality Theory	12
2	**Some Ethological Concepts**	15
	Evolutionary Basis of Behavior	15
	Ethology's Objectives	17
	General Methods	17

Fixed Action Patterns (FAPs)	18
Unvarying Form	18
Coordination of Several Muscle Groups	19
Environmental Influences	19
Genetic Factors	22
Brain Stimulation	22
Conflict Behaviors	22
Redirection	24
Displacement	25
Intention Movement	27
Alternation	27
Ambivalence	27
Compromise	28
Conflict in Psychiatry	28
Vacuum Activity	30
Ritualization and Displays	30
Stereotypy	31
Typical Intensity	31
Association with Conspicuous Morphological Features	31
Threshold Changes	32
Motivational Changes	32

3 Some More Ethological Concepts   37

Sign Stimuli, Releasers, and Innate Releasing Mechanisms	37
Perception versus Attention	37
Definitions	38
Properties of Sign Stimuli and Releasers	38
Modifying Influences	41
Supranormal Stimuli	43
Advantages and Disadvantages of Innate Releasing Mechanisms	43
Programmed Learning and Imprinting	44
Programmed Learning	44
Imprinting	45
Drive or Motivation	48
Illustrative Model for Thirst	49
Drive Models	51

4 Assessment of Hereditary Influences   55

Indirect Methods	57
Presence in Geographically Isolated Populations	57
Presence in Individuals Raised in Isolation from Conspecifics	58
Phylogenetic Comparisons: Presence in Closely Related Species	58

Association with Specialized Morphological or Physiological Traits	58
Studies on Human Twins	59
Direct Methods	59
Artificial Selection	60
Inbreeding	60
Hybridization	64
Molecular Changes	64

5 Behavioral Endocrinology: Gonadal Hormones ... 67

Synthesis and Major Sites of Production	69
Estrogens and Progestins	69
Androgens	70
Transport	71
Mechanisms of Action	71
Hormone Receptors	71
Effects on Tissues	71
Organizational Effects During Development	74
Sexual Differentiation: Somatic Modifications	74
Sexual Differentiation: Behavioral Modifications	75
Sexual Dimorphism in Brain Structures	77
Activational Effects in Adults	77
Breeding Seasonality	78
Activational Effects in the Female	78
Activational Effects in the Male	83

6 Behavioral Endocrinology: Stress and Adrenal Hormones ... 89

Definition	89
Types of Stress	89
The Stress Responses of the Body	90
The Adrenal Medulla and Sympathetic Arousal	91
The Hypothalamus and the Adrenal Cortex	91
Corticosteroid Production	92
Corticosteroid Metabolism	93
Habituation to Stress	95
Functions of Stress	96
Psychosomatic Medicine	96
Two Psychiatric Syndromes	97

7 Biological Rhythms ... 99

Functions of Biological Rhythms	99
Circadian Rhythms (23–26 Hours)	101

	Studying Circadian Rhythms	102
	Location of the Internal Clock	105
	Circadian Rhythms in Humans	108
	Circatidal Rhythms (12.4 Hours)	111
	Circalunar Rhythms (14.8 Days)	111
	Monthly Rhythms (29.5 Days)	112
	Circannual Rhythms (365 Days)	113
	Rhythms in Human Disease	113

8 Orientation and Navigation ... 117

Orienting Responses ... 117
    Kinesis ... 118
    Taxis ... 118
Navigation ... 121
    Navigational Mechanisms ... 121
    The Compass Mechanism ... 124
    Navigational Cues ... 125
Migration ... 128

9 Feeding, Foraging, and Predation ... 131

Feeding Behavior ... 132
    Social Learning and Facilitation of Feeding ... 132
Foraging ... 134
    Optimal Foraging ... 135
    Constraints on Optimal Foraging ... 138
    Coping with Changes in Food Supply ... 139
    Feeding in Humans ... 144
    Some Physiological Aspects of Feeding ... 145
Predatory Techniques and Antipredator Defense ... 146
    Somatic Adaptations ... 146
    Predatory Techniques ... 148
    Antipredator Defense ... 149

10 Social Behavior ... 151

Social Systems ... 151
    Coelenterate Colonies ... 152
    Eusociality in Insects ... 152
    Vertebrates ... 154
Benefits of Sociality ... 155
    Reduction in Predator Pressure ... 155

    Improved Foraging and Hunting Efficiency 156
    Improved Defense of Limited Resources 156
    Improved Care of Offspring 157
Costs of Sociality 157
    Increased Competition between Conspecifics 157
    Increased Risk of Infection 157
    Increased Risk of Mating Interference and Parental Exploitation by
        Conspecifics 159
    Increased Risk That Offspring Are Killed by Conspecifics 159
Philopatry and Dispersal 159
    Dispersal Hypotheses 160
Evolution of Cooperative Behavior 162
    Cooperation (Mutualism) 163
    Reciprocity (Reciprocal Altruism) 163
    Altruism (Kin Selection) 164
Mechanisms of Kin Recognition 165
    Location 165
    Familiarity 165
    Phenotype Matching 166
    Allele Recognition 167
Environmental and Cultural Influences in Primates 167
    Nonhuman Primates 167
    Humans 168

11 Communication **171**

Definition 171
Functions of Communication 171
"Honesty" and "Deception" in Communication 172
Communicatory Signals 174
Sensory Channels of Communication 176
    Visual Communication by Reflected Light 177
    Auditory Communication 180
    Chemical Communication (Olfaction) 188
    Tactile Communication 195
    Electrical Communication 197

12 Agonistic Behavior **199**

Interspecific Agonism 200
    Predatory Aggression 202
    Antipredatory Agonism 203
Intraspecific Agonism 203
    Individual Distance 203

Intraspecific Aggression	204
Categories of Intraspecific Aggression	206
Intraspecific Submission and Flight	215
Comparisons between Interspecific and Intraspecific Agonism	216
Human Aggression	216

13 Sexual Selection ... 221

Asexual and Sexual Reproduction	221
Sex Determination	222
Sex Ratio (SR)	223
Theoretical Considerations	224
Bateman's Principle	224
Trivers's Theory of Parental Investment	224
Trivers–Willard Hypothesis	226
Intrasexual Selection	227
Competition among Males	227
Competition among Females	233
Intersexual (Epigamic) Selection	233
Desirable Male Qualities	233
Markers of Male Qualities	233
Evaluation of Male Qualities	234
Evolution of Male Traits and Female Preferences for Them	234
Mate Choice by Males	236

14 Courtship and Mating ... 237

Factors Important for the Onset of Courtship and Mating	238
Seasonal Factors	238
Hormonal Stimulation	238
Social Stimulation	239
Functions of Courtship	239
Species (and Strain) Identification	239
Gender Identification	240
Aggression Reduction between the Male and Female	240
Individual Recognition	241
Behavioral and Physiological Synchronization between the Male and Female	241
Signaling Competitive and Parental Abilities	242
Mating Categories	243
External and Internal Fertilization	243
Copulatory Patterns in Mammals	243
Bisexual Behavior	244
Nonhuman Animals	244
Humans	245

15 Parental Behavior and Mating Systems 249

Models of the Parent–Offspring Relationship 250
  The Parental Provision Model 250
  The Mutual Benefit (Symbiosis) Model 250
  The Conflict Model 250
Evolution of Parental Care 252
  Sex Differences in Parental Care 252
  Selection Pressures for Parental Care 255
Mating Systems 258
  Definition of Mating Systems 258
  Classification of Mating Systems 259

16 Nonhuman Primates 265

Ethology 265
  Fixed Action Patterns (FAPs) 267
  Conflict Behaviors 268
  Ritualization 268
  Releasers 269
  Sensitive Periods, Imprinting 272
Sociobiology 273
  Social Systems 273
  Mating Systems 277
  Mate Competition and Mate Choice 277
  Parental Investment 278
  Dispersal and Inbreeding Avoidance 279
  Hormonal and Seasonal Influences 282
Language in Apes 286

17 Humans 289

Human Ethology 289
  Fixed Action Patterns (FAPs) 290
  Conflict Behaviors 291
  Ritualization 292
  Releasers 294
  Ethology in Clinical Settings 298
  Sensitive Periods, Imprinting 298
Human Sociobiology 299
  Mating Systems 299
  Mate Competition and Mate Choice 299
  Parental Investment 303
  Incest Avoidance 305
  Hormonal and Seasonal Influences 308

References 313

Author Index 323

Subject Index 327

CHAPTER 1
The Study of Behavior
History

A BRIEF HISTORY

The attempt to understand and classify the behavior of animals, and in particular that of humans, has engaged scholars for many centuries, at least since the time of the Greek philosophers Aristotle and Plato. Ethology developed from zoology, that is, from the study by naturalists of animals in their natural habitat, whereas psychology developed from philosophy. Much of the difference between the theoretical and methodological approaches taken by ethologists and psychologists can be traced to their different origins. The following sections briefly summarize the principal branches of the behavioral sciences, ethology, and comparative psychology, the controversies between them during the 1950s and 1960s, and the subsequent synthesis that has developed into the modern study of animal behavior. This may be the most difficult chapter in the book to master.

Classical Ethology

The term ethology comes from the Greek word *ethos*, meaning habit or manner, and was once applied to what we now know as ecology. From the turn of the twentieth century it gradually came to be used to describe the study of the naturalistic behavior of animals, and has been the generally accepted term for this branch of natural history since 1951.

Ethology received an impetus from scientists who recognized the importance of Charles Darwin's groundbreaking insights into evolution (Darwin, 1859) that, for the first time, linked the behavior of animals with their environment. The basic tenet of ethology is that animals evolve not only morphologically but also behaviorally, due to the selection of traits that allow animals to interact successfully, if not optimally, with their environments to survive and reproduce. Ethologists are therefore concerned with why animals perform certain behaviors in their natural habitats, and seek to understand how animals interact and cope with their environment, that is, how they feed, avoid predation, seek and court a mate, and raise their young. One of the earliest scientists that we would now call an ethologist was Douglas Spalding, a tutor to the family of Bertrand Russell, a twentieth-century philosopher and nuclear disarmament advocate. He conducted experiments with birds, mostly chickens, showing that various adaptive behaviors, those increasing the probability of survival and of producing young, are instinctive (inborn or innate and coded by the genes), although a period of maturation might be necessary for their appearance (Spalding, 1873). At the time, the prevailing view was that an animal or person is born as a *tabula rasa* (a clean slate), and that all behaviors are then learned more or less rapidly during life. The work of Spalding, and others like him, remained unrecognized and mostly forgotten, and some 60 years passed before several European naturalists brought the field of ethology into prominence and eventually into public recognition with the joint award of the 1973 Nobel Prize for Physiology or Medicine to its three main pioneers, Konrad Lorenz, Nikolaas Tinbergen and Karl von Frisch.

Lorenz (Fig. 1-1), generally recognized as the founder of ethology, which he named, was born into the family of a prominent and wealthy Austrian surgeon in 1903. From an early age, he surrounded himself with many different wild animals, especially birds. He obtained a medical degree but soon concentrated on animal studies, and in 1940 took the Chair of Philosophy once held by Emmanuel Kant (psychology was still subsumed under philosophy at that university) at Königsberg University in Germany, until called up as an army doctor in 1941 to serve on the Russian front. He resumed his work in Germany after the war and returned to his ancestral home near Vienna after retirement. Although unaware of Spalding's studies, Lorenz was influenced by the work of earlier naturalists and also that of Sigmund Freud. Lorenz discovered most of the phenomena of ethology on the basis of his keen observational powers, intuitive insights into animal behavior, and inductive reasoning. He relied on studying animals in their natural environments and rarely performed an experiment, although he collaborated with others who did and encouraged his students to do so. The Dutch ethologist Tinbergen, considered the founder of experimental ethology, began his work in Holland with studies on insects and fishes, and with the advent of World War II moved to Oxford University in England, where he remained for the rest of his life working mostly with birds. He shared Lorenz's concern for observing animals in their natural environments, which he combined with elegant experiments conducted mostly in the field. Based on Lorenz's observations and his own experiments, Tinbergen helped to formalize some general principles of ethology, including fixed action patterns, releasers, supranormal stimuli, innate releasing mechanisms, and imprinting. The Austrian zoologist von Frisch

FIGURE 1-1. Konrad Lorenz, considered the founder of ethology. (Photo courtesy of Anne Kirchbach)

was primarily an experimentalist, and began his famous, lifelong work on honeybees by wondering why flowers were so colorful, although, at the time, the insects feeding on them were thought to be color blind. He used conditioning techniques that had recently been demonstrated by Pavlov (see below) to show that bees could both discriminate colors from shades of gray and see ultraviolet (UV) light. From this he went on to characterize the sensory world and communicatory systems of honeybees, including the "dance" that signals the location of a food source to bees in the hive.

Ethologists differed from some zoologists (Bierens de Haan, 1940) and psychologists (McDougall, 1936; Tolman, 1932; Russell, 1938), who were called "vitalists" because they postulated inexplicable instincts as the final determinants of behavior, and they differed also from "mechanists," namely, behaviorists who believed the internal motivational state of an animal to be irrelevant for its behavior (see below). Ethologists held the view that an animal's behavior is intimately related to its motivational states, and that the physiological mechanisms underlying both could eventually be determined by research.

Comparative Psychology

Comparative psychology occupied a place in the United States that was filled by ethology in Europe, insofar as it was concerned with studying the behavior of several different species. Unlike ethology, it emphasized controlled laboratory experimentation and the quantification of data, and discouraged any assumptions about an animal's mental state. These ideas were formalized in Morgan's Canon (1894): "In no case may we interpret an action as the outcome of the exercise of a higher psychical faculty if it can be interpreted as the outcome of the exercise of one which stands lower in the psychological scale." This was espoused and perhaps taken to extremes by behaviorism (see below). The emphasis on laboratory experimentation led naturally to studies that were almost exclusively on learned behavior.

Operant and Classical Conditioning

Operant conditioning, also known as trial and error learning, was pioneered by Thorndike, who devised various "problem boxes" from which animals had to escape in order to gain access to a reward such as food. He found that the time it took an animal to escape from a box and attain the reward decreased with repeated trials. He studied many species, and concluded that the process of learning was the same in all species, although animals differed in what, and how fast, they learned (Thorndike, 1898). Classical conditioning, now generally known as associative learning, was discovered some years later by the noted Russian physiologist Pavlov. While working on his Nobel Prize–winning research on the physiology of digestion, he noticed that hungry dogs began salivating at the sight of food. If he repeatedly paired the sight of food, the unconditioned stimulus for salivation, with an irrelevant stimulus, for example, a light or the sound of a bell, the dogs would eventually respond with salivation to the light or sound, the conditioned stimulus, even when no food was present (Pavlov, 1927). In contrast to behaviorists, Pavlov assumed that recognition of the unconditioned stimulus, the sight of food, as well as the unconditioned response to it, salivation, were innate. The techniques of operant and classical conditioning, widely used today to study motivation and learning in animals, have become useful tools in ethological studies.

Behaviorism

The school of comparative psychology known as behaviorism dominated all other schools for the first half of the twentieth century, and was espoused by those scientists who never accepted Darwin's proposition that behavior, like morphology, is shaped by evolution. Developed in the United States by its chief proponent, Watson (1919), and subsequently elaborated and popularized by Skinner (1953), behaviorism was founded on the premise that each animal and human comes into the world as a clean slate, uninfluenced by predispositions to act and respond in certain ways, and that the organism is subsequently molded entirely by its environment. It was believed that environmental stimuli elicit reflexes that are then linked together and built by conditioning into complex behavior patterns. Skinner emphasized the role of positive and negative reinforcing stimuli, which he considered sufficient to shape an animal's behavior. There seemed no reason to suppose that genetic influences were needed, even for modulating that which can and cannot be learned. Consequently, it was thought that laboratory animals such as inbred rats and pigeons could serve as models for humans in elucidating laws of behavior believed to be common to all birds and mammals. Since behavior was thought to depend exclusively on environmental variables, it could be understood by examining the effect of one variable after another under controlled laboratory conditions. In essence, the animal was regarded as a black box, with an input (an environmental variable) and an output (the animal's response); it seemed unnecessary to examine the workings inside the box because they were regarded as irrelevant. In the 1950s, some psychologists found that there were clear limits to these approaches (Chapter 3, Programmed Learning), and the more extreme views advocated by behaviorists have ceased to carry as much weight as they formerly did.

Physiological Psychology

Proponents of a subgroup of comparative psychology known as physiological psychology were deeply concerned with the physiological mechanisms responsible for behavior, notably, for learning and reproductive behavior. Lashley attempted to localize the areas of the brain important for learning various tasks by cutting or removing regions of cerebral cortex after animals had learned a task and then retesting them. The finding that most of the cerebral cortex was generally necessary for complex problem solving (Lashley, 1950) did not harmonize with the prevailing view that learning depended on the growth or functional enhancement of neural connections between task-specific, circumscribed regions of the cerebral cortex. Sperry (1958) examined the neural mechanisms responsible for locomotion and other motor patterns in several vertebrates. He emphasized the role of neural plasticity, in which there is great interest today, and also found that many movements developed independently of learning and are modified little, if at all, by a learning process (Chapter 2, Fixed Action Patterns). Other physiological psychologists, foremost among them Beach and Lehrman, made important contributions to our understanding of the effects of hormones on reproductive and related behaviors in several species of birds and mammals (Beach, 1948; Lehrman, 1961).

Emergence of Modern Behavioral Research

Because of the impact of Darwin's work, ethological research was directed almost exclusively toward innate behavior, especially that of insects, fishes, and birds, giving rise to the still common misconception that ethologists consider all behavior to be innate and minimally affected by experience or learning, which is not so. The research of behaviorists as well as some comparative psychologists, on the other hand, focused on learned behavior and the effects of experience. This dichotomy contributed to a nature–nurture controversy during the 1950s and 1960s, as a consequence of which an influential book was written to harmonize the two disciplines (Hinde, 1970), some of the more implausible and extreme views of both factions were put aside, and there emerged a general consensus that both nature and nurture are involved in most naturalistic behaviors. In other words, there are genetic predispositions to perform certain behavioral patterns and to learn certain things, while experiential factors contribute to a greater or lesser degree in shaping behavior and the context in which it occurs. This consensus seems sensible.

Subsequently, three other factors helped to change the study of behavior into what it is today. The first was a critical reexamination of evolutionary theory in the 1960s. Although Darwin (1859) had concluded that natural selection acts on the individual, the assumption that selection acts on groups or on the species as a whole, at least where social and altruistic behavior is concerned, was common in the mid-twentieth century. The view that individuals acted "for the good of the species" was argued formally by Wynne-Edwards (1962). This elicited a response by Williams (1966) in which he reasoned convincingly that selection acting on individual differences will generally have stronger effects than those produced by the selection of differences between groups. Williams' view has gained acceptance and was taken further by Dawkins in his book *The Selfish Gene* (1976). The problems concerning the evolution of altruism had been elegantly solved by Hamilton (1964) (see below, Fitness and Inclusive Fitness). The second factor was the impact of a new discipline, called *sociobiology* or *behavioral ecology*, which emphasized the application of evolutionary theory to the study of social behavior. Earlier workers had suggested that ecological variables might be better predictors of social behavior than phylogenetic relationships, but these ideas did not have their full impact until Wilson published his book *Sociobiology: The New Synthesis* (1975). Whereas ethology tended to concentrate on differences between species, sociobiology typically focuses on differences between conspecifics (those of the same species) to examine how individuals cope with social and environmental variables, including social rank, predation, and availability of food and shelter, and how this affects their survival and reproduction. The third factor influencing behavioral research was the rapid development of technologies that permitted the measurement of a variety of different physiological variables, including hormone levels in blood and the activity of individual neurons in the brains of freely behaving animals. These developments led to a greater understanding of the physiology underlying ethological concepts such as the motivational state and the releaser, and contributed to a cross-fertilization between the original disciplines and a renaissance of interest in the study of behavior itself.

SOME THEORY AND TERMINOLOGY

Due largely to the influence of sociobiology which, unlike ethology, tends to be hypothesis-oriented, many concepts and mathematical models have been developed that need brief explanations.

Fitness and Inclusive Fitness

As noted previously, modern evolutionary theory holds, as did Darwin originally, that natural selection typically acts on the individual, and not on the species as a whole, by favoring individuals with phenotypic (somatic and behavioral) traits, which are encoded in their alleles, enabling them to survive and reproduce. Over time, individuals with these traits (and alleles) will increase in the population at the expense of individuals with less favorable traits and alleles. An extreme view is that the organism can be regarded as a machine driven by its alleles to behave in such a way that these alleles replicate themselves into the future to the maximum extent possible. As stated by Dawkins (1976), organisms "are survival machines—robot vehicles blindly programmed to preserve the selfish molecules known as genes." This statement would not have been made before 1964 because it could not account for the evolution of well-documented behaviors, for example, predator alarm calls and behavior that lures a predator away from others, that appeared altruistic since they put the individual in harm's way. Nor could it account for the evolution of eusocial species, namely, those with sterile castes, such as honeybees. Indeed, the mostly unspoken acceptance of group selection had gradually developed until Hamilton (1964) showed how evolution can act on "selfish genes" to produce altruism and eusociality. He drew attention to the implications of the fact that an individual shares at least some of its genes with other kin as well as its offspring, the proportion of shared alleles depending on the degree of relatedness. If an animal helps its kin to survive and reproduce, then this will enable some of its genes to be passed on, even if it should die without offspring. Consequently, a mutation that results in its bearer's altruistic behavior could gradually spread through a population if the benefits of altruism exceed its costs (Chapter 10, Social Behavior). The benefits and costs of a behavior are typically measured by its effects on (1) the animal's fitness, defined by the number of alleles passed on to offspring that survive and reproduce, and (2) the animal's inclusive fitness, defined by the total number of its alleles in its offspring and in other kin that survive and reproduce due to the help given (and harm prevented) by the animal (Fig. 1-2).

Benefit–Cost Ratios

Cost–benefit analysis is a tool used by economists to predict the net profit of companies and the effectiveness of manufacturing processes. They typically refer to financial costs and benefits. The technique has been applied to estimate the costs and benefits of the behavior of animals, often in relation to different environmental variables, where the "currency" is the animal's fitness or, more commonly, its inclusive fitness. In behavioral studies, the ratio is typically reversed and expressed as a benefit–cost (B/C) ratio. The assumption is that for a behavior to have evolved, the fitness benefits must equal or exceed the fitness costs (Fig. 1-3; B/C = 1 or higher).

Optimality Theory

An organism compromises between opposing selection pressures; for example, attracting a mate may also happen to attract a predator. The advantages and disadvantages of a trait are estimated by its B/C ratio. Optimality theory holds that a trait is not just reasonably well adapted in these terms, but optimally adapted. The phenotype of a population of animals will therefore change during the course of evolution as a consequence of different and sometimes opposing selection pressures on many traits, a process called dynamic selection, in such a way that these selection pressures eventually become balanced and maintained by stabilizing selection. Long-term studies involving optimality theory are hampered by difficulties in measuring inclusive fitness, but optimality theory is frequently invoked in short-term studies, notably on foraging and

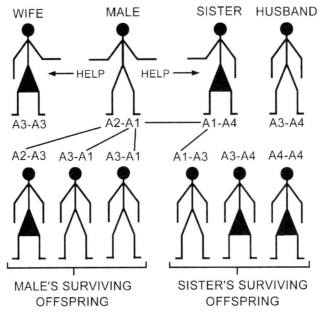

FIGURE 1-2. The fitness of the male is measured by the number of his alleles (A1, A2) passed on directly to his surviving offspring. His inclusive fitness is measured by the total number of his alleles in his surviving offspring and in all of his other relatives (sister and sister's son) who survive because of his help.

feeding (Chapter 9), where benefits and costs are typically measured both in energy intake and in time and energy expenditure.

Game Theory

Game theory is so named because its mathematical basis is similar to that of games such as chess or checkers. It is used as a tool to predict evolutionarily stable strategies (see below). Here, the term *strategy* simply describes a set of behaviors and does not imply conscious choice or conscious planning in any way. Individuals are treated as opponents in a game, and account is taken of all existing behavioral strategies, the frequency of each strategy in the population, and the payoff in fitness for each possible combination of strategies. Game theories have names such as "Hawk and Dove" (aggressive behavior) and "Prisoner's Dilemma" (cooperation). The latter has been used to develop a model for the evolution of cooperation between nonkin. Two players, prisoners arrested for committing a crime and unable to communicate with each other, have a choice between cooperating (not giving each other away) or defecting (accusing the other of the crime in return for a lighter sentence). In a single game, it is better to defect no matter what the other player does, but if the game is played repeatedly and both players defect, they do worse than if both cooperate. When the two players meet repeatedly, the most effective strategy overall was found to be the "tit for tat" strategy:

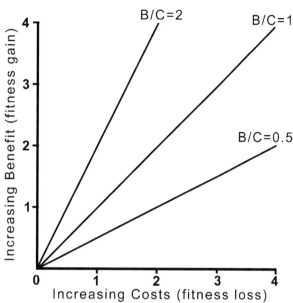

FIGURE 1-3. The benefit–cost ratio (B/C) of an evolved behavior would be expected to be one or greater.

cooperation on the first encounter and subsequently doing what the opponent did last time, namely, defect if he or she defected and cooperate if he or she cooperated (Axelrod & Hamilton, 1981; Axelrod & Dion, 1988). Game theory has been useful for testing predictions from evolutionary hypotheses because it facilitates the application of mathematical models to behavioral data in computer-generated simulations.

Evolutionarily Stable Strategies

Game theory shows that the fitness benefits and costs of a behavioral strategy often depend on the behavior of other conspecifics in the population. By definition, an evolutionarily stable strategy (ESS) is the optimal strategy that cannot be bettered by an alternative strategy that provides greater reproductive success once most members of the population have adopted it (Maynard Smith, 1976). A *pure* ESS is a single strategy employed by all individuals. For example, Tinbergen found that the white interior of eggshells from newly hatched black-headed gull chicks attracted predators (herring gulls and crows) to the nest, but the parents did not remove them for several minutes or even hours. Further observation showed that neighboring black-headed gulls might cannibalize wet, newly hatched chicks if the parents were momentarily inattentive or away. He concluded that the delay in eggshell removal until chicks were dry (fluffy and harder to swallow) reduced the high incidence of cannibalism, although it increased predation by other species, whose incidence is lower than cannibalism (Tinbergen *et al.*, 1962). A *mixed* ESS is where two or more strategies exist, with a stable equilibrium between the two within the population. This may involve individuals with different phenotypes, that is, groups of individuals using different strategies, or it may involve individuals using them at different times. Dawkins (1980) used a hypothetical example of fish-hunting birds. If a bird has a mutation that conserves energy (increases its fitness) by stealing fish from others, the mutation will initially increase in the population. At some point, the benefits of stealing will decrease as the thieves are increasingly likely to meet other thieves or birds that have just been robbed of their catch. This gives an edge to the fish-catching birds, which will multiply at the expense of the thieves. The proportion of hunters and thieves in the population will therefore remain at a stable equilibrium; the two forms of feeding are a mixed ESS. The same situation would apply if individuals switched between the two feeding strategies.

General Life Strategies: *K* and *r* Selection

Some habitats are stable or vary *predictably* over time, and others are unstable and subject to *unpredictable* changes. These ecological differences have produced two extremes in lifestyle called "*K*-selected" and "*r*-selected." The terms are derived from the carrying capacity (K) of the habitat (the maximal size of a population that can be maintained by the habitat) and the intrinsic population growth rate of the population (r). *K*-selected species are adapted to stable, predictable habitats. Populations initially increase at the intrinsic growth rate, but then growth slows and the population typically remains just below the carrying capacity (Fig. 1-4). With *K*-selection there is keen competition between conspecifics for resources, especially for those that determine and limit the carrying capacity, such as food, water, shelter, or nest sites. Body size is larger (giving an edge in competition) and development is therefore slower; there are fewer offspring in each of several reproductive efforts per lifetime, and there is greater parental care investment in offspring (Table 1-1, left; Gould, 1982). *K*-selected species are typically specialists occupying different niches of a particular habitat, which reduces competition with other species in the habitat. A good example is the food specialization,

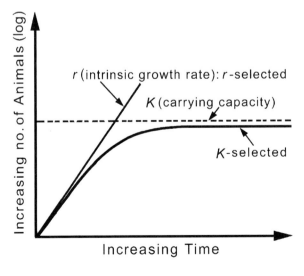

FIGURE 1-4. Schematic of population growth curves of *K*-selected and *r*-selected species.

Table 1-1. General Trends in Life Strategies

	K-selected species	r-selected species
Population	At carrying capacity	Usually below carrying capacity
Environment	Constant or predictable	Unstable
Intraspecific competition	High	Low
Body size	Large	Small
Development	Slow	Fast
Number of offspring	Few	Many
Investment per offspring	Large	Small
Niche	Specialists	Generalists

SOURCE: James L. Gould, *Ethology: The Mechanisms and Evolution of Behavior.* Copyright © 1982 by James L. Gould. Used by permission of W.W. Norton & Company, Inc.

reflected in beak morphology, of the Galapagos finches (Chapter 12, Fig. 12-2), which probably evolved from a single finch species. In contrast to K-selected species, r-selected species are niche generalists and adapted to unstable, unpredictable habitats. Populations increase explosively at the intrinsic growth rate and, if favorable conditions remain long enough, the carrying capacity is exceeded and the population crashes. Typically, however, populations remain well below the carrying capacity, and there is little intraspecific competition for resources. Adaptations for exploiting short periods of favorable conditions include small body size, rapid development, many offspring in a single reproductive effort per lifetime, and little investment in offspring (Table 1-1, right). All animals fall somewhere on the continuum between extreme K-selection and extreme r-selection. An example of extreme K-selection would be some large, long-lived birds of prey that rear one chick every few years (Chapter 15). Mosquitoes are a good example of r-selection. As the weather warms, the few aquatic larvae that have survived the winter hatch into a world full of potential prey, and the survivors rapidly reproduce through many generations, until the population boom abruptly ceases as cold weather returns.

WARNINGS, FALLACIES, AND PITFALLS

The attempt to understand behavioral phenomena, and especially social organizations, in terms of the evolutionary selection pressures determining them is both valuable and intellectually attractive but contains pitfalls for the unwary.

Correlational Studies

Many behavioral studies, especially those conducted in the field, do not involve experimental interventions because they are either not feasible or not ethical. The relationship between two behaviors, between hormone levels and behavior, or between habitat and social organization, is deduced from whether or not the variables are correlated. This is problematic for two reasons: Correlations between two variables cannot prove causation because there may be a cause common to both, and even where a causal relationship exists, the direction of causality often cannot be ascertained. For example, increasing testosterone levels in many male mammals increases their sexual activity, but sexual activity itself can also increase testosterone levels; because of feedback mechanisms, it is not always clear whether a hormonal change is the cause or the effect of a behavioral change. Many studies in clinical medicine rely on correlations, for example, those between certain diets and the incidence of heart disease or cancer.

Genes: What They Do and Do Not Do

It is easy to overlook the fact that genes can code for the anatomy and physiology of the organs and limbs of an individual, for example, the wing of a pigeon and the wing of a penguin, but not for their functions, namely, flying by the pigeon and underwater swimming by the penguin. Note that we do not use the term "function" here in its physiological sense, but in the sense of serving an adaptive purpose. Genes determine the size of an individual, large or small, but not the size difference between two individuals, one being larger and the other being smaller. Thus, genes code for traits within the individual, while natural selection favors individuals with traits whose functions are adaptive. In these two examples, the pigeon wing and large size can be regarded as the structures of the two traits, while flying and "being larger than" are their respective functions. Since there is no function (e.g., flying) without the necessary structure (wing), the structure is the cause and the function is its effect. We have used morphological traits as examples, but these considerations also apply to the structure and function of behavioral traits, and are particularly important when attempting to understand social systems. Furthermore, gene mutations occur by chance and not on demand: Giant pandas might avoid extinction if they could switch from bamboo to grass consumption, but it does not follow that the necessary mutations will occur. If the foregoing seems self-evident, the following quotation from the literature should give one pause, as both errors (an effect can produce its own cause, and mutations can occur on demand) are contained in it: "Obviously, when male post-parturitional effort is adaptive, selection can provide an appropriate machinery to ensure its emergence." (Cited by Bernstein, 1987). Male postparturitional effort (the effect) cannot produce an appropriate machinery (the cause), nor do "designer" mutations occur just when they are needed, but the statements are written so speciously that the fallacies are not easy to detect.

Understanding Optimality Theory

Optimality theory, as described earlier, does not imply perfect adaptations of a trait to different and often opposing selection pressures. Kummer (1971) put it succinctly: "Discussions of adaptiveness sometimes leave us with the impression that every trait observed in a species must by definition be ideally adaptive, whereas all we can say with certainty is that it must be tolerable, since it did not lead to extinction. Evolution, after all, is not sorcery." It can never be proven that a trait is optimally adapted, and sometimes for the reasons given below, it is impossible even to know if it has an adaptive function.

It takes time for adaptions to a selection pressure to occur, certainly many generations. Time delays may be a few weeks or months in rapidly reproducing organisms such as viruses and bacteria, but will be hundreds or thousands of years for larger vertebrates. Environmental changes have accelerated during the last few centuries in many regions of the world due in part to human influences. When we investigate today the selection pressures that may have brought about a species' behavioral and morphological traits, we are correlating traits that evolved many thousands of years ago, in response to conditions back then, with conditions prevailing now, which are likely to be very different. Most organisms have the same problem: At no point in time can they be totally adapted to the world they live in, although cockroaches come close. Phylogenetic inertia can result in the retention of traits that were adaptive once but are no longer. Unless they are maladaptive (and some maladaptive traits are retained because the genes coding for them have multiple effects) and therefore selected against, they will remain but no longer have any adaptive function—the well-known example of the human appendix. Random processes can have similar effects, producing traits that are neutral or even maladaptive. It is also worth remembering that some traits might be a concomitant or consequence of a trait that is adaptive, without being adaptive themselves. The human navel has no adaptive value itself but is an unavoidable consequence of an umbilical cord and placenta. This example emphasizes the point made by Tinbergen when he advocated the need to ask four questions in order to understand a behavior fully: its mechanism, ontogeny, function and phylogeny (Chapter 2). In the initial enthusiasm to implement the sociobiological approach to the study of behavior, there was a tendency to neglect the mechanism question and to adopt an extreme "adaptationist" approach by assuming that all behavior must have an adaptive function. Optimality theory, then, may be understood as optimal adaptation to selection pressures given the constraints described earlier and those imposed by an organism's phylogenetic history.

CHAPTER 2

Some Ethological Concepts

EVOLUTIONARY BASIS OF BEHAVIOR

The theoretical foundation of ethology is that behavior, like morphology, is adaptive and subject to selection pressures, and that it may have a genetic basis and phylogenetic history. As with morphological and anatomical features, a behavioral trait can best be understood by determining (1) the physiological and environmental factors regulating it (its immediate or proximate mechanisms), (2) its development in the individual (ontogeny), (3) its function in terms of the advantages it confers on the animal, and (4) its evolutionary development (phylogeny).

Tinbergen (1951) emphasized that these four questions should be asked about a behavior pattern in order to understand it: questions that can be couched in terms of "how" or "why." The first two questions are often "how" questions: (1) How does a behavior come about in terms of the animal's physiology and response to the environment? and (2) How does it develop during ontogeny? The third and fourth questions are often "why" questions: (3) Why does the animal perform the behavior in terms of its immediate consequences and long-term advantages? and (4) Why did it evolve as it did? All this is arbitrary because the questions are closely interrelated and could be put either way. The first two questions mainly involve what are called proximate processes, those processes operating during the animal's life. The last two questions generally involve ultimate processes, those processes subject to evolutionary selection pressures. None of this should be taken too rigidly. For example, the third question can address both proximate and ultimate functions. In the case of eating, these two functions are nearly

15

identical: The animal eats to obtain the energy to keep alive (survival value–proximate function), which is necessary for it to reproduce and pass its genes into the next generation (adaptive function–ultimate function). In other cases, especially with altruistic behaviors, the proximate function may be at odds with the ultimate function: The proximate function may be to attract a predator *toward* the animal but away from its young, which will not do much for its own survival but the ultimate function may be the payoff in terms of the survival of the animal's genes via its offspring and other kin. These two examples help to emphasize that we are not only concerned merely with semantics but also with fundamental biological processes at different levels of analysis.

Like morphological traits, behavioral traits may have undergone divergent evolution resulting in different forms and functions of homologous traits in different species (Fig. 2-1). Alternatively, they may have undergone convergent evolution resulting in similar forms and functions of analogous traits in different species (Fig. 2-2). Both homology and analogy tell us much about ultimate function, selection pressures, and phylogeny.

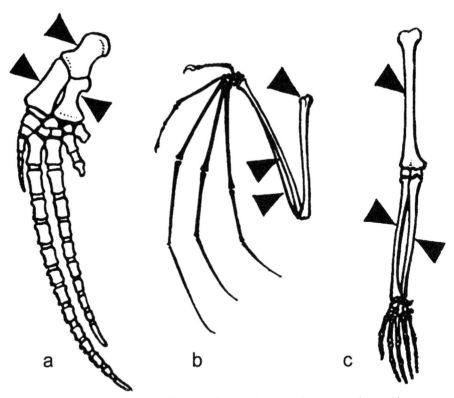

FIGURE 2-1. Homology illustrated by the whale's flipper (a), bat's wing (b), and human arm (c). (SOURCE: Lorenz, 1965b, with permission)

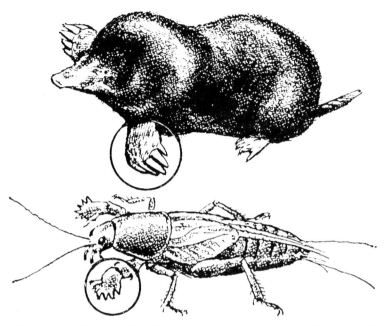

FIGURE 2-2. Analogy illustrated by the digging claws of the mole (above) and the mole cricket (below). (SOURCE: Lorenz, 1965b, with permission)

ETHOLOGY'S OBJECTIVES

Studying behavior using an ethological approach is akin to figuring out the rules of a sport (such as cricket) one does not understand without the benefit of a commentary by someone who understands the sport. To determine its rules, one must watch the game for a long time in order to recognize and group fragments of behavior or "plays" into different categories. Typical sequences emerge from analyses of which fragments of behavior precede or follow other fragments so that patterns become apparent. The frequencies of these sequences can then be measured and subjected to statistical analyses, and finally studied during experimental interventions. This is the first task of ethology, namely, to obtain a complete description of the behavioral repertoire of a species in these terms, called an ethogram. When this has been accomplished, we can ask the four questions already described.

GENERAL METHODS

To develop an ethogram, the species is observed in its natural environment and all behavioral sequences including postures, movements, facial expressions, and vocalizations are recorded and broken down into the smallest units with independent functions. Lorenz emphasized the value of trusting one's Gestalt perception (appreciation of the overall shape and pattern) for classifying behaviors into related categories; subsequently, these can be checked out as correct or incorrect on the basis of numerical analyses. Human perception is quite remarkable in this respect. It would be difficult to explain to someone who had never seen any dogs which features a Chihuahua, a Dachshund and a Great Dane have in common that categorize them as dogs, but a baby just a few months old has no difficulty with the Gestalt and says "bow-wow" for all three. Many of Lorenz's creative insights came from simply observing animals without any theoretical preconceptions. This method makes the replication of data difficult and runs counter to today's scientific needs and philosophy, which require the rigorous testing of *a priori* hypotheses.

When the ethogram has been compiled, the following basic strategies are used to answer Tinbergen's four questions:

1. *Physiological and environmental regulation* are investigated by conducting experiments to determine the effects of changing one variable while holding all others constant.
2. Ontogeny is studied by determining when a behavior pattern first appears during development and identifying any subsequent changes in form, mechanism, and function.
3. *Adaptive function* is examined by analyzing sequences of behavior patterns, the events eliciting them, and their consequences.
4. *Phylogeny* is likely to be involved if the behavior occurs in geographically isolated populations and in social isolates; it must be involved if the behavior has specific morphological and physiological correlates, and phylogenetic pathways can be traced if the behavior occurs in closely related species (Chapter 4).

This way of studying insects, fishes, birds and mammals led Lorenz, Tinbergen, von Frisch and the other founders of ethology to develop a number of concepts, some of which have subsequently been refined and modified, but all retain utility for conceptualizing a wide range of behavioral phenomena: among these are fixed action patterns, conflict behavior, and ritualization.

FIXED ACTION PATTERNS (FAPs)

Fixed action patterns (FAPs) are generally characterized by (1) an unvarying form; (2) the coordination of several muscle groups; that is, they are more than a simple neuromuscular reflex; (3) a form that is largely independent of environmental stimuli and feedback, although they may be needed to elicit or orient the FAP; (4) being species-typical and under genetic control; and (5) being elicited in some cases by electrical stimulation of discrete regions of the central nervous system, for example the spinal cord, hypothalamus, or limbic system.

Unvarying Form

The behavioral repertoire of most animals contains movements that are easily recognizable because they have an unvarying form or stereotypy; that is, they are form-constant across all members of the species. These movements do not need to be learned by the animal but are innate. Stereotypy per se does not mean that a movement is an FAP, because stereotyped movements may be acquired during an animal's life. For example, cage stereotypies were once prevalent among animals constrained in small cages in circuses and zoos. Cage stereotypies, unlike FAPs, are form-constant in a single individual only. FAPs, which fit so smoothly into the animal's normal life, appear inappropriately when the animal is reared in the wrong environment. Red squirrels raised in captivity exhibit their species-typical nut-hoarding behavior in the fall and will try unsuccessfully to bury nuts in the hard floor of the room. This shows that there is a strong genetic component to the behavior that is little influenced by learning or by the environment. FAPs cannot be broken down into smaller behavioral components that themselves can be elicited by different types of stimulus. For example, when we walk (a FAP) we lift up one foot and place it in front of us, shift our weight onto it, then lift the other foot to bring it forwards, and so on. We can lift and bring forward a foot (to tie a shoelace or kick a ball) and shift our weight onto the other foot (to begin a jump) for different reasons. Neither component is a FAP, but in combination with other movements these components produce the human FAP of walking.

Coordination of Several Muscle Groups

von Holst was one of the first to propose that locomotor behavior in fishes, namely, the undulating body movements of eels and the fin movements of other fishes, are controlled by an endogenous production of excitatory electrical potentials that act as a type of generator. This generator is thought to be located in the anterior spinal cord. Spinal preparations deprived of information from the muscles, joints, and skin by excision of the dorsal root ganglia and dorsal columns of the spinal cord resumed their species-typical locomotor movements shortly after recovery from surgery, despite the absence of any proprioceptive feedback. These locomotor behaviors are not simply chains of interconnected reflexes.

Environmental Influences

When Environmental Stimuli Act as Modulators. Although the form of an FAP is largely independent of environmental stimuli, the latter may be important in two ways: They may elicit an FAP and also orient it. Many ground-nesting birds have an FAP whereby if an egg has rolled off a nest, they retrieve it by placing their beak on the far side of the egg, then drawing it back into the nest. In his studies on egg retrieval by geese, Tinbergen showed that the sight of a goose egg or similar object just outside the rim of the nest elicits retrieval on the part of the adult incubating the eggs, and that the movement continues to completion even when the egg is removed. However, the way the egg rolls to the right or to the left determines the exact orientation of the head and beak of the goose as the egg is retrieved.

When Environmental Stimuli Do Not Act as Modulators: Learning versus Maturation. When a stereotyped, species-typical behavior does not appear until several weeks or months after hatching or birth, it is tempting to assume that it has been learned from other conspecifics and is not really innate and under direct genetic control. This assumption may be incorrect; the delay can be due to a maturational process, which is obvious in many courtship displays that do not occur until sexual maturation is reached. Maturation is also necessary for locomotion. In precocial species, such as the ungulates, this may occur *in utero* because such species must be able to run with the herd immediately after birth. In altricial species, which are born relatively immature, time is essential for locomotor development. Spalding (1873), in an ingenious experiment on swallows, demonstrated this by raising chicks in tubes that did not allow them to flap their wings. In the absence of any opportunity for learning by trial and error, these juveniles flew immediately upon release from their tubes at an age when swallows raised normally begin to fly. More recently, the development of flight by domestic chickens has been investigated (Provine, 1981). Several morphological and flight parameters were measured in naive chicks removed from incubators at various ages posthatching. They were placed in a standard test setting in which flight was evoked by dropping them from a height of 1.7 m onto a soft surface. At 13 days posthatching, distance of lateral flight (Fig. 2-3), wing area to body ratio (Fig. 2-4), and wing-flap rate measured by stroboscopic photography all reached their maxima. Identical maturational changes in flap rate were observed in mutant chicks born without feathers and in those whose wings were taped to the body immediately after hatching until testing. It appears from these and other experiments that flight is centrally patterned and coordinated, and that maturation of both the neural mechanism and the physical effectors is responsible for the time

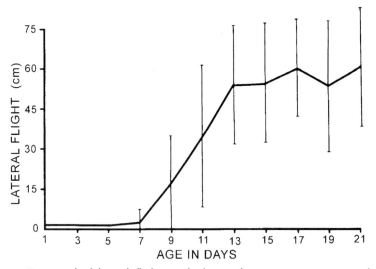

FIGURE 2-3. Drop-evoked lateral flight in chicks reaches a maximum at 13 days after hatching. (SOURCE: Provine, 1981, Development of wing-flapping and flight in normal and flap-deprived domestic chickens. *Devel. Psychobiol.* **14**, 279–291. Copyright © 1981 by John Wiley & Sons, Inc. Reprinted with permission of John Wiley & Sons, Inc.)

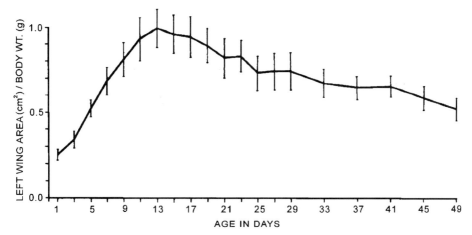

FIGURE 2-4. Ratio between wing area and body weight in chicks reaches a maximum at 13 days after hatching. (SOURCE: Provine, 1981, Development of wing-flapping and flight in normal and flap-deprived domestic chickens. *Devel. Psychobiol.* **14**, 279–281. Copyright © 1981 by John Wiley & Sons, Inc. Reprinted with permission of John Wiley & Sons, Inc.)

delays often attributed to learning. There is convincing evidence that locomotor behaviors in other species, including the human, are also FAPs. Premature infants (7 months) are much lighter in body weight than term infants and are capable of hanging by their hands and feet and also walking with minimal support (Chapter 17).

FAPs may occur in chains producing organized sequences of behavior in which the performance of the first FAP acts as the eliciting stimulus for the next element in the chain. Experimental disruption of the sequence causes resumption at an earlier FAP than the one currently in progress. The sequence in which spiders spin their webs is a chain of FAPs; when the spider is interrupted, it resumes by repeating earlier sequences before completing the web. Chains of FAPs may give the appearance of goal-directed behavior but have a stereotyped, unvarying form. They are also largely independent of the environment. Naive red squirrels given a supply of nuts in an empty room will perform the entire sequence of behaviors involved in nut hoarding (Eibl-Eibesfeldt, 1963). They run to a prominent vertical marker (normally a tree but in the experimental setting the corner of a room), scratch on the bare floor (normally resulting in a hole in the earth), tamp down the nut (normally into the hole), scratch sideways on the ground (normally replacing earth over the nut), and pat down on the floor (normally tamping down the replaced earth). Earlier FAPs in a chain can often be elicited more frequently than later ones, as demonstrated for predatory behavior in cats, in which the initial components of stalking and crouching continue long after the subsequent components of pouncing, killing, and eating have disappeared following satiation. This phenomenon is presumably an adaptation to the fact that, in the natural environment, earlier phases of hunting are less likely to result in a meal than the later ones and are therefore repeated more frequently.

Genetic Factors

Genetic programming is implicated when the behavior pattern is species-typical, that is, found in all conspecifics of the appropriate age–sex category irrespective of the geographical location of individual populations. The assumption is strengthened when the behavior appears in individuals raised in isolation from other conspecifics or cross-fostered to a different species (Chapter 4). A chick hatched by a duck will immediately scratch the ground and peck at seeds, and shake itself when wet, just like chicks hatched by a hen. Conversely, a duckling hatched by a hen will run to water, swim, dive, feed under water, and oil its feathers, just like ducklings hatched by a duck. This shows that there is a very strong genetic component to these behaviors and rather minimal environmental influences, be they direct or indirect via learning mechanisms. Even stronger evidence for genetic control is provided when the behavior pattern is shared by several closely related species. Lorenz's comparative studies with ducks and geese demonstrated that if FAPs are common to several species, then they can be as useful as any morphological character in tracing phylogenetic pathways and taxonomic relationships. In Figure 2-5, the vertical lines represent 20 different genera or species of ducks and geese, while the horizontal lines connecting two or more taxonomic groups represent courtship displays and other behaviors that are common to the groups involved. For example, "EVP," the monosyllabic piping vocalization of the duckling or gosling that has lost its mother, is common to all ducks and geese, but "Is," the conflict behavior (see below) of displacement shaking as a display, is common to ducks but absent in the ruddy sheld duck and geese, and so on. Similar taxonomic relationships can be traced among primates, including the side-to-side movements of the human baby searching for the mother's nipple, which is shared by all primates, and smiling and laughing in humans, which have close parallels in the "play-face" of chimpanzees (Chapter 17).

Brain Stimulation

It has been shown that some single or chained FAPs can be elicited in complete form by electrical stimulation of certain regions of the brain, notably, the hypothalamus and limbic system. Furthermore, both the duration of the electrical stimulus and the quality of the visual stimulus modulate the intensity of the domestic hen's defense against a ground predator (Fig. 2-6). At a constant brain site, constant voltage and stimulus duration, a human fist elicits mild unrest, whereas a stuffed polecat (ground predator) elicits threat, attack, and screeching and, if the stimulus duration is increased, eventually flight. von Holst elicited other sequences of behavior, such as crowing by cockerels, by stimulating other hypothalamic and limbic sites.

CONFLICT BEHAVIORS

In the normal environment, many different and conflicting stimuli impinge on an animal simultaneously, and stimuli arising internally, such as hunger, may conflict with those arising externally, such as threat from a predator. If it is to survive, the animal must resolve such conflicts rapidly and respond adaptively. It is not generally understood that

SOME ETHOLOGICAL CONCEPTS

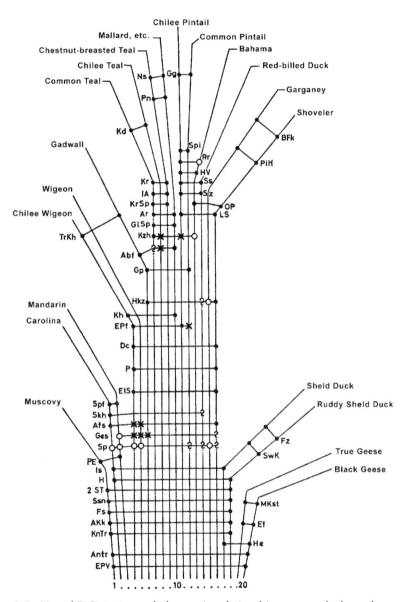

FIGURE 2-5. Use of FAPs to trace phylogenetic relationships among ducks and geese (see text). Crosses indicate the absence of a behavior, and circles indicate unusual differentiation of the behavior. (SOURCE: Modified from Lorenz, 1965c. Copyright © 1965 Piper Verlag, GmbH, München)

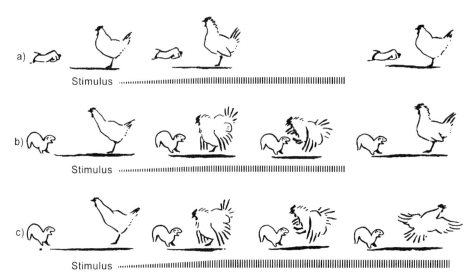

FIGURE 2-6. Brain stimulation in a domestic hen elicited mild unrest in response to a human fist (top) and predator defense in response to a stuffed polecat (middle and bottom). Shaded horizontal bars give the onset and duration of the electrical stimulus. (SOURCE: Fig. 32 in von Holst & von Saint Paul, 1960. Copyright © 1960 by Springer-Verlag)

the initial approaches between two conspecifics of almost any species are frequently associated with much ambivalence and conflict; the human is certainly no exception. Many courtship rituals in vertebrates are derived from the conflict between sexual attraction on the one hand and either fear or aggression on the other. This is particularly obvious in seasonally breeding species that become territorially aggressive during the mating season but must avoid driving away a prospective mate that initially arouses both aggressive and sexual behavior.

Behavior patterns arising from the simultaneous activation of incompatible motivations are classified into six categories: (1) redirection, (2) displacement, (3) intention movement, (4) alternation, (5) ambivalence, and (6) compromise.

Redirection

Redirected behavior occurs when a single stimulus simultaneously arouses two conflicting tendencies such as attack and flight or attack and sexual activity. The animal performs the behavior appropriate to one tendency, but redirects it *away* from the animal that elicits it. For example, a female may at first elicit an attack from a territorial male, but the attack is redirected away from the female onto another conspecific or even onto an inanimate object in the environment. This redirection occurs during the early phases of courtship in many different species, including fishes, birds, nonprimate mammals, and macaques. If courtship is to proceed successfully, it is essential that the initial attacks of the male are redirected and do not end up in an attack on the female. The

FIGURE 2-7. Redirected threats during greeting in a male–female pair of blackheaded gulls. (Photo by N. Tinbergen, from Eibl-Eibesfeldt, 1975)

potential mate may facilitate this by joining in with the redirected attacks and threats, so that the male and female threaten together in parallel side by side (Fig. 2-7). This joint redirection can now, of course, become directed at a third individual or at some irrelevant feature of the environment. It appears to have a strong bonding effect and also helps to keep other conspecifics at bay. If aggression cannot be redirected for some reason, courtship breaks down and direct aggression occurs, which, in captivity, can prove fatal in certain species if escape distances are too short. Those familiar with cichlid fishes know that some species do not breed if a male–female pair has no visual access to other conspecifics in the same or neighboring tank. In the absence of this, the male continues to attack the female and eventually kills her. Redirection occurs in many contexts; children, and also adults, will punch or kick an inanimate object when angered by an authority figure they dare not attack directly. Redirected aggression travels down the dominance hierarchy; scapegoating of religious minorities and of women and children (whipping boys) are familiar examples.

Displacement

Animals in conflict situations perform abbreviated segments of a behavior that seems totally irrelevant to the situation in which they find themselves. Displacement activity involves the sudden appearance of a completely different type of behavior, often a self-maintenance activity such as scratching, preening, self-grooming, feeding, or drinking. During a dog fight, one dog may suddenly sit down and scratch itself quickly before resuming the fight; the scratching seems to have nothing to do with either of the underlying tendencies for attack and flight. There was once a theory that displacement activities were the result of a "sparking-over" of energy from both of the thwarted tendencies into a third, irrelevant tendency (Fig. 2-8, top). When attack (Motivation A) and flight (Motivation B) are mutually inhibitory, their combined energy was thought to spark over to activate both the motivation (Motivation C) and hence the behavior (Displacement activity C) of scratching. It was later observed that the type of displacement activity occurring depended on the presence of its normal eliciting stimulus; for example, fighting turkeys in approach–avoidance conflict may drink in the presence of water but may feed in the presence of food. The current view is that lower priority maintenance behavior (e.g., scratching in the dog) appears as a displacement activity when the mutual inhibition of the two conflicting higher-priority tendencies (attack and flight) decreases the inhibitory effects of both on the low priority behavior (Fig. 2-8, bottom). Human examples include elements of self-grooming, for example, the inevitable tie straightening or hair smoothing by a speaker stepping onto the podium to give an address. It is quite common for people to snack in stressful situations, and this, just like lighting a cigarette, may be a form of displacement behavior.

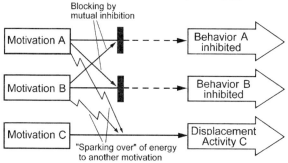

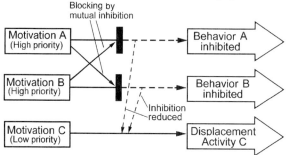

FIGURE 2-8. Original displacement hypothesis (top) and more recent disinhibition hypothesis (bottom) of displacement behavior.

(Attention must be drawn to a semantic matter that could cause confusion: Displacement is the term used by psychiatrists and psychoanalysts for what ethologists term redirection.)

Intention Movement

Intention movements are abbreviated, initial components of an entire behavioral sequence, for example, a forward lunge as the first component of the intention to attack, and opening the jaws as the first component of a bite. Intention movements occur in conflict situations when one behavioral tendency partially overrides the other, such as fear partially overriding aggression in the examples just given. A job applicant seated in an armchair during an interview may lean forward and place his or her hands on the chair's arms in an intention movement to depart, but does not do so because of the conflict between leaving a stressful interview and staying to secure the job. In addition, intention movements may appear when motivation is too low to elicit the entire sequence. For example, someone seated at home in an armchair late at night may show the same intention movement when halfhearted about rising to turn off the television and go to bed. The way to distinguish between an intention movement due to conflict and that due to low motivation is by the context in which it occurs.

Alternation

Alternation occurs when first one and then another competing tendency briefly overrides the other, so that the animal alternates between the two appropriate behavioral responses, for example, between approach and avoidance. A classic example is the zigzag courtship "dance" of the male three-spined stickleback (*Gasterosteus aculeatus*) (Fig. 2-9), thought to be derived from an alternation between swimming toward the female in his territory to attack (aggressive component) and swimming toward his nest, where he leads the female for spawning (sexual component). Similar directional changes in locomotion may be observed in someone torn between seeing the end of a mystery movie on television and departing for the dining room when called for supper.

Ambivalence

Ambivalent behavior results when intention movements appropriate for each of two conflicting tendencies are combined into a single pattern. Analyses of facial expressions and body postures in several mammalian species have documented this derivation from components of behavior of varying intensities. Figure 2-10 illustrates differences in the body postures (top) and facial expressions (bottom) of the domestic cat. In each case, aggressive tendencies increase from left to right, and submissive tendencies increase from top to bottom. A young dog in training may approach its master, wagging its tail vigorously (pleasure), but at the same time adopt a crouching posture, with flattened ears (fear of reprimand). As in nonhuman animals, a characteristic of human ambivalence is that opposing impulses are expressed almost simultaneously. In this way, an intended caress becomes an inadvertent scratch and an affectionate nibble or kiss may degenerate into a bite.

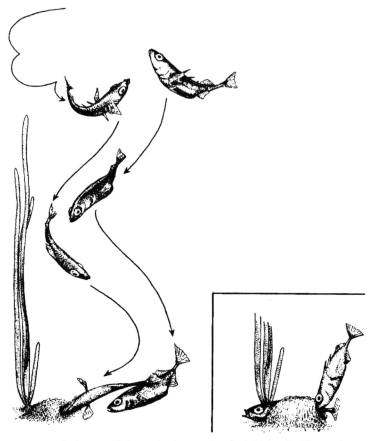

FIGURE 2-9. Mating behavior of the male three-spined stickleback. Alternation occurs during the zigzag dance, indicated by the line in top-left corner of the drawing. (SOURCE: Reprinted from *The Study of Instinct* by N. Tinbergen (1951) with permission of Oxford University Press)

Compromise

Compromise occurs when an animal shows behavior that is only partially appropriate for each conflicting behavioral tendency. For example, a bird may raise and lower its body on a branch while extending its wings, the intention movements for taking off, which are appropriate for flying either toward a conspecific to attack or away from it to flee.

Conflict in Psychiatry

Unacceptable sexual and aggressive thoughts and impulses are usually dealt with successfully by repression. But the repressed material may continue to seek expression,

SOME ETHOLOGICAL CONCEPTS 29

FIGURE 2-10. Ambivalence between aggression and fear in the postures (top) and facial expressions (bottom) of the cat. (SOURCE: Leyhausen, 1973, with permission)

and this can result in considerable emotional discomfort and anxiety, and even disruptive symptoms. The need to discharge this affect results in the unconscious conflict becoming preconscious and more accessible in fantasies and dreams. There are several well-recognized defense mechanisms used to reduce the pain of conflicts and, as in nonhumans, we see displacements, redirections, and ambivalence.

VACUUM ACTIVITY

Vacuum activity is a behavior that occurs spontaneously in the absence of an eliciting stimulus. Lorenz coined the term when he saw his tame starling, which always lived in a large room in his home and was fed all its life from a bowl containing boiled eggs, fly up to the ceiling, snap at a nonexistent fly and swallow it in the unmistakable species-typical sequence. When deprived of the natural stimulus, an innate behavioral response may eventually be performed anyway. The owners of dogs will be familiar with the way in which they often run to the end of the yard on being let out and bark vigorously, as if an intruder (dog or human) were in their territory, although neither is present. The owner is left to wonder what the fuss is about. Another example is the threat behavior of caged rhesus monkeys. If one observes them in a large observation cage, they periodically interrupt their activities to threaten at the observer. For this reason, observers are usually seated behind an angled, one-way vision mirror to prevent animals from seeing the observers and their own reflections, but the threat behavior, now oriented in different directions, persists. The animal, even when alone, will fixate its gaze on some point outside the cage and then begin to threaten at it. At first one might think there must be something there to elicit these threats, but this is not so. If monkeys in observation cages are monitored by closed-circuit television, the same threat behavior builds up and occurs at intervals; it is a vacuum activity.

RITUALIZATION AND DISPLAYS

Many conflict behaviors, as well as autonomic responses such as piloerection (hair raising), have become specialized in form and frequency during phylogeny and are then said to have become ritualized into displays. Ritualization is an evolutionary process. The selection pressures responsible for ritualization are partly understood and partly speculative. Three major possibilities are mentioned here: (1) Ritualization may decrease the ambiguity of the signal and enhance communication, the original view; (2) it may hide the true motivation of the signal's sender (Krebs & Davies, 1978); and (3) it may facilitate the comparison of the qualities of competitors in sexual selection (Zahavi, 1975). It is probable that different selection pressures have shaped displays in different species. Ritualization may involve several or all of the following specializations: (1) stereotypy, (2) typical intensity, (3) association with conspicuous morphological features, (4) threshold changes, and (5) motivational changes.

Stereotypy

Displays can become inflexible and stereotyped (sometimes termed "formalized"). This involves (1) the loss of some components of the original movements, (2) a change in the rate of performance, the rate of rhythmic repetition, or both. In ducks, "inciting" is a female courtship behavior consisting of threats redirected away from the male partner onto a neighboring pair. The female sheld duck threatens directly toward another pair (Fig. 2-11, top), but the female mallard always turns her head back over her wing while threatening, whether the other pair is behind, to one side, or even in front of her. Close inspection shows that the female mallard always fixates on the pair being threatened, and that her sideward head movements encompass a larger angle if the other pair is farther to the rear (Lorenz, 1963) (Fig. 2-11, bottom). The stereotypy involved in evolutionary ritualization, unlike cage stereotypy, is characterized by the fact that displays are shared by all conspecifics of the same age and sex class, and meet other criteria for FAPs.

Typical Intensity

In general, the intensity of a display parallels changes in the strength of the tendency to perform it. By analogy, during a meal, one is more likely to eat more of a favored food than of one that is disliked, and more likely to eat a disliked food 24 hours after than 4 hours after the last meal. This type of relationship is almost eliminated in a behavior having a typical intensity, in which the form of the behavior changes little whether the tendency to perform it is high or low (Fig. 2-12). By analogy, without typical intensity, our vocal response to mild discomfort, moderate pain, or unbearable agony would be virtual silence, low moaning, or screaming, respectively, whereas with typical intensity, it would be screaming in all three circumstances. At the expense of some loss of informational content, the standardization brought about by a typical intensity enhances the signal value of the display and minimizes its chances of being misunderstood.

Association with Conspicuous Morphological Features

Displays may have become associated during phylogeny with very conspicuous morphological features that exaggerate and draw attention to the behavior; there has been the coevolution of mutually enhancing behavioral and morphological traits. A courtship display by cocks of all species of game birds, including domestic fowl, pheasants, turkeys, and peacocks, has evolved from pecking at food on the ground, which attracts the female. Pecking exposes the back and the spread tail of the cock to the hen, which stands immediately in front of him. In all species, the tail feathers of the cock are more elaborate and colorful than those of the hen, and the peacock exhibits the most remarkable and famous specialization of its tail feathers (Fig. 2-13). Fiddler crab males attract females by waving one claw up and down, and this signal is greatly enhanced by the fact that the claw that makes the signal is very much larger than the other.

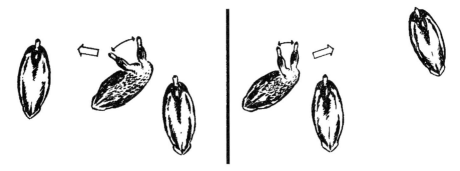

FIGURE 2-11. Inciting female sheld ducks (top) threaten directly at the other pair. The female mallard (bottom), however, always threatens over her wing, whether the other pair is to her left, behind her (left), or virtually in front of her (right). The open arrows indicate the female's direction of gaze, and the two-headed arrows point at the two extremes of her alternating head positions. (SOURCE: Konrad Lorenz, 1963, *Das Sogenannte Böse*. Copyright © 1983 by Deutscher Taschenbuch Verlag, München)

Threshold Changes

Ritualization may also involve changes in the threshold of homologous displays in closely related species; that is, the display may come to require a higher intensity of motivation in some species than in others. This may be so extreme that the display disappears entirely in some species, only to reappear in hybrids. For example, the "down-up display" of many species of duck during courtship is absent in both pintail and yellowbilled ducks, but the display reappeared in hybrids between the two.

Motivational Changes

Ritualization may also result in a complete change in the motivations that elicit the behavior. For example, a display based on an intention movement to attack might come to be expressed almost entirely in a sexual situation. A consort pair of male and female rhesus monkeys frequently expresses redirected aggression toward other individuals, and this redirected aggression also occurs as a vacuum activity in the absence of other

SOME ETHOLOGICAL CONCEPTS 33

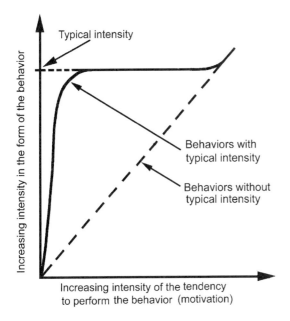

FIGURE 2-12. Graph illustrating typical intensity.

FIGURE 2-13. Gallinacious cocks with elaborated tail feathers. From top left: domestic rooster, ring-necked pheasant, Impeyan pheasant, peacock pheasant, and peacock. (After Schenkel, 1956)

FIGURE 2-14. A male and female rhesus monkey in a cage redirect their threats onto an environmental feature outside the cage ("threatening-away").

conspecifics just before and during a copulatory sequence (Fig. 2-14). The form of the behavior is indistinguishable from that of ordinary, direct aggression, but now the context is a sexual one. The sexual motivation of both partners was manipulated by treating the ovariectomized female with different combinations of estradiol (OEST) and progesterone (PROG) (Fig. 2-15). When untreated, females were unattractive to males, which made few mounting attempts (top panel), and were unreceptive to male mounting attempts, as shown by high number of female refusals (bottom panel). In general, OEST treatment reversed these effects, increasing the sexual motivation of both males and females, while additional PROG treatment decreased the sexual motivation of both partners again. However, male sexual motivation remained high when females were given systemic estrogen together with PROG, but the PROG decreased female sexual motivation, as shown by the increase in refusals (third histograms from left). In all five treatment conditions, male threatening-away frequencies were positively correlated with male mounting attempts, while female threatening-away was negatively correlated with refusals. In other words, in both sexes threatening-away (Fig. 2-15, middle two panels) increased with increasing sexual motivation and decreased with decreasing sexual motivation, from which it can be concluded that the behavior is related primarily to sexual rather than aggressive motivation.

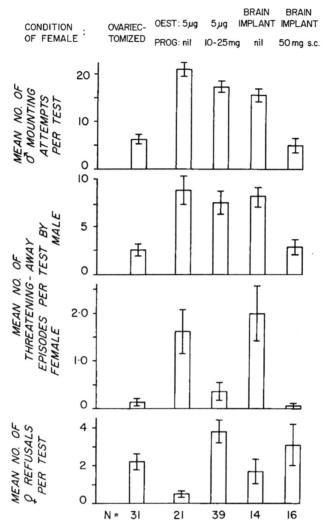

FIGURE 2-15. In both male and female rhesus monkeys, "threatening-away" was correlated with measures of sexual motivation that were altered by manipulations of the female's hormonal status. (SOURCE: Zumpe & Michael, 1970, with permission)

CHAPTER 3

Some More Ethological Concepts

SIGN STIMULI, RELEASERS, AND INNATE RELEASING MECHANISMS

Chapter 2 dealt with specializations in the sender of signals, but there are also genetically programmed specializations in the receiver of signals coming from the environment or from social companions. So there can be a genetic basis for receiving as well as sending signals with adaptive significance. This is important because of the many hundreds of bits of sensory information with which animals are bombarded every moment of the day and night. Specialization in the receiver requires information reduction brought about by selective filtering, which helps to separate relevant from irrelevant information. This filtering can be considered on two separate levels: perception and attention.

Perception versus Attention

Early in this century, von Uexküll (1921) pointed out very graphically that each species experiences its environment in a distinctive way. We look at our town in spring and may see green grass, tall trees with white or pink blossoms, colorful azaleas, and other flowers. Our dog, however, perceives a very different world. Dogs are color-blind but live in a fascinating environment of scents: the urine of a bitch in heat, squirrel tracks, the odor of the person who rang your doorbell that morning, and so on. Filtering occurs at the level of the perceptual capabilities of the animal, determined by its sense organs and central nervous system (CNS); these change little during life. Further filtering depends on what the animal is paying attention to at a particular time; this changes from moment to moment. von Uexküll emphasized these distinctions by contrasting an animal's total environment (*Umwelt*) with its perceptual environment (*Merkwelt*) and its significant environment (*Wirkwelt*). All three differ for different species but are generally similar for all members of the same species. Clarifying a species' *Merkwelt* is a task for physiologists, while understanding its *Wirkwelt* is a task for ethologists.

Definitions

Cues or signals, to which all members of a species respond with the same FAP the first time in their lives they are exposed to them, are known as sign stimuli or releasers. Some authorities reserve the term *sign stimulus* for environmental cues or signals from other species, whereas the term *releaser* is reserved for signals from conspecifics. Together, the signal and the response are known as an innate releasing mechanism or IRM, a concept developed by Lorenz and Tinbergen (Lorenz, 1981). The term was defined by Tinbergen as a "special neurosensory mechanism that releases the reaction and is responsible for its [the reaction's] selective susceptibility to a very special combination of sign stimuli." An analogy would be a key-and-lock mechanism, the sign stimulus or releaser being the key that "unlocks" a very specific FAP.

Properties of Sign Stimuli and Releasers

A sign stimulus may be a very simple environmental feature, such as the edge of a precipice that naive kittens and chicks will hesitate to step over, even when the chasm is covered by a glass plate. Or it may be a simple characteristic of another species, identifying it as a prey or a predator; for example, any moving object of appropriate size and speed elicits the prey-catching response of dragonfly larvae, frogs, toads, and various fishes. Frogs have neurons in the striate (visual) cortex specialized for this type of detection (fly detector neurons). A releaser may also be a simple morphological or behavioral characteristic, often an FAP, that identifies a conspecific, a parent, an offspring, a potential competitor, or a mate. The classic example of this comes from Tinbergen's (1951) experiments on the courtship behavior of sticklebacks. During the breeding season, males become territorial and develop a red belly. By using various models, Tinbergen found that the red belly of the male elicits attacks from other males (Fig. 3-1), but the swollen belly of the gravid female elicits courtship (Fig. 3-2).

Sign stimuli or releasers may comprise a configurational stimulus, such as a particular ratio between head and body areas of cardboard models of the parent, independent of absolute size. This ratio must be maintained to elicit the food-begging response of blackbird chicks (*Turdus merula*) (Fig. 3-3). In other cases, what was first thought to be a configurational stimulus turned out to be the summation of two or more independent stimuli, called the law of heterogeneous summation. Thus, food-begging by herring gull chicks depends on two independent signals: the red spot near the tip of the parent's bill and its rate of horizontal movement.

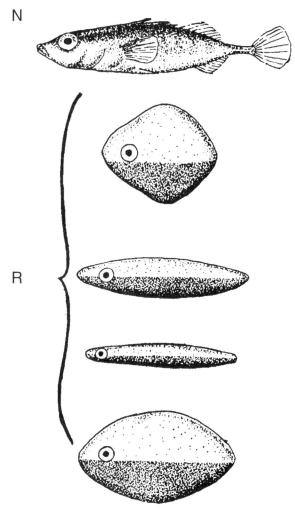

FIGURE 3-1. All models with red underneath (dark shading) elicited aggression by male sticklebacks, but a naturalistic model without red underneath (top) did not. (SOURCE: *The Study of Instinct* by N. Tinbergen, 1951, with permission of the Oxford University Press)

These are simple bits of information that can be encoded in the responses of individual neurons, and there are compelling data suggesting that more complex releasers also depend upon neuronal specialization. Many primates, including young humans, perceive prolonged eye contact as a threat. Between 3 and 4 months of age, rhesus monkeys reared in isolation suddenly show decreased lever pressing to see pictures of monkeys making threats. Infrared corneal refraction studies have shown that monkeys, like humans, pay great attention to faces, especially to the eye region. Single unit recordings of the activity of neurons in the brains of monkeys looking at pictures and schematics have helped elucidate the neurophysiology of this behavioral specialization. Individual neurons in the upper and lower banks of the superior temporal sulcus of the cerebral cortex respond selectively to different facial configurations, for example, either to full face or to half profile (head averted 45°) or to faces making eye contact or to

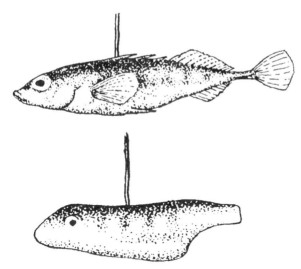

FIGURE 3-2. A crude model with a swollen belly (bottom) elicited courtship by male sticklebacks, but a naturalistic model without a swollen belly (top) did not. (SOURCE: *The Study of Instinct* by N. Tinbergen, 1951, with permission of the Oxford University Press)

those with eyes averted 45° (Fig. 3-4). We now know that there are two distinct populations of neurons. The first, mainly in the upper and lower banks of the superior temporal sulcus, is primarily responsive to facial expressions of different emotions (calm vs. threatening), independent of the identity of the individual monkey expressing them. The second group of neurons, in the inferior temporal cortex, is primarily

FIGURE 3-3. Configurational stimulus elicits food begging by blackbird chicks if the proportion between the "head" and "body" of the model is naturalistic. Model a elicited begging when the smaller appendage acted as the "head", while model b elicited begging when the larger appendage acted as the "head". (SOURCE: Modified from Tinbergen & Kuenen, 1939, with permission)

responsive to the identity of the other monkey, independent of its facial expression (Hasselmo et al., 1989) (Fig. 3-5). Clearly, monkeys have the neural apparatus to perceive different facial features, enabling them to recognize their social companions individually and also changes in the facial expressions that reflect altered motivational states such as attack or flight. Humans may have similar neuronal specializations. Those who have suffered trauma in these brain regions are unable to recognize their own faces and the faces of relatives and friends, although they recognize familiar people by gait and voice; this distressing condition is called prosopagnosia.

Modifying Influences

Very early learning, as well as habituation, can quickly modify the responses to a sign stimulus or releaser. Associative learning may be involved. For example, a freshly metamorphosed frog needs to snap only once or twice at a wasp to learn that it stings and tastes bad. Habituation was found to affect the responses of turkey chicks to a cardboard silhouette of a bird passing over them. When moved in one direction, the silhouette resembled that of a bird of prey, a natural predator, but when moved in the opposite direction, it resembled that of a goose or duck (Fig. 3-6). Working outdoors, where ducks and geese often flew over, Lorenz found that 9-week-old turkey chicks either fled into cover or ignored the model depending on its direction of movement and the relative

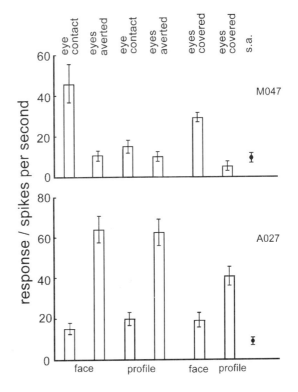

FIGURE 3-4. Selective activation of two neurons in the temporal cortex of a rhesus monkey being shown different facial features. With eyes visible, neuron M047 was activated by full face and eye contact, while neuron A027 was activated by averted eyes. With eyes covered, both neurons still showed a corresponding effect of head orientation. The dots on the right give the neurons' mean baseline activity, and the vertical bars give standard errors of means. (SOURCE: Perrett et al., 1985, with permission)

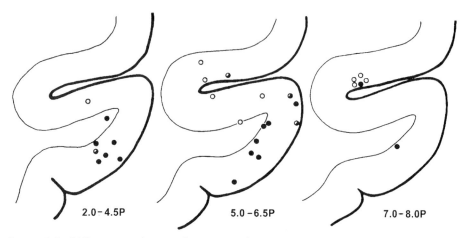

FIGURE 3-5. Different populations of neurons in the temporal cortex of rhesus monkeys were responsive to facial expressions of emotion (open circles) or to the identity of the monkey showing them (solid circles). (SOURCE: Modified from Hasselmo et al., 1989. The role of expression and identity in the face-selective responses of neurons in the temporal visual cortex of the monkey. *Behav. Brain Res.* **32**, 203–218. Copyright © 1989, with permission from Elsevier Science)

speed of travel (diameters/time unit). But working indoors in the same location, Schleidt (1961) found that direction was not a critical factor in newly hatched turkey chicks, and that they quickly habituated to the configuration presented most frequently. Outdoors, the chicks had already become habituated to the overflights of ducks and geese. Habituation to "goose" silhouettes occurred more quickly than habituation to "hawk"

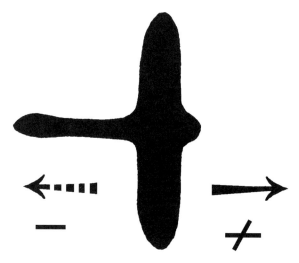

FIGURE 3-6. A cardboard model resembling a duck or goose when pulled overhead to the left was ignored by turkey chicks, which fled for cover when the model was pulled to the right and resembled a raptor. (SOURCE: *The Study of Instinct* by N. Tinbergen, 1951, with permission of the Oxford University Press)

silhouettes, indicating that the habituation effect is genetically programmed. In contrast to the flexibility shown to environmental inputs, other IRMs in the same species can be rigid and not easily modified. Turkey hens will brood practically anything that emits the calls of turkey chicks, including a stuffed polecat (a ground predator) equipped with a loudspeaker playing back chick calls. Those same hens, when experimentally deprived of hearing, will kill their own chicks. The flexibility of the response to bird silhouettes is adaptive because different habitats may have differently shaped raptors, but it is difficult to imagine any selection pressures against the rigidity of the brooding response, since mutant deaf or mute turkeys are unlikely to survive.

The motivational state of the animal also modifies its response to a releaser. This is clearly seen in seasonal breeding species, which do not perform or respond to courtship displays unless primed by hormones during the mating season. When there are small, progressive changes in the motivational state, IRMs can appear to be rather flexible, as if the response to a given releaser varies. But this may not be so when the changes in motivational state are taken into account. In male guppies, color patterns change with increasing sexual motivation, probably under the influence of hormones, and larger females are more effective releasers of male courtship displays than are smaller ones. For each of several intensities of the male guppy's courtship display, there was a direct relationship between male sexual motivation and the body size of the female eliciting the display: As male sexual motivation increased, progressively smaller females evoked the display response (Baerends et al., 1955).

Supranormal Stimuli

In the case of intraspecific IRMs, that is, when the releaser is provided by members of the same species, the releaser and the response have probably coevolved, and until there are selection pressures operating against this, the positive feedback effects of one upon the other can lead to extremes such as the immense claw of male fiddler crabs, and the large and elaborate tail of the peacock. Absence of negative selection pressures on IRMs can result in supranormal releasers, in which an exaggerated version of the normal releaser has greater stimulus value than the normal releaser itself. For example, ground-nesting birds such as geese and oystercatchers retrieve eggs that are just outside the rim of the nest and will attempt to retrieve a giant egg model in preference to their own natural egg (Fig. 3-7). Many humans find exaggerations of the "infant schema" more appealing than naturally occurring infantile characteristics, including rounded heads and small faces (Chapter 17), and this is used in advertising. The tendency for exaggerations to be more effective than naturally occurring releasers seems inherent in most intraspecific IRMs and provides the impetus for evolutionary changes in releasers until they are halted by physiological or anatomical constraints.

Advantages and Disadvantages of Innate Releasing Mechanisms

IRMs have advantages in that they enable naive individuals to make fast responses when learning processes may be too slow to assure survival or reproductive success, and also when the ability to learn is limited by a primitive CNS. Disadvantages are based on

FIGURE 3-7. The supranormal stimulus of an oversized egg model releases egg retrieving by an oystercatcher, which ignores its own, natural egg (foreground) and a herring gull's egg (left). (SOURCE: *The Study of Instinct* by N. Tinbergen, 1951, with permission of the Oxford University Press)

the rigidity and inflexibility of IRMs that leave animals open to mimicry (Chapter 9) and parasitism by others. The young of certain birds that are brood parasites, which includecuckoos and cowbirds, have almost identical markings inside the bill and throat as the nestlings of their host species. The host parents' rigid feeding response to this releaser, especially when the parasitic chicks outgrow the host chicks, can severely reduce their reproductive success. In humans, the preference for fatty, salty, and sweet-tasting foods, which were difficult to obtain during most of human evolution, can be regarded as IRMs, since they are consumed to excess (with a negative impact on health) in societies having access to great quantities of them.

PROGRAMMED LEARNING AND IMPRINTING

We now consider just two types of learning that show particularly clear evidence of genetic constraint, namely, programmed learning and a specialized form of programmed learning called imprinting.

Programmed Learning

American experimental psychology had its roots in the school of behaviorism, which held that behavior can be understood in terms of conditioned reflexes built upon unconditioned reflexes, so that the behavior of any individual can be shaped at will by appropriate positive and negative reinforcements (Chapter 1). But some researchers began to recognize what had long been familiar to animal trainers, namely, that species differ profoundly in what they can and cannot learn, independent of anatomical differences and the size and complexity of their brains. For example, it was noted that rats could be conditioned to avoid the smell and taste of certain foods by making them ill after ingestion of such foods, and that they could be conditioned by an auditory stimulus (a click) to avoid pain (electric shock). But they could not be conditioned by a click to avoid food that would make them ill, or by the smell and taste of nauseating food to avoid pain. So rats could associate smell and taste with subsequent sickness, and a sound with subsequent pain—associations that make sense in the natural world—but could not associate a sound with nausea or a smell with pain, associations that rats would be unlikely to encounter in their real world. Many of us have experienced a long-lasting distaste for certain foods that, by chance, were ingested just before an unpleasant bout of nausea and vomiting, whatever the latter's cause (seasickness, gastric flu, or hangover). Animals are predisposed (programmed) to learn certain things in certain contexts and are constrained from learning them in others, hence the term programmed learning. Humans learn well with both positive and negative reinforcement (carrot and stick); pigs do not. Pigs learn well with positive reinforcement in the form of food rewards but very poorly with punishment.

An example of naturally occurring programmed learning comes from three closely related species of seabirds that differ with respect to their nesting habits. Guillemots nest in the cracks and ledges of steep cliffs, so their nests are crammed close together. Royal terns nest on the ground in very dense colonies so that their nests are also very close together. Both species are capable of learning to recognize their own eggs, so they do not incubate the eggs of others when they return from the sea after feeding. Herring gulls, on the other hand, nest on the ground in colonies whose nests are quite far apart. They recognize their nest location but not their eggs; however, they do recognize their own chicks and do not feed others that might wander near or into the nest.

Imprinting

This phenomenon is now regarded as a specialized form of programmed learning, usually occurring early in life. It was discovered and named by Lorenz in the 1980s when he hand-reared birds (jackdaws, ducks, and geese) and found that they subsequently treated humans as parents and, on maturity, as prospective mates. Numerous studies have shown that imprinting has a number of characteristics, some of which are shared with other forms of learning: (1) a sensitive (critical) period for the acquisition of information; (2) the acquisition of information for use later in life; (3) retention of the information for life; (4) the acquisition of supraindividual, species-specific information; and (5) response specificity. There are three major categories of imprinting: filial, sexual, and parental.

Sensitive or Critical Period

This is a restricted period, usually very early in life, when special information is acquired. The information identifies the parent and, subsequently, appropriate mates. In effect, the animal learns to which species it belongs. A goose or duck that is not imprinted on its own species is destined not to reproduce. In ground-nesting birds with precocial young, the imprinted chick follows the imprinting object, normally the mother, and the sensitive period occurs within hours of hatching, between 5 and 22 hours, with a peak at 13 hours in the domestic chick. After the sensitive period, imprinting is usually no longer possible (but see zebra finches below). Laboratory studies have shown that chicks may be imprinted on the most unlikely objects, for example, cylinders with spiral stripes, as long as these are the only moving objects within sight during the sensitive period. It is also very important that the imprinting object emits sounds similar to the maternal call.

Retention of Information for Life

Information acquired during imprinting is retained for life; this stands in contrast to other forms of learning in which the acquired information tends to be lost or "forgotten" with the passage of time. Male wood ducks that were imprinted on other male wood ducks continued to court each other every spring, although females were available; each male showed male courtship behaviors but none showed female courtship behaviors. These males courted and mated females when housed with them in the absence of other males but immediately reverted to courting males when allowed access.

Acquisition of Information for Use Later in Life

The acquisition of information about the parent is termed filial imprinting, and the acquisition of information about future mates is termed sexual imprinting. In ducks and several other species, filial imprinting occurs within hours of hatching, and sexual imprinting takes place shortly thereafter, during the first few days of life. The information acquired during sexual imprinting, unlike that of filial imprinting, is not actually used until maturation occurs some months later. Similarly, song imprinting in sparrows and certain other birds occurs well before the chick itself starts to sing.

Supraindividual, Species-Specific Information

The information acquired by imprinting other than parental imprinting (see below) is not restricted to characteristics unique to the individual upon which an animal is imprinted (the imprinting object) but generalizes to all members of that individual's species. Thus, a goose or duck sexually imprinted on Lorenz would court any available human, while a male duck imprinted on a hen would court any available hen.

Response Specificity

Generally, only very specific responses are affected by imprinting. A hand-raised jackdaw directed filial and sexual responses toward humans but flew with hooded crows as flight companions and directed parental behavior toward young jackdaws.

The third category of imprinting, parental imprinting by the parent on its offspring, differs from filial and sexual imprinting in several respects. The sensitive period occurs in the adult, is closely linked to the time when the parent gives birth (see below), and may occur more than once in the lifetime of the individual, with each new litter or clutch. The imprinted information is acquired at a time when it can be used. Whether or not the information is retained for life is questionable.

Originally, the imprinting process was conceptualized as a sort of patchwork depending on a mixture of innate predispositions and learned ones—an instinct-learning intercalation. Subsequent work on song learning in male zebra finches has suggested that imprinting is best understood as a finely tuned and surprisingly flexible developmental process. When male zebra finches were raised by mixed pairs of zebra and Bengalese finches, they showed marked sexual preferences for female zebra finches, suggesting an innate preference for conspecifics. But it was observed that, due to a parental preference, the zebra finch parent interacted more than the Bengalese finch parent with the zebra finch chicks, and experiments demonstrated that sexual preference for Bengalese finches depends on the amount of interaction with other Bengalese finches during the sensitive period (Fig. 3-8). While the normal sensitive period lasts from about day 35 to 65 days posthatching if a song "tutor" is present, it can be delayed to 6 months of age if a song "tutor" is withheld until then. There is also evidence that the imprinting individual influences the acquisition of information. Male zebra finches raised in the presence of adult male Bengalese finches, whose song is softer and more complex than that of zebra finches, were observed to show intense listening behavior after day 35.

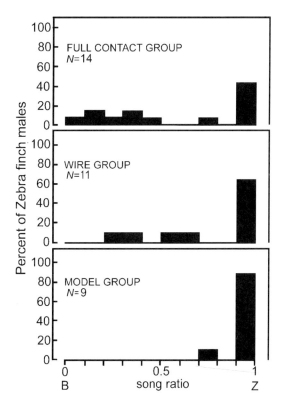

FIGURE 3-8. Sexual preference of male zebra finches for Bengalese finches, measured by the proportion of Bengalese finch song components in their songs, was greatest when male zebra finches had full contact with a family of Bengalese finches (top), less when they were separated from it by a wire-mesh barrier (middle), and virtually absent when they were exposed to a group of stuffed Bengalese finches (bottom). B = 100% Bengalese finch song components; Z = 100% zebra finch song components. (SOURCE: ten Cate, 1989, with permission)

Chicks and ducklings will press a lever more vigorously to obtain access to a more effective imprinting stimulus than to a less effective one.

Imprinting in birds generally relies on auditory and visual cues. In mammals, olfactory stimuli appear to be very important for maternal imprinting in goats and ewes, which depends on the odor of neonates. The mother imprints on the odor during a sensitive period, lasting a few hours after birth, under the influence of increased oxytocin levels and stimulation of the uterine cervix by expansion during the birth process (Kendrick *et al.*, 1997). There is some evidence that a maternal imprinting-like process occurs in the human. Some 90% of new mothers exposed to their newborns for 10–60 minutes following parturition could reliably discriminate between the odor of their own infant's garment and the odor of garments from two other neonates (Kaitz *et al.*, 1987).

In summary, imprinting appears to be a finely tuned, somewhat flexible, yet genetically constrained developmental process rather than the all-or-nothing, fixed type of learning, as originally conceived. The study of filial imprinting in birds has led to the proposition that there is a neural network involving three systems: The first is concerned with analyzing the features of novel stimuli, the second with the recognition of familiar stimuli in terms of their features, and the third with the execution of the behavioral response (Bateson & Horn, 1994). This concept of imprinting makes it more, rather than less, likely that certain kinds of programmed learning in the human, such as the acquisition of language and gender identity, may also be imprinting-like processes.

DRIVE OR MOTIVATION

The terms *drive* and *motivation* lost popularity at one time in the 1930s, because so-called "vitalists" such as McDougall, Tolman, and Russell postulated unfailing and inexplicable instincts as the final causes of behavior—a position that effectively bars the path to further research and understanding, a hindrance rather than a help. Nevertheless, under certain circumstances, they can be useful hypothetical constructs and help us to define relationships between changes in environmental and behavioral variables. The idea of drive, although an abstraction, can lead to working hypotheses about the possible physiological mechanisms involved.

The probability of engaging in a certain type of behavior is rarely constant; usually, it is changing. If a change is unidirectional, it is generally attributed to learning. Once a frog catches a bee or wasp and experiences the consequences, the probability of its doing so again decreases sharply and does not increase thereafter. Usually, the tendency to perform a behavior fluctuates back and forth, increasing at one time and decreasing at another, and the concept of drive or motivation can help relate changing behavioral tendencies to changes in environmental or physiological variables. The following observations led to the development of drive models: (1) The same stimulus elicits responses of different intensities at different times; (2) the frequency of a response changes under constant external conditions; (3) the strength of a stimulus necessary to elicit the response (threshold) varies with time; (4) an animal responds to different stimuli at different times; (5) the strength of an aversive stimulus, such as electric shock, that is needed for avoidance varies at different times; (6) the amount of work performed

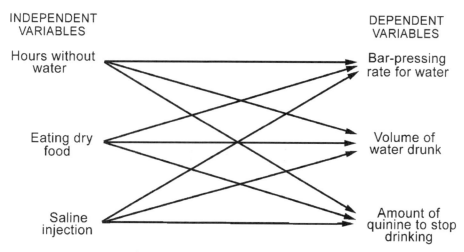

FIGURE 3-9. Interactions between three independent and three dependent variables of drinking behavior in rats. (SOURCE: Modified from Miller, 1959, in S. Koch, ed., *Psychology: A Study of Science.* Copyright © 1959, with permission of the McGraw-Hill Companies)

(operant behavior) to obtain access to a rewarding stimulus also varies; and (7) vacuum activities.

In all these cases, one can postulate changes in a hypothetical internal variable, first to assess the relationship between the stimulus and the response, and second to formulate hypotheses as to what the actual physiological mechanisms might be. This can be illustrated with drinking behavior in rats. It has been shown that three independent variables—the duration of water deprivation, the consumption of dry food, and the administration of saline—all affect three dependent variables of drinking behavior. These are the rate of lever pressing for access to water, the volume of water drunk, and the amount of quinine (bitter taste) in the water needed to stop the rat drinking. Without the concept of a hypothetical intervening variable, namely, thirst (drive), a total of nine relationships need to be considered (Fig. 3-9), whereas the inclusion of thirst reduces relationships to six and simplifies conceptualization (Fig. 3-10) (Miller, 1959). When it is found that the dependent variables are not tightly correlated (Fig. 3-11), it becomes clear that a unitary thirst drive is an oversimplification and that several partly independent physiological processes are involved.

Illustrative Model for Thirst

Much is known about the physiology of thirst. Because of perspiration and excretion, we continually lose water and electrolytes. Exercise is dehydrating, and the problem for humans and other animals is to maintain the total volume of blood and fluid in the extracellular and intracellular spaces. This requires the conservation and replacement of electrolytes, principally sodium and potassium ions. When this replacement is

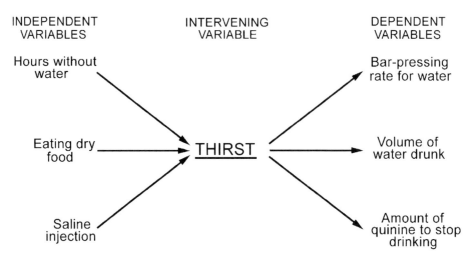

FIGURE 3-10. Postulating a thirst drive as an intervening variable simplifies the relationships and helps to generate testable hypotheses. (SOURCE: Modified from Miller, 1959, in S. Koch, ed., *Psychology: A Study of Science.* Copyright © 1959, with permission of the McGraw-Hill Companies)

needed, we experience mild to extreme thirst, which can be regarded as an expression of the drive or motivation to drink. Locating fluid and taking it into the mouth are appetitive behaviors involved in finding the releasing stimulus for the consummatory response of swallowing, which results in stomach distension and satiates thirst. But it may take 30 minutes to an hour for the fluid volume in the different body spaces to be restored and to equilibrate. To attain the homeostasis that the drinking drive restores, there are osmo-

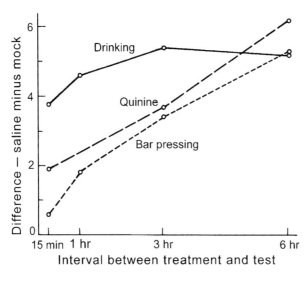

FIGURE 3-11. The three dependent variables of thirst are not tightly correlated in the rat. Saline = test following intragastric saline administration; mock = control test. (SOURCE: Miller, 1956, with permission)

receptors and sodium receptors in the supraoptic nucleus (SON) of the hypothalamus that, through the supraoptico-hypophyseal tract, regulate the secretion of antidiuretic hormone (ADH) by the pars nervosa (posterior lobe) of the pituitary gland, and its release severely restricts water loss by the kidneys. This occurs after exercise. In the condition of diabetes insipidus, on the other hand, in which ADH cannot be released, the patient secretes up to ten times the normal volume of very dilute urine daily and drinks ten times more than normal. Just anterior and lateral to the SON is a small brain region, regarded as the thirst center, from which drinking can be induced by both electrical stimulation of the area and direct microinjection of hypertonic saline. Also important for thirst, drinking, and fluid balance is the renin–angiotensin–aldosterone system. Renin is released from the juxtaglomerular cells of the kidney when blood volume is reduced and blood flow through the kidneys is lowered. This results in an enzymatic conversion to angiotensin I and II from a protein precursor. The former have direct effects on the kidney's nephrons by decreasing salt and water secretion and, separately, by stimulating aldosterone secretion from the zona glomerulosa cells of the cortex of the adrenal gland. Aldosterone, a mineralocorticoid, increases sodium reabsorption by the kidney tubules and so conserves sodium ions in extracellular fluid; it is also important for the control of extracellular potassium ion concentrations. We have considered thirst in some detail because it exemplifies the interrelation between the drive concept and the physiology underlying it. Another example, sexual motivation, will be considered similarly because so much is known about the physiology of reproduction (Chapter 5).

Drive Models

It is of historical interest that both Freud, with his neurological and psychiatric background, and Lorenz, the ethologist and student of animal behavior who was influenced by Freud's ideas, saw the drive concept of motivation in terms of a simple hydraulic model, along the lines of the old-fashioned toilet cistern (Fig. 3-12). The readiness to perform a behavior was thought to be produced internally (endogenously). This idea arose from the observation that the strength of the stimulus needed to elicit a behavior decreases when the animal has been prevented from performing it for a long time; this is known as a deprivation effect. An extreme case of this is a vacuum activity that occurs without a stimulus. In the simple model (Fig. 3-12, A), the continuous endogenous release (ER) of what was termed "action-specific potential" (Asp) gradually raises its level in the cistern. This empties when the animal encounters an appropriate sign stimulus or releaser (SR) and the behavior is performed. In the absence of an appropriate SR, Asp in the reservoir increases to a maximum, which also "opens the spring valve," resulting in a vacuum activity. Once the reservoir empties, the behavior cannot be triggered at all, or only with difficulty, until the Asp again reaches a critical level. The somewhat more physiological model (Fig. 3-12, B) is modified to take into account the fact that other factors (AR) can increase the readiness to perform a behavior, and that these differ from SR only in the time required to elicit the behavioral response. The opening of the valve in this model depends exclusively on the level of Asp in the reservoir.

While this model had its origins in what was known at the time about the "all-or-nothing" firing of neurons, it did not gain much popularity. This was in part because

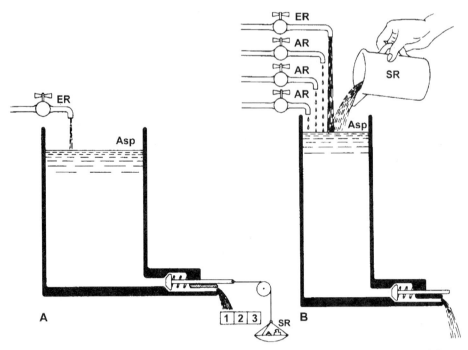

FIGURE 3-12. Original (A) and modified (B) versions of Lorenz's hydraulic drive model (see text for explanation). Asp = action-specific potential, ER = endogenous release of Asp, SR = sign stimulus or releaser, AR = other factors increasing the tendency to perform the behavior. (SOURCE: Lorenz, 1981, with permission)

studies by physiological psychologists on feeding and drinking in rats failed to yield results compatible with the model. Drinking studies measured drive intensity by the amount of water drunk, namely, by the consummatory response. This did not take into account that the consummatory response can be affected by the performance of the appetitive behaviors themselves. This is illustrated by infant feeding. When human babies are fed from bottles of milk whose nipples have a large hole that requires little suckling effort, they quickly empty the bottle but continue to be restless and to cry. When fed from bottles whose nipples have a small hole that requires much more suckling effort, they do not empty the bottle completely and, despite a lower food intake, fall contentedly asleep. Observations showing that both appetitive and consummatory responses need expression led Tinbergen to develop the hydraulic model into a more generalized one, namely, the hierarchical organization of behavior. In the reproductive behavior of the male stickleback, for example, Tinbergen regarded the spring migration to shallow, warm water, elicited by hormonal changes induced by environmental factors such as increasing day length and temperature (Marshall, 1936, see Chapter 5), as a first-order behavior, territory establishment, elicited by the releaser of other red-bellied males, as a second-order behavior, and fighting, nest building, courtship, mating, and parental behavior, all elicited by their respective releasers, as third-order behaviors. The individual behaviors in each third-order behavior, for example, displaying, chasing, and biting, in the case of fighting, were regarded as fourth-order behaviors, and so on down to the activity of individual neurons and muscle fibers of the rays of one fin. He was mistaken in his view that we would find neural centers and mechanisms corresponding to these different levels of control, an expectation that has not been realized.

Despite its limitations, we have seen that the drive concept is helpful for defining relationships between independent and dependent variables. It is also useful for classifying and relating different dependent variables, for example, when constructing an ethogram. As noted earlier, many courtship displays are ritualized conflict behaviors with aggressive components. A behavior may appear to be aggressive but occur in a sexual context. In these circumstances, lack of a correlation between the unknown behavior and a known aggressive behavior (threat) but a positive correlation with a known sexual behavior (copulation) can resolve the ambiguity and lead to a hypothesis that the behavior may depend on a physiological mechanism such as gonadal hormone secretion. This is illustrated in the example of threatening-away by rhesus monkeys (Chapter 2, Ritualization).

CHAPTER 4

Assessment of Hereditary Influences

The following discussion assumes that the reader has taken a course in introductory and Mendelian genetics and has a basic understanding of its terminology, principles, and role in evolution. Briefly, within the nucleus of every cell except the gametes, most species have pairs of homologous chromosomes, each consisting of a long strand of the molecule deoxyribonucleic acid (DNA), with a centromere near one end. One chromosome of each pair is derived from the father and the other from the mother, and both contain the same gene sequence. Perhaps the most important discovery in biological science during the twentieth century, for which the 1962 Nobel Prize for Physiology or Medicine was awarded to Watson, Crick, and Wilkins, was the structure of DNA, a double helix consisting of two intertwined sugar-phosphate strands connected by a series of four bases, adenine pairing with thymine, and cytosine with guanine (Watson & Crick, 1953). This structure showed how chromosomes replicate themselves, and thereby the genetic code they contain, during cell division. Chromosomes consist of many genes, each being a section of chromosome at a specific locus that contains instructions for producing a particular protein that, together with others, helps to determine the development, structures, and physiological functions of the organism. In addition to the autosomes, there are two sex chromosomes that are heterologous in one sex, for example, XY in male mammals and WZ in female birds. Humans have 46 chromosomes, 22 pairs of homologous autosomes, together with a pair of sex chromosomes.

A gene at a given locus can be polymorphic in that it produces somewhat different effects depending on which variants are present on the two homologous chromosomes. For example, a gene for hair color may have one allele coding for black, a second for brown, a third for chestnut, and a fourth for blonde. One allele can be dominant over another that is recessive for the trait. If black hair is dominant and blonde hair is recessive, an individual with an allele for black and an allele for blonde is heterozygous for that gene and will have black hair, while a sibling with two alleles for blonde is homozygous for that gene and will have blonde hair. The genetic makeup is termed the genotype, while the traits produced by the genotype make up the phenotype of the individual. In sexual reproduction, the diploid gonadal cells, having pairs of chromosomes derived from each parent, divide by reduction division (meiosis) to produce haploid gametes with half the diploid number of chromosomes. These contain only one of each pair of chromosomes; the full complement is restored when the male and female gametes fuse during fertilization. The chromosomes in the gonadal cell are first replicated, giving rise to chromatids, and those from homologous chromosomes pair up together, at which time they may exchange sections, a process called recombination, before the cell divides. Each gamete therefore contains just one of each homologous chromosome pair, while another gamete may contain a chromosome that carries different alleles of some of its genes. The diploid zygote resulting from fertilization contains chromosomes from both parents, so in the absence of any spontaneous or induced mutations (structural change of a chromosome), offspring will share half of their alleles with each parent and also with their full siblings (Chapter 1, Fig. 1-2) unless the zygote splits to produce monozygotic twins, which are genetically identical. Consequently, there is much variation in the phenotype of members of sexually reproducing species. Further phenotypic variation may result from deletions and duplications of chromosome sections and from extra chromosomes, known as aneuploidy, for example, XYY in some tall men.

One of the two causes of evolutionary change thought to be most important is natural selection acting on phenotypic differences; individuals with a phenotype that is advantageous at a given time and place will survive and produce more offspring than will those with a less advantageous phenotype. The other main cause of evolutionary change is genetic drift, resulting from random fluctuations in allele frequencies in small (but not large) populations, because some individuals may by chance produce more offspring than others with more advantageous alleles, resulting in underrepresentation of the latter in the next generation. Thus, genetic drift can be regarded as the antithesis of the trends usually characteristic of natural selection. An extreme case is the founder effect, in which a new population originates from a very few individuals, so that the new population possesses a small fraction of the alleles present in the original population. Examples of founder effects are Darwin's finches on the Galapagos Islands or the human intersex condition in the Dominican Republic (Chapter 5, Clinical Syndromes). Other causes of evolutionary change are mutations, which occur infrequently and are often deleterious, and gene flow, when individuals moving from one population into another with rather different allele frequencies gradually change the latter during subsequent generations.

Genes produce proteins that are either the structural building blocks of the cell or enzymes that facilitate chemical reactions. Since all somatic cells contain the same genes, a major function of genes is to turn other genes on or off in different tissues and in response to physiological or environmental changes. In the special case of genomic imprinting, either the maternal or the paternal allele of a gene is expressed in the offspring; the other is silenced. Genes may have several different effects when they are said to be pleiotropic; they may mask the effects of other genes when they are said to be epistatic, and several genes may be necessary to produce a trait, in which case it is called a polygenic trait. Moreover, organisms do not develop and live in a vacuum, and many if not all traits are products of the interaction between the effects of genes and of environmental influences. For example, fishes in small aquaria and young rhesus monkeys in cages do not grow as large as conspecifics maintained in large environments; in honeybees, the amount of royal jelly fed to the larva determines whether it will develop into a sterile worker or a sexually reproducing queen; a mammal not allowed to see with one eye during infancy will be blind in that eye as an adult; and humans growing up in the absence of talking conspecifics cannot, as adults, learn to speak normally. For all these reasons, the connection between genes and behavior is both indirect and very complex. Nevertheless, there are both indirect and direct methods for examining the hypothesis that differences in the behavior of organisms are in part a consequence of the degree of genetic difference between them.

INDIRECT METHODS

Indirect methods for assessing hereditary influences on behavior are based on comparisons between individuals of similar genetic makeup subjected to different environmental conditions. Behavioral similarities reflect strong hereditary but weak environmental influences, whereas behavioral differences reflect strong environmental but weak hereditary influences. By behavioral similarities, we mean the form of a behavior as well as the context in which it occurs, for example, during feeding, fighting, courtship or parental care, and its proximate function, such as hunting prey, protecting the territory, attracting mates, or nourishing offspring. This is important because, while the individual can possess the genes for performing certain behaviors, the behavior may be activated only under specific conditions, for example, in the case of courtship at a certain time of year, under the influence of gonadal hormones, and in the presence of an appropriate mate. The first four of the major categories of indirect methods listed below have been used to characterize many ethological concepts described in Chapters 2 and 3, so they are summarized only briefly here.

Presence in Geographically Isolated Populations

Many species live in populations whose members do not intermingle or interbreed because they are separated by geographical barriers such as oceans, rivers, or high mountains. There are, for example, populations of mallard ducks, starlings, gray squirrels, and black rats both in North America and in Europe. Any behavior patterns shared by both populations are likely to have been inherited from their ancestors before the populations separated; in fact, the behavioral similarities are so close that individuals from one population readily interbreed with one from another. Among humans, certain communicatory behaviors such as laughing, smiling, crying, and the "eyebrow" flash during greeting (Chapter 17), occur in identical form throughout the world, even in populations that have until recently had few contacts with any others.

Presence in Individuals Raised in Isolation from Conspecifics

The offspring of many invertebrate and some vertebrate species develop in the absence of parents and other conspecifics that can protect them and serve as behavioral models. The role of behavioral evolution is obscured in species with parental care and social organizations, and in such cases, experiments in which an individual is raised in isolation from conspecifics, either on its own or fostered onto another species, have been informative. However, these experiments demand knowledge about the life history and normal behavior of the species and extraordinary care on the part of the experimenter, especially when total isolation is involved. For example, naive rats raised in bare metal cages showed the species-specific nest-building behavior when supplied with straw or crepe paper in their home cages but did not when moved to an unfamiliar test cage in which a sleeping area had not yet been established, and a knothole at a duckling's eye level in the wooden enclosure separating fledglings from adult conspecifics actually vitiated an experiment on sexual imprinting. Fostering experiments are preferable because total isolation perturbs many aspects of development and, consequently, not just the behavior of interest. Many fostering or cross-fostering studies have been conducted with birds because fertilized eggs can be introduced for incubation and subsequent hatching into the nests of suitable foster species. These experiments have documented species-specific FAPs in ducklings and chickens (Chapter 2), as well as the importance of imprinting in the development of sexual preferences of male zebra finches (Chapter 3).

Phylogenetic Comparisons: Presence in Closely Related Species

If behavior patterns have a similar form and occur in similar contexts in related species, they are likely to be homologous and inherited from a common ancestor. Reproductive behavior patterns occurring during courtship, mating, and parental care are more likely to be evolutionarily conserved than others, for example, those involved in feeding. Comparisons of courtship displays and parental behaviors in related species with phylogenetic trees based on morphological traits have been made in several taxonomic groups including, among birds, the *Anseriformes* (ducks, geese, and swans, Chapter 2, Fig. 2-5) and *Pelecaniformes* (pelicans, cormorants, and gannets). In general, it has been found that related species sharing more morphological features also share more behavioral displays and vice versa. With modern techniques for assessing phylogenetic relationships, it should be possible to obtain more accurate information about the evolutionary history of behavior.

Association with Specialized Morphological or Physiological Traits

Many courtship and aggressive displays involve, or are released by, specialized morphological features. As indicated in the previous section, a close association between morphological and behavioral traits argues for the heritability of both. This is especially the case when a morphological trait appears to serve no function other than to elicit a very specific behavioral response from another individual, for example, the release of a male stickleback's attack by the red belly of another male (Chapter 3, Fig. 3-1), or to emphasize and exaggerate a specific behavior pattern by its bearer, such as the peacock's huge tail spread during courtship (Chapter 2, Fig. 2-13) and the large claw that male fiddler crabs wave during courtship. As discussed in Chapter 13, both the courtship behaviors and their morphological correlates may result from sexual selection and in some cases seem to have coevolved with female preferences for them.

Studies on Human Twins

Because monozygotic twins result from the splitting of a single fertilized egg and are therefore genetically identical, with 100% of their alleles in common, they provide a unique opportunity for assessing the heritability of morphological, personality, behavioral, and IQ traits, as well as various medical conditions. Comparisons are often made (1) with fraternal, dizygotic twins, a result of the implantation of two fertilized eggs, that have only 50% of their alleles in common but share the intrauterine environment and subsequent rearing environments unless reared apart, (2) with siblings, which also have 50% of their alleles in common and share the rearing but not the intrauterine environment, and, to minimize all shared environmental influences as much as possible, (3) *between* identical twins reared apart from birth. As might be expected, such studies have shown that identical twins, even when raised in different environments, generally share more traits than do fraternal twins or siblings. There are also differences between identical twins even when reared together. These differences may result from environmental influences that are probably perceived, and responded to, in ways that are subtly and uniquely different for every human being. The incidence of male homosexuality in the human is by most estimates between 4% and 10% of the general population (Chapter 14). In one family study on sexual orientation by men, for example, 52% of monozygotic twins, 22% of dizygotic male twins, and 11% of unrelated, adoptive brothers of homosexual or bisexual men were also homosexual (Bailey & Pillard, 1991). In another study, increased rates of homosexuality were found among maternal uncles and male cousins but not among fathers or paternal relatives of homosexual men in 114 families, implicating maternal transmission via the X chromosome. DNA-linkage analyses in a subset of 40 families demonstrated an association between sexual orientation and inheritance of polymorphic markers on the X chromosome in about 64% of the pairs of homosexual brothers tested; the genetic loading was statistically highly significant (Hamer *et al.*, 1993). Although this is an important first step in locating a region containing several hundred genes that may contribute to homosexual orientation in a subpopulation, the work needs further confirmation.

DIRECT METHODS

Direct methods for assessing hereditary influences on behavior use an opposite approach to that used in indirect methods. Under the same environmental conditions, the behavior of individuals whose genetic constitution has been changed is compared with that of unchanged individuals of the same species or strain, so that any behavioral differences can be attributed primarily to the genetic differences.

Artificial Selection

Humans have domesticated a variety of animals for thousands of years and, in the process, used artificial selection to maximize desirable traits and minimize undesirable ones, and to produce different strains or "breeds" of the same species that have different functions, for example, beef cattle and dairy cattle, or cart horses and racehorses. Dogs are believed to have descended from wolves and to have been domesticated for thousands of years, and we are familiar with many different breeds, from Chihuahuas to Great Danes, whose morphological and behavioral traits vary enormously. For example, Dachshunds were bred for chasing badgers out of their burrows, retrievers for finding and bringing back game birds killed by the hunter, Sheepdogs for herding sheep, and Pit Bull Terriers for fighting. The morphological differences among these breeds are associated with equally distinct behavioral differences that suit their different functions.

Artificial selection involves mating together those individuals in a population that happen to possess the trait of choice, and doing so for many generations. If the trait has a genetic basis, it will, with time, become more frequent in the population. This has been shown in laboratory studies in which environmental influences can be held far more constant than was the case during the thousands of years of domestication. In a classic study on rats, for example, all rats in a population were tested for their ability to navigate a particular T-maze containing 17 blind alleys, as measured by the time taken to complete the run and the numbers of errors and errorless runs. The "bright" rats, making few errors, were mated with each other, as were the "dull" rats that made many errors. Their offspring were tested in the same maze and, again, the bright rats were mated with each other, as were the dull rats. This process was repeated for several generations; by the seventh generation, there was virtually no overlap between the performances of the bright and dull rats (Fig. 4-1). It should be noted that it is not clear which ability was selected out in this study. It was thought to be learning ability or intelligence, but subsequent research has found that while the progeny of the two rat strains performed as expected in the original maze, there were no differences in their performance when tested in a different type of maze.

Inbreeding

Another experimental approach for separating genetic and environmental influences on behavior is to compare different inbred strains in the same environment and the same strains in different environments. While inbreeding can be used to select for a particular trait, its purpose when used to assess behavioral heritability is to produce several different populations, each composed of individuals that are genetically almost identical but different from those in other inbred populations. Even if the first male–female pairs chosen as founders for each strain are selected on the basis of a particular trait, for example, in laboratory rodents often for high or low activity levels or for low aggressiveness, they are likely to differ in many other traits, which depend solely on the alleles that the founders happened to carry. The homozygosity produced by inbreeding

ASSESSMENT OF HEREDITARY INFLUENCES 61

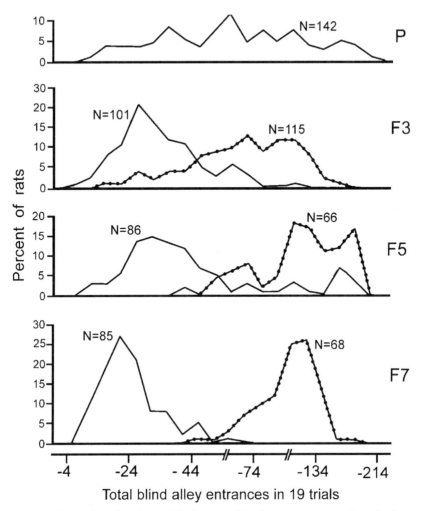

FIGURE 4-1. Selective breeding for "bright" rats (making few errors by entering blind alleys in a maze) and for "dull" rats (making many errors) from a heterogeneous parental sample (P) during the third (F3), fifth (F5) and seventh (F7) generations. N = number of rats. (SOURCE: Modified from Tryon, 1940, with permission)

may, of course, cause deleterious and even lethal recessive traits to come to the surface, but the alleles responsible for them will eventually be lost from the strain if they prevent survival or reproduction by the individuals carrying them. The offspring of each founder pair are then mated with their siblings, whose offspring are again mated with their siblings, and so on.

Genetic diversity decreases because only the alleles present in the founder pair will be retained in a given strain, and because heterozygosity decreases by about 25% in every subsequent generation of offspring from brother–sister matings. It has been estimated that by the twentieth generation, all individuals will have identical alleles for

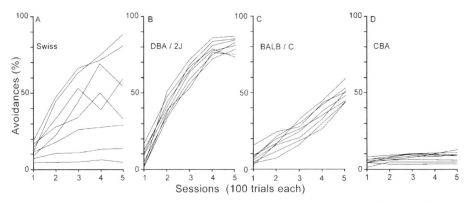

FIGURE 4-2. Individual differences in the acquisition of a foot shock avoidance task were large in heterogeneous Swiss Webster mice (A) but very small in each of three inbred mouse strains (B–D) that, between them, represented the highest, intermediate, and lowest acquisition rates observed in the heterogeneous population. Each line gives the percentage of avoidances by one mouse during successive test sessions. (SOURCE: Reprinted with permission from Bovet *et al.* (1969), Genetic aspects of learning and memory in mice. *Science* 163, 139–149. Copyright © 1969 by the American Association for the Advancement of Science)

98% of their genes. Differences between the genotypes of several different inbred strains derived from the same original population will be random but large. Figure 4-2 compares the acquisition of an operant avoidance task by individuals from a heterogeneous population of Swiss Webster mice with that of individuals from each of three inbred strains of mice (Bovet *et al.*, 1969). A mouse in a two-compartment shuttle box had to learn that it could avoid a light electrical shock from the floor of its compartment if it moved to the other compartment within 5 seconds of a light signal. The data showed large variability in the performance of individual Swiss Webster mice and also *between* the three inbred strains, but there was very little variance *within* each inbred strain. This, again, implicated the involvement of one or more genetically determined traits in the differences between the performance of the animals in this artificial task.

However, selective breeding and inbreeding studies with rats (Fig. 4-1) and mice (Fig. 4-2) should be interpreted with caution because mammals, by necessity, associate closely with their mothers until weaning, so that a mother's interactions with her pups may result in behavioral changes in her offspring that they in turn pass on to the next generation, producing data that seem to imply genetic transmission. A cross-fostering study (see Indirect Methods) in rats has demonstrated that differences in maternal care, resulting in differences in the stress reactivity of pups, can be transmitted nongenomically across generations. When visiting her nest, a mother rat typically collects the pups beneath her, licks and grooms them, then nurses them while continuing to lick and groom occasionally; some mothers do so more than others. Pups from "high" and "low" groomers were cross-fostered. When adults, pups of low-groomer mothers fostered to high-groomer mothers readily explored new surroundings, indicating a low level of fear, and showed high levels of grooming and licking toward their own pups, whereas high-groomer pups fostered to low-groomer mothers were fearful in new surroundings and showed low levels of grooming toward their pups. This demonstrated that the mothers' behavior, rather than genetic factors, determined mothering style and fearfulness. Human manipulation of pups can compensate for reduced maternal grooming and licking, and a follow-up study examined the effects of handling or not handling the pups of both high- and low-groomer mothers. While unhandled daughters of low-groomer mothers showed low levels of licking and grooming, those that were handled showed high levels of licking and grooming, similar to those in unhandled and handled daughters of high-groomer mothers (Fig. 4-3, top left), and the same pattern was observed in the next generation without pup handling (Fig. 4-3, bottom left). These behavioral differences were associated in the third generation with differences in behavioral (Fig. 4-3, top right) and physiological (Fig. 4-3, bottom right) measures of

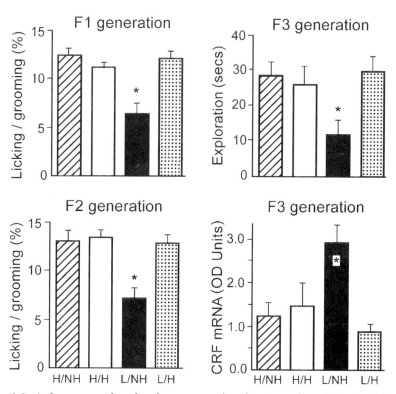

FIGURE 4-3. *Left*: compared with other groups, female rat mothers (F1) raised by low-groomer mothers and not handled (L/NH, black bars) showed significantly lower grooming and licking of their own pups, and this difference persisted into the next generation. *Right*: in the third generation, L/NH progeny explored less in an open field test and had more messenger ribonucleic acid (mRNA) for corticotropin-releasing factor (CRF) in the brain, indicating higher stress. H = high-groomer mother; L = low-groomer mother; H = F1 handled as pup; NH = F1 not handled as pup; * = significant difference. (SOURCE: Modified with permission from Francis et al. (1999), Nongenomic transmission across generations of maternal behavior and stress responses in the rat. *Science* 286, 1155–1158. Copyright © 1999 by the American Association for the Advancement of Science)

stress, both of which were greater in the progeny of unhandled low-groomer females than in any other group (Francis *et al.*, 1999). Similar profound and permanent effects on stress physiology have been found in the adult offspring of bonnet monkey (*Macaca radiata*) mothers subjected to a variable foraging demand to which they were unable to adjust (Chapter 9), so this phenomenon is possibly quite widespread among mammals, including humans.

Hybridization

Hybridization between different strains or closely related species can reveal behavioral traits that are genetically coded, for example, the appearance of the down–up courtship display in Lorenz's hybrid of pintail and yellow-billed ducks, which is absent in both parents (Chapter 2). Blackcap warblers (*Sylvia atricapilla*) are migratory European songbirds; those breeding in Germany migrate to West Africa via a southwesterly route, while those breeding in Austria take a southeasterly path to Ethiopia and Kenya (see similar migratory differences between West German and Baltic storks, Chapter 7). The migratory directions of the parental strains and their hybrids were revealed by scratch marks left by the birds, during their fall migratory restlessness, on sensitive paper in their funnel cages, whose tops provided a view of the night sky. The migratory orientation of the hybrids averaged almost due south, intermediate between the two parental strains (Helbig, 1991).

Hybridization between two Australian species of cricket, *Teleogryllus oceanicus* and *T. commodus*, has shown that characteristics of the courtship "song" of males, which vary between species and attract conspecific females but not females of other species, are genetically controlled. The songs consist of a chirp comprising 5 or 6 pulses in both species, followed by about 9 trills of 2 pulses each in *T. oceanicus*, and about 2 trills of 10–13 pulses each in *T. commodus*. Differences in song characteristics can be identified in three modes of interpulse interval: (1) the intratrill interval with the shortest duration, (2) the intrachirp interval with intermediate duration, and (3) the intertrill interval with the longest duration. Figure 4-4 compares these three modes of interpulse interval in the two parental species (top), and in the offspring of female *T. oceanicus* and male *T. commodus* (bottom left), and those of female *T. commodus* and male *T. oceanicus* (bottom right). In general, the song characteristics of the hybrids tended to fall between the two parental songs, suggesting a role for several genes, and backcrossing hybrids to each of the paternal species supported this view. However, the intertrill interval of hybrids clearly resembled that of the mother, indicating that this particular trait is influenced by genes on the X chromosome.

Molecular Changes

Spontaneous or induced genomic changes can in some cases also provide strong evidence for the genetic control of behavioral traits. As noted in Chapter 2, the finding that the development of wing flapping in mutant chicks without feathers was identical to that in normal chicks supported the view that flying is genetically programmed and not a consequence of learning from proprioceptive feedback. With the development of molecular biology, experimental manipulation of specific genes has become possible. One

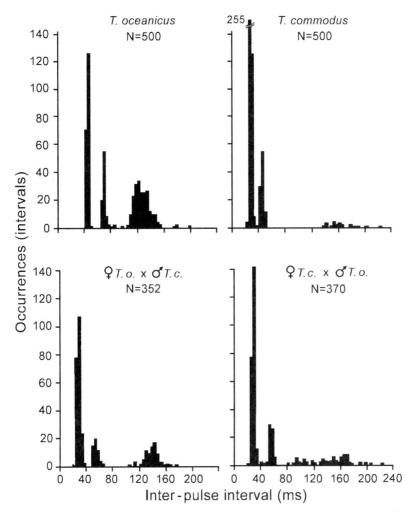

FIGURE 4-4. Distributions of intratrill (shortest), intrachirp (intermediate) and intertrill (longest) intervals in male crickets from the two parental species (top two panels) and in hybrid males (bottom two panels). The intertrill intervals of hybrids more closely resembled the maternal than the paternal species, suggesting an X-linked trait. (SOURCE: Bentley & Hoy, 1972, with permission)

approach is to delete a gene from the animal's DNA to produce a so-called "knock-out" animal; this is often done in mouse studies examining the behavioral role of a hormone or neurotransmitter by eliminating the gene or genes that express their receptors. This sometimes produces unexpected results, perhaps because of unknown interactions with other hormones or neurotransmitters, or because the deleted gene has pleiotropic effects.

The other, most powerful approach is to exchange alleles of a gene between two strains or species that differ in their phenotypic expressions of a particular behavioral trait. Like crickets, male fruitflies (*Drosophila*) attract conspecific females with a

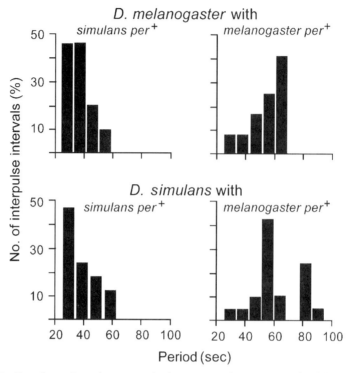

FIGURE 4-5. Transformation of mutant arhythmic *D. melanogaster* and wild-type *D. simulans* with the *per* gene from the latter resulted in predominantly short courtship song cycles, whereas transformations with the *per* gene from wild-type *D. melanogaster* resulted in predominantly long song cycles. (SOURCE: Reprinted with permission from Wheeler et al. (1991), Molecular transfer of a species-specific behavior from *Drosophila simulans* to *Drosophila melanogaster*. *Science* **251**, 1082–1085. Copyright © 1991 by the American Association for the Advancement of Science)

courtship "song" produced by vibration of their wings. Both the interpulse intervals and the average duration or period (Chapter 7) of cycles of interpulse intervals are species-specific, the latter being shorter (30–40 seconds) in *D. simulans* than in *D. melanogaster* (50–65 seconds). Mutant alleles of the *period* (*per* for short) gene on the X chromosome (together with the *time* gene) shorten, lengthen, or abolish circadian rhythms of eclosion (hatching from egg or pupa) and locomotor activity. Wheeler et al. (1991) demonstrated that it also controls the period of interpulse interval cycles in courtship songs. Germlines of mutant *D. melanogaster* without rhythmicity, and those of *D. simulans*, were each transformed by inserting *per* genes from wild-type *D. melanogaster* or from *D. simulans*. The results (Fig. 4-5) showed that male courtship song periods in both species resembled those typical for the donor species, being shorter in males transformed with *per* from *D. simulans* (Fig. 4-5, left) and longer in males transformed with *per* from wild-type *D. melanogaster* (Fig. 4-5, right).

CHAPTER 5
Behavioral Endocrinology
Gonadal Hormones

Hormones, particularly gonadal hormones, affect behavior by influencing (1) the development and activity of the CNS, for example, changing sexual and aggressive motivation; (2) sensory or perceptual mechanisms, for example, changing sensitivity to tactile and olfactory cues; and (3) effector mechanisms important for the elicitation and execution of the behavior, for example, the production of secondary sexual characteristics and the growth of antlers in deer.

Many hormones are produced by endocrine glands situated in one part of the body and transported via the blood stream to distant areas, where they bring about physiological changes in hormone target tissues. The *steroid hormones*, and also thyroid hormone, which is nonsteroidal, meet these criteria well. Steroids, derived from cholesterol, are fat-soluble, so they can pass readily through cell membranes, which are made of lipids. They can act on the cell membrane but for the most part act on the cell nucleus. The steroid hormones are closely related to each other and include the gonadal hormones (estrogens, progestins, and androgens) produced primarily by the ovary and testis, and the adrenal hormones (mineralocorticoids and glucocorticoids) produced by the adrenal cortex. In addition to the steroids, another category of hormones, the *peptide hormones*, are water-soluble and range in size from as few as three amino acids (thyroid-stimulating hormone) to those with over a hundred. The peptide hormones act primarily on the cell membrane. Examples are the gonadotropins such as follicle-stimulating hormone (FSH) and the decapeptide luteinizing hormone (LH), prolactin, and adrenocorticotropic hormone (ACTH), all of which are produced by secretory cells in the anterior pituitary gland when stimulated by the appropriate hypothalamic releasing factors. The anterior pituitary produces other important hormones, including growth hormone and melanophore-stimulating hormone. The gonadal hormones are the topic of this chapter and the adrenal hormones are dealt with in Chapter 6. Other hormones, first recognized by their neurosecretory granules, the *neurohormones* produced by the endoplasmic reticulum and Golgi apparatus in the bodies of neurons in the hypothalamus, travel along axons by axoplasmic flow directly to their targets. Neurons can use peptides, as well as neurotransmitters such as norepinephrine and dopamine, at the synapse and other sites; they are therefore referred to as peptidergic neurons. The majority of invertebrate hormones are produced by nerve cells and are not considered further. We restrict ourselves here to gonadal hormones in mammals. The pivotal role of the hypothalamus and pituitary gland, together with the feedback effects of the secretions of target glands, on neuroendocrine integration is illustrated in Figure 5-1.

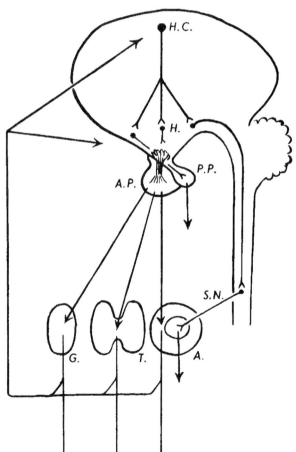

FIGURE 5-1. Diagram illustrating the reciprocal relationship between the central nervous system and the endocrine system. A. = adrenal gland; A.P. = anterior pituitary gland; G. = gonads (male and female); H. = hypothalamus; H.C. = "higher centers"; P.P. = posterior pituitary gland; S.N. = splanchnic nerves; T. = thyroid gland. (SOURCE: Harris, 1955, with permission)

SYNTHESIS AND MAJOR SITES OF PRODUCTION

As mentioned earlier, the gonadal hormones are closely related to each other structurally and all are derived from cholesterol (Fig. 5-2). Cholesterol is converted to progesterone, the main progestin in mammals, which in turn can be converted to the androgen testosterone. Testosterone may then be metabolized either to another androgen, 5α-dihydrotestosterone (DHT), or to the major estrogen, 17β-estradiol (estradiol) (see also Fig. 6-4).

Estrogens and Progestins

In females, estradiol is secreted by the granulosa cells of a developing Graafian follicle within the ovary, resulting in a gradual increase in the levels of estradiol in blood. Before ovulation, an increase in estradiol secretion by the granulosa cells brings about the preovulatory surge of LH secretion and a further increase in estradiol secretion, the so-called positive feedback effect of estrogen on pituitary function. This slow effect, taking 1–2 days, may be contrasted with the more familiar negative feedback effect of estrogen, which inhibits gonadotropin output rapidly in minutes or hours. These hormonal surges result in the rupture of the Graafian follicle and the release of the ovum from the ovary. Immediately thereafter, the granulosa cells become luteinized and

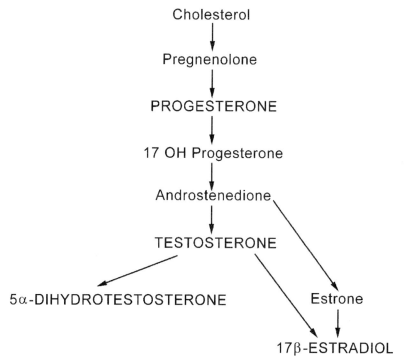

FIGURE 5-2. Simplified metabolic pathway of synthesis of the gonadal hormones.

increase in number to form a corpus luteum (yellow body), which starts to secrete progesterone, and this is associated with an abrupt decline in estradiol production. The ovary also produces a small amount of testosterone, which can become important clinically. The ovaries are largely inactive during embryonic life and do not secrete their hormones until the approach of puberty. In addition to the cyclic production of gonadal steroids by the ovary, there is a smaller, noncyclic production of estradiol, progesterone, and testosterone by the cortex of the adrenal gland (Chapter 6). It is known that cyclic ovarian function, which is responsible for the estrous cycle, depends upon the phasic secretion of gonadotropin-releasing hormone (GnRH) by the hypothalamus and FSH and LH by the anterior pituitary gland. It is not certain if there are separate FSH and LH releasing hormones, so the umbrella term GnRH is often used. These same hormones determine the menstrual cycle, which occurs only in Old World monkeys, apes, and humans. The menstrual cycle refers to the periodic breakdown of the inner layer of the uterus (endometrium), which desquamates about once a month and is then shed during menstruation. In normal health, the estrous (ovarian) cycle and menstrual (endometrial) cycle are interlocked, but there can be menstrual cycles without ovulation (anovular cycles) and also ovulation without menstruation. The length of the ovarian cycle, from ovulation to ovulation, varies widely in different mammals, from a few days to over a month. In addition to these cycles, there are more rapid pulses of gonadotropin secretion lasting a few minutes and occurring at approximately hourly intervals (circhoral rhythm). These pulses are not, in themselves, associated with ovulation. Changes in the frequency of these pulses over the 24-hour cycle cause the diurnal (circadian) rhythm (Chapter 7) in plasma hormone levels.

Androgens

In males, the Leydig cells of the interstitial tissue (lying between the seminiferous tubules) of the testis are one of the main sources of testosterone and other androgens. These cells become active early in embryonic development and the fetal testes play an important role in male sexual differentiation (see Organizational Effects). The Leydig cells secrete androgens during fetal life, unlike the granulosa cells of the fetal ovary, which do not secrete estrogens at this stage. The Leydig cells cease androgen production immediately after birth and then remain dormant until prepuberty. As in females, the adrenal cortex of males is the source of small amounts of testosterone as well as estradiol and progesterone, and both genetic abnormalities and adrenal tumors can result in virilization of the fetus by adrenal androgens (Chapter 6). It was thought until recently that the production of androgens by the testis was the consequence of the steady or tonic, acyclic secretion of hypothalamic and pituitary hormones, but we now know that testicular secretion is sustained by a series of rapid LH pulses occurring throughout the 24-hour cycle. The frequency of these pulses varies throughout the day and night, giving rise to a circadian rhythm of androgen production that, in men, results in higher blood androgen levels during the early morning. It has been established clearly in the rhesus monkey that the pulse generator for LH production resides in the arcuate nucleus region of the mediobasal hypothalamus, and pulsatile LH in the human is essential for testicular development and activity—an observation important for the treatment of hypogonadal males.

TRANSPORT

It is the free or unbound steroid moiety in blood that is the metabolically active form because it is available to diffuse into the tissues. Much the larger fraction of the steroid pool in plasma is, however, protein-bound in a way that conforms to a reversible binding equilibrium. The steroid is not metabolically active until the steroid–protein complex disassociates. Some high-affinity binding of estradiol and testosterone is to sex hormone binding globulin (SHBG). In addition, steroids bind to serum albumen with low affinity, but albumen is present in large amounts in plasma, so this binding is quantitatively important for estradiol. Estrogen is also bound to alpha-fetoprotein present in the blood of pregnant rats and in the blood of neonatal rats; alpha-fetoprotein declines to low levels before puberty. Gonadal steroids are excreted by the kidneys into the urine and by the liver into the bile, part of which is reabsorbed by the intestine, the enterohepatic circulation. Excretion is mostly in the form of sulfate and glucoronide conjugates. The most important steroids used by women in hormone replacement treatment are estrogens in a preparation called Premarin. This biological preparation comprises a mixture of several different estrogenic substances conjugated mainly as water soluble sulphates, all of which are present in large amounts in the urine of pregnant mares.

MECHANISMS OF ACTION

Hormone Receptors

It is now widely accepted that steroid hormones act by first binding to macromolecules called receptors in the target tissues where steroids exert physiological effects. The receptors are large protein molecules with specific binding sites for a particular steroid, and they are present in both the cytoplasm and the cell nucleus. Having bound the steroid, the steroid–receptor complex then binds to other sites on both the nuclear membrane and within the cell nucleus. By binding to acceptor sites near certain DNA (double strand) sequences, gene expression is altered and mRNA (single strand) transcription and processing result in translation and new protein formation and enzyme synthesis in the cytoplasm, a process in which the ribosomes take part. As a consequence, the metabolism of target tissues changes, tissue growth occurs, and differentiation into new structures is stimulated. The steroid is eventually inactivated, partly by metabolism to a steroid with lower affinity for the receptor, and then diffuses from the target cell. Gonadal hormone receptors are present throughout the body, including many brain regions, where hormones are thought to mediate changes in the motivation of the animal for sexual, aggressive, and other adaptive behaviors.

Effects on Tissues

Estrogens and Progestins

Estrogens, primarily estradiol, and progesterone are responsible for the growth and development of the female genital tract, most importantly, the growth of the uterus.

During the first minutes or hours after estrogen treatment, the uterus enlarges because of water retention and hyperemia (increased vascularity); there is influx of calcium ions and eosinophils, and an increase in RNA and protein synthesis. Subsequently, there is cell proliferation, true uterine hypertrophy, and growth. Estrogens have other effects: The epithelium of the vaginal wall increases in thickness, becoming multi-layered, and cells next to the vaginal lumen become cornified. Estrogens are also responsible for laying down the distribution of subcutaneous fat characteristic of the human female and act on the mammary glands to cause their development. In humans, these changes signal puberty. Estrogens have other widespread metabolic effects on motor activity, bone metabolism, and texture of the skin and hair, and they exert a "protective" effect on the cardiovascular system, which is to some extent lost at menopause and treated with hormone replacement therapy.

Progesterone secretion inhibits any additional estrogen-stimulated endometrial development, producing a secretory type of endometrium by causing development and proliferation of the tubular glands; there is also edema of the uterine stroma. Progesterone rapidly decreases estrogen receptor levels and also the level of its own receptor, which is called "down" regulation. All these changes prepare the uterus for pregnancy and successful implantation of the blastocyst. The effects of progesterone on the uterus depend on estrogen priming, without which the uterus remains insensitive to progesterone administration. The estrogen priming stimulates the production of the cytosolic progesterone receptor. Originally thought to antagonize the action of estrogen, progesterone is now more accurately regarded as a modulator.

Androgens

Androgens differ from estrogens because they may undergo metabolic changes and activation before receptor binding. Testosterone can act directly on the androgen receptor, just as estrogen acts on the estrogen receptor, but it may also undergo reduction by the enzyme 5α-reductase to DHT, which binds to the androgen receptor but may also have a receptor of its own. DHT has major actions on the skin, on bodily and facial hair, and acts very importantly on differentiation of the male genital tract, which is derived embryonically from the Wolffian duct system. Furthermore, testosterone is converted in the brain and elsewhere by aromatization to estradiol, which then binds to the estrogen receptor. This may seem somewhat paradoxical, but it is very important for sexual differentiation. Testosterone and DHT are responsible for male secondary sexual characteristics and male differentiation of Wolffian duct structures in the embryo. The development and growth of the scrotum, penis, prostate gland, and seminal vesicles depend on the androgens. Acting in concert with pituitary gonadotropins, testosterone promotes spermatogenesis by an action on the Sertoli cells of the testis. DHT acts more specifically on the skin to promote the male-type distribution of bodily and facial hair as well as on genital tract tissues. DHT is therefore essential for full male sexual differentiation.

In the brain and pituitary gland there are estrogen, progestin, and androgen receptors involved in the control of gonadotropin secretion and ovulation, as well as sexual and other types of behavior (described below).

Antagonists

It is worthwhile mentioning steroid antagonists, the antiestrogens and antiandrogens, because they now have importance in the treatment of disease. Most have both agonistic and antagonistic actions depending on dosage and duration of action. The best known antiestrogens are triphenylethylene derivatives such as clomiphene and tamoxifen. They bind to the estrogen receptor and act in part by competing for this substrate. Tamoxifen is important in the treatment of certain types of cancer in women. If a breast cancer contains estrogen receptors, it is likely to be hormone-sensitive and susceptible to antiestrogen treatment with tamoxifen as an adjunct to surgery and radiotherapy. The best-known antiandrogens are cyproterone acetate and flutamide, which bind to the androgen receptors and act by competitive inhibition. Benign prostatic hyperplasia, a troublesome condition prevalent in older men, is thought to be due to excessive conversion of testosterone to DHT locally within the gland and, in many cases, responds to antiandrogen treatment. Some types of prostate cancer also respond to antiandrogen administration.

Illustrative Clinical Syndromes

Two clinical syndromes in the human male exemplify the importance of enzymes and receptors, and both are genetically determined. The first, steroid 5α-reductase deficiency, is an autosomal recessive trait. Patients must be homozygous for this genetic defect to be expressed and they cannot produce DHT. They are usually classified as female at birth for the following reasons: Tissues and structures derived from the embryonic urogenital sinus and tubercle are female in appearance because there is no penis or scrotum, although structures derived from the Wolffian (male) duct system are present internally. There is partial virilization at puberty, when large amounts of testosterone are produced, but the main metabolic defect is the absence of DHT formation from testosterone because the reductase enzyme is missing. Without DHT, development of the scrotum, penis, male urethra, and prostate gland cannot occur. Although regarded initially during infancy and childhood as girls and treated as normal genetic females would be treated by family and friends (including clothing and hairstyles), affected individuals are masculine in orientation, and this identity develops more firmly both psychologically and somatically after puberty. Several examples of this condition were identified first in the Dominican Republic and subsequently in other parts of the world, and are due to a genetic founder effect. The second condition, the androgen insensitivity syndrome, formerly called testicular feminizing syndrome, occurs in genetic males whose androgen receptors are defective due to mutation of an X-linked gene specific for androgen receptors. The androgen receptors simply do not "see" the circulating androgens. There is a deficiency in the nuclear binding of androgens because cytosolic androgen receptors are absent. The tissues of these cases are completely unresponsive to androgens throughout uterine development and subsequent life, and because the basic mammalian somatotype is female (see Organizational Effects), the result is a completely feminized body and temperament; superficially, these cases are indistinguishable from normal females. These two syndromes demonstrate that if there is an enzyme deficiency preventing normal metabolic conversions, or if the receptor lacks functionality, very major changes in sexual differentiation result. For genetic defects in the female, see Chapter 6.

ORGANIZATIONAL EFFECTS DURING DEVELOPMENT

A useful contrast is sometimes made between the early effects of gonadal hormones during embryonic development, the so-called organizational effects which are generally permanent and irreversible, and the stimulatory effects of gonadal hormones occurring after puberty and during adult life, the so-called activational effects which are transient and generally last only as long as the hormones are circulating in the blood stream (see below). The organizational effects of gonadal hormones are critically important in the sexual differentiation of both structural and behavioral characteristics.

Sexual Differentiation: Somatic Modifications

In mammals, sexual differentiation is genetically determined by XX chromosome pairs for the female and XY chromosome pairs for the male. But in birds, the situation is reversed, the female being XY and the male XX, usually designated ZW and ZZ to prevent confusion. The basic somatotype in each vertebrate class is therefore homozygous for sex. By basic somatotype, we mean here the bodily form and characteristics that would develop solely under the influence of genes and without any additional influence of hormones. This, of course, occurs mostly in experimental conditions; it has been shown, for example, in hypophysectomized (pituitary removed) and gonadectomized (gonads removed) fetal rabbits that both genetic males and genetic females survive and develop with the female somatotype. The hormonal situation is therefore not "symmetrical" between the male and the female. In mammals, all anhormonal development results in female somatotypes. To produce males, the Y chromosome is needed to induce embryonic testes, which are critical for determining the male sex of the primordial germ cells and, eventually, the sex of the embryonic testes. Primordial germ cells have been identified in the human embryo 3–4 weeks postconception, before the woman may even be aware that she is pregnant. They arise outside the embryo in the endoderm of the yolk sac near the allantoic origin and migrate into the genital ridges of the embryo. In future males, through the action of the testis-determining factor (TDF) on the Y chromosome, primordial spermatogonia and primordial Sertoli cells develop and aggregate more centrally, resulting in primitive testes, while in genetic females, the peripheral or cortical region of the genital ridge develops more strongly as oogonia begin to develop in it, resulting in primitive ovaries (Fig. 5-3). The early Sertoli cells secrete Müllerian inhibiting factors (MIFs), sometimes called Müllerian regression factors (MRFs), which suppress development of the Müllerian (female) ducts, and the primitive Leydig cells begin to secrete testosterone at about 63–65 days gestation in the human. The testosterone causes differentiation of the Wolffian duct system, the urogenital sinus, and the external genitalia. One might emphasize again that the ovarian developmental pathway occurs by default in mammals in the absence of the Y chromosome and of any hormonal influences.

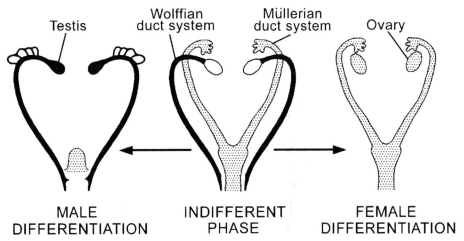

FIGURE 5-3. In early embryonic life, both duct systems are present—the indifferent phase (middle). When an embryonic testis develops, Wolffian duct structures are enhanced and Müllerian structures regress (left). In the absence of a fetal testis, Wolffian duct structures regress and female duct differentiation proceeds (right).

True hermaphrodites, which carry the gonads and germ cells of both sexes, are very rare. The duct systems and the tissues derived therefrom are relatively sensitive to hormonal influences, and pseudohermaphrodites, those with a mixture of male and female characteristics, are not rare. Changing the sex of the gonads themselves is a more difficult proposition. One of the best examples is the naturally occurring freemartin condition in calf twins. The normal twin is genetically male, while the freemartin twin is genetically female, having somatic cells that are chromatin-positive (Barr bodies). The external genitalia and mammary glands of the freemartin are female but there may be an enlarged penis-like clitoris. Internally, male duct structures are well preserved, while female duct structures are rudimentary. The gonads are either sterile testes or poorly developed ovotestes. For this condition to occur, it is essential that there be an anastomotic communication between the placental circulations of the twins, and the extent of the vascular interconnection determines the extent of the virilization. It is thought that the testosterone formed by the embryonic gonads of the male twin produces the masculinization of the female twin because the blood supply of the twins is comingled. There is also a freemartin condition in goats but it is different, being genetically determined. In rats, which produce large litters, it has been observed that there may be some degree of masculinization of the female fetus when it is situated in the uterine horn between male fetuses.

Sexual Differentiation: Behavioral Modifications

In addition to the structural or somatic modifications produced by hormones during development, it is now clear that hormones exert equally profound effects on behavior and on the neural mechanisms underlying it. It seems that there are sensitive periods during embryonic life when the developing fetus is particularly vulnerable to these effects, which do not occur when the sensitive period is over; this is reminiscent of imprinting. In species with relatively long gestations (guinea pig, 68 days; rhesus monkey, 164 days), the sensitive period occurs entirely *in utero*, but in species with relatively short gestations (rat, 22 days) the sensitive period may be extended several days postpartum. Testosterone administered either to the pregnant mother or directly to the fetus has masculinizing effects on behavior that may not become obvious until after puberty: Measures of male sexual behavior are increased, while those of estrous behavior are decreased.

In the primate also, the female fetus can be masculinized by injections of testosterone to the pregnant mother during the middle third of pregnancy and, at birth, the vagina is absent, and the phallus is developed but the scrotum is empty. These pseudohermaphroditic rhesus monkeys subsequently show behavior patterns (rough-and-tumble play, chasing play, threatening, and mounting) during the first few years of life that are intermediate between those of normal males and normal females (Fig. 5-4), although these differences tend to lessen as animals mature. After puberty, treated females show more mounting behavior than untreated females. These findings in primates prompted speculation about similar effects in the human female. We know that in congenital adrenal hyperplasia (Chapter 6) diagnosed in neonatal girls, both the neural and genital tissues are exposed to very high concentrations of adrenal androgens before and after birth. Follow-up studies have shown that the sexual orientation of these girls is not

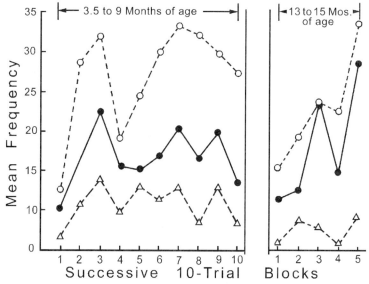

FIGURE 5-4. Rough-and-tumble play frequencies by pseudohermaphroditic female rhesus monkeys (closed circles) exposed to androgens during fetal life (approximately 40–90 days postconception) were intermediate between those of normal males (open circles) and normal females (open triangles). (SOURCE: Reprinted with permission from R. W. Goy in *Endocrinology and Human Behaviour*, edited by R. P. Michael, 1968. Copyright © 1968 by the Oxford University Press)

markedly changed in the male direction, and they do not have gender identity disorders. Nevertheless, some behaviors are masculinized to a mild degree: In childhood, they are regarded as tomboys, with high energy expenditure, more rough outdoor play, and a preference for boys rather than girls as playmates.

Another relevant situation is the medically induced (iatrogenic) condition resulting from treating pregnant mothers threatened by spontaneous abortion with synthetic progestins in combination with estrogens in an attempt to maintain the pregnancy. Some daughters showed masculinization of the clitoris and some degree of labial fusion. A second iatrogenic condition that caused considerable public consternation occurred when diethylstilbestrol (DES), the first orally active synthetic estrogen, was also used similarly to maintain pregnancy. A few of the daughters of mothers given DES during pregnancy developed vaginal and cervical cell dysplasia (abnormal cytology) and, in very few cases, adenocarcinoma. It was suggested that some of these DES daughters showed increased levels of bisexual behavior and were less maternal than controls, but the behavioral findings were not very convincing.

Sexual Dimorphism in Brain Structures

There are clear-cut and convincing structural differences between the brains of male and female rats and other mammals. A group of densely staining neurons, now termed the sexually dimorphic nucleus of the preoptic area (SDN-POA), is much larger in males than in females. Neonatal male castration results in smaller SDNs, while androgen administration to newborn females causes enlarged SDNs. So not only is there a morphological brain difference, but it is one that is sensitive to hormonal manipulations. This brain nucleus is present in the males of several mammalian species, including rats, gerbils, guinea pigs, and ferrets, but absent in others, and its precise function is not clear. Other neural structures showing sexual dimorphism include the spinal nucleus (motor) of the bulbocavernosus muscle, which subserves erection and is present in male rats but absent in females. Also, in birds such as canaries and zebra finches, the brain nuclei responsible for song by males are much larger than the corresponding structures in females. The corpus callosum is a large structure containing a mass of nerve fibers (white matter) connecting the left and right sides of the brain. It has been reported that the posterior part, or splenium of the corpus callosum, in women is larger and more bulbous than in men when measured in the midsagittal plane, although the overall size of the corpus callosum is similar in the two sexes. It is well known that cognitive function is lateralized in the human, speech being mainly controlled by the left cerebral hemisphere and spatial reasoning by the right hemisphere. It appears there is less lateralization of function in the female than in the male, and this has important clinical implications for the effects of strokes, brain tumors, and head injuries. Perhaps the greater communication between the left and right sides of the brain in females than in males is reflected in the larger size of the splenium of the female's corpus callosum.

ACTIVATIONAL EFFECTS IN ADULTS

As noted earlier, activational effects of gonadal hormones occur both during and after puberty, and in the adult; they are transient, generally lasting for as long as the hormones are circulating in the bloodstream, for example, throughout the mating season in males of many seasonally breeding mammals, and during the short periods of estrus in females. However, there are some exceptions; mating behavior and the gonadal hormones are not invariably associated in several species of fishes, reptiles, and birds. An example of dissociation in mammals are hibernating bats, in which mating occurs in the fall and sperm is stored in the female's genital tract until ovulation occurs in spring, after the winter hibernation.

There are two approaches to examining the effects of gonadal hormones on behavior: (1) correlating changes in behavior with changes in hormone levels in blood, and (2) removing the main source of hormones, the gonads, to examine whether behavior is changed, and subsequently administering hormones to examine whether the behavioral change is reversed. Both approaches are useful, but we should remember that a correlation between two variables cannot itself establish a causal relationship between them, because each variable may be regulated by other variables not being monitored. Even if a causal relationship between two variables is established, a correlation does not reveal the *direction* of causality. This problem is particularly relevant for hormone–behavior relationships, because we now know that hormonal changes affect behavior and also that behavioral changes can affect hormone levels.

Breeding Seasonality

Like other species, many mammals, including some primates, show breeding seasonality, the timing of which depends on changes in exteroceptive factors, a term introduced by Marshall (Marshall, 1936; Amoroso & Marshall, 1960) for extrinsic environmental stimuli, including day length, temperature, humidity, rainfall, as well as social stimulation by conspecifics, all of which influence an animal's physiology and consequently behavior. The timing of mammalian breeding seasonality depends primarily on day length but also on other exteroceptive factors, including temperature and rainfall (Chapter 14). For the majority of light-entrained cycles, whether they be short (circadian) or long (circannual), it appears that the suprachiasmatic nuclei of the hypothalamus play a crucial role, and another important role is played by melatonin secreted by the pineal gland (Chapter 7). Carnivores and rodents tend to breed in the spring, when daylength is increasing, while ungulates such as red deer, sheep, and goats tend to breed in the fall, when daylength is decreasing (see below). If animals are transported across the equator, sheep and rhesus monkeys, for example, the timing of the breeding season may be reversed by 6 months after a few seasons. In the nonbreeding season, the gonads of both sexes are quiescent but become active and secretory as the breeding season approaches.

Activational Effects in the Female

Female Cycles

Some mammalian species, including the rat, primate, and human, ovulate rhythmically, while others ovulate reflexly upon mating. In species such as rabbit, cat, and ferret, the coital stimulus reaches the brain of the female via the pudendal nerves and spinal cord, and thence to the hypothalamus, causing a surge of FSH and LH release. The precise role of the ovarian hormones varies much with different species. In the rat, progesterone acts synergistically with estradiol after estrogen has primed the tissues. In the ferret, progesterone antagonizes the actions of estrogen and terminates estrous (heat) behavior. In the sheep, progesterone priming is required before estradiol can have its stimulatory effect (the reverse of the rat), so the first estrus of the sheep breeding season is behaviorally "silent." Heat behavior in the ewe does not occur until the second cycle, when estradiol can act on the basis of the progesterone secreted by the corpus luteum of the first cycle. In the guinea pig, progesterone priming is also needed for estradiol to have its stimulatory behavioral effects. In the cat, progesterone has little effect either way on estrogen-induced behavior.

We have mentioned that the estrous cycle refers to ovarian events such as the ripening and rupture of Graafian follicles at estrus and corpus luteum formation. This cycle can be studied quite easily by collecting vaginal smear samples using pipettes or cotton swabs. After appropriate staining, vaginal smears give a microscopic indication of the status of the epithelium: This is really the "Pap" smear used to detect abnormal vaginal cytology in women. Based on the cytology, vaginal smear appearance can be divided into four phases: diestrus, proestrus, estrus, and metestrus. Diestrus is devoid of estrogenic stimulation, proestrus and estrus reflect increasing estrogen levels and increasing cornification, while the last phase, metestrus, is characterized by a marked infiltration of leukocytes into the vaginal wall and epithelium. A behavioral cycle is associated with the vaginal cycle. The majority of nonprimate mammals are behaviorally receptive to the male only during estrus, when females are maximally attractive to him. In proestrus, although not fully receptive, the female shows approach and invitational behavior, and is said to be proceptive; she also becomes sexually more attractive to the male. During diestrus and metestrus, the female is neither receptive nor attractive. The period of ovarian quiescence characterizing the nonbreeding season in seasonally breeding species is usually termed anestrus. If mating is sterile and fertilization does not occur, there may follow a period of pseudopregnancy during which the female remains unreceptive. In the absence of pregnancy, some species show consecutive estrous cycles and are said to be polyestrous, while other species have but one or two cycles in a breeding season and are said to be monestrous.

In rats and other species the timing of estrus is determined by the photoperiod (day length), and ovulation usually occurs every 4 or 5 days, depending on the strain of rat, at around 2 A.M., during the physically active phase of this nocturnal species, and sexual receptivity lasts about 2 hours. The coordination between the cycle of behavioral changes and the cycle of tissue changes in the vagina and uterus is exemplified well by the female cat (Figs. 5-5 and 5-6). In Old World primates, in contrast, vaginal cytology is much less clear-cut and more difficult to interpret. Mating may occur throughout the cycle, although there are peaks and troughs (Chapter 16), and the term *estrus* or heat is not generally so useful in these primates. (For clarification, *estrus* is the noun and *estrous* is the adjective, and the terms were first introduced to describe behavior.)

Estrous Cycles. The estrous behavior of the rat has been studied most extensively. When in proestrus, the female shows several courtship behaviors, which are now sometimes termed proceptive behaviors. These include hopping and darting about in

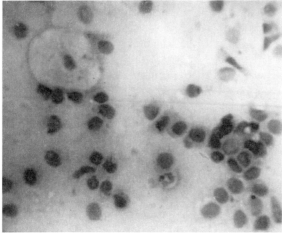

FIGURE 5-5. During diestrus, the female cat responds aggressively to the male's sexual approach (top). The vaginal smear at this time (bottom) shows only basal cells and leukocytes, with no signs of epithelial cornification. (SOURCE: Michael, 1961)

front of the male, and very rapid "ear wiggling" brought about by very fast head shaking. Increased motor activity at this stage of the cycle is also evident in increased wheel running in the absence of a male. If receptive during estrus and mounted by the male from the rear, the female will stay put and show some degree of lordosis response. This response is characterized by hyperextension of the lower spine, which brings the introitus into a more dorsal position to facilitate the entry of the male's penis. The spinal reflexes involved in this response are potentiated by a combination of estradiol and progesterone, although very high doses of estradiol alone may have the same effect. The flanks of the female are stimulated by the male during mounting, and it has been shown that estradiol increases the area of flank skin from which the lordosis response can be elicited.

The diestrous cat reacts aggressively to male sexual interest; Figure 5-5 (top; frame from a moving film) shows the female lashing out at the male, which is in midair jumping clear. Under the influence of estradiol, female aggression changes to female courtship behavior. The proestrous cat shows proceptive or courtship behavior by

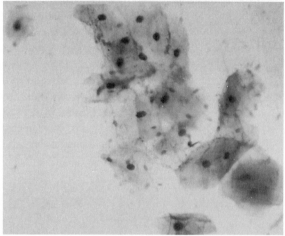

FIGURE 5-6. During estrus, the female cat assumes a receptive posture (lordosis) that facilitates male mounting (top); the majority of cells in the vaginal smear (bottom) are strongly cornified but they retain pyknotic nuclei, not seen in rodents. Small dots are sperm heads. (SOURCE: Michael, 1961)

approaching the male and emitting soft vocalizations while squirming and rolling around in front of him. These responses are evoked by the smell, sound, and sight of the tomcat. When in full estrus the cat, like the rat, adopts an extreme lordosis posture (Fig. 5-6), and the hind paws move backwards, pushing the vulva even more dorsally, with alternate treading movements of the hind legs. There is brisk lateral deviation of the tail and, at the moment of intromission, which lasts only some 5 seconds, a piercing scream is emitted by the female as she frees herself from the neck bite of the male and immediately goes into the "after-reaction." This is a violent rolling and rubbing, accompanied by a display of aggression; but the male has already jumped well clear. Several identical matings may ensue over the next 12–24 hours, until the female goes into metestrus and becomes unreceptive. Lions and tigers have an almost identical behavioral pattern and breed readily in zoos. Despite the drama of these events in carnivores, estrous behavior for many if not the majority of mammals merely requires the female to stand still so that the male can catch up and mount. The estrous sow has an immobilization reflex that is activated by the odor of the boar, and the rigidity induced in all four limbs helps to support the boar's great weight during mating. In all these species, ovariectomy abolishes not only the female's proceptive and receptive behaviors but also her attractiveness to the male, and appropriate hormone replacement treatments restore them. During diestrus, pregnancy, and, in some species, during lactation, the female remains unreceptive and unattractive to the male. Some species have a postpartum estrus, so that the female immediately conceives again, before the offspring of the previous mating are weaned.

Menstrual Cycles. As noted earlier, only Old World monkeys, apes, and humans have true menstrual cycles, and the behavioral changes associated with them differ somewhat from those associated with estrous cycles in nonprimate species. Most studied is the rhesus monkey, and most observations have been made upon animals in captivity. Under these conditions, it appears that the female will accept the copulatory attempts of the male at any time during the menstrual cycle but does so with more enthusiasm around midcycle, the expected time of ovulation (Chapter 16, Fig. 16-9). Certainly the restricted period of estrus seen in nonprimate mammals no longer applies, and it has been proposed that, with increasing development during phylogeny of the brain's neocortex, the overriding role of hormones is correspondingly diminished. Nevertheless, it has been shown both experimentally and in the wild that the hormonal milieu drives both the bonding patterns and the copulatory frequencies of the male–female pair. As in women, menstrual cycles in rhesus monkeys vary in length from cycle to cycle and between individuals, but the relationship between changes in individual hormones (Fig. 5-7) and also between hormones and behavior becomes clearer when data are aligned by

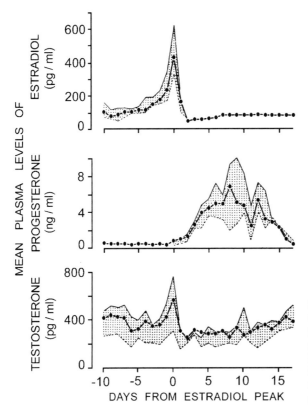

FIGURE 5-7. Plasma hormone changes (means and ranges) in 33 menstrual cycles from 5 female rhesus monkeys. (SOURCE: Reprinted with permission from Bonsall et al., 1978. Copyright © 1978 by the American Psychological Association)

the midcycle estradiol peak; this occurs 24–48 hours before ovulation. Female rhesus monkeys show several types of proceptive, sexual initiating behaviors (Chapter 16) during the follicular and ovulatory phases of the cycle, and changes in sexual motivation can also be measured by the speed with which they perform a task to gain access to a male. For example, pressing a lever 250–500 times to open a door giving access to a male employs an operant technique that provides a rather precise measure of motivation. The time taken to achieve the chosen criterion varies quite dramatically with the phase of the female's menstrual cycle; some females do not lever-press at all in the luteal phase. These techniques demonstrate a synchronization between maximal female sexual motivation (shortest access times) and maximal mating activity, which occurs 1 day after the midcycle estradiol peak at the expected time of ovulation (Fig. 5-8). During much of pregnancy, and until the infant is weaned, the female is unreceptive and unattractive to males. After ovariectomy, proceptive and receptive behavior as well as female attractiveness may continue, but only at low levels, and show no rhythmic changes; this differs from nonprimate species. Constant daily doses of estradiol restore high levels of female proceptivity, receptivity, and attractiveness. Additional progesterone treatment providing physiological levels in blood has little effect on behavior, but at high doses, progesterone clearly antagonizes the effects of estradiol.

Artificial Menstrual Cycle. Rhythmic behavioral change can be restored in ovariectomized females when they are given artificial menstrual cycles. This is an experimental hormone technique useful for primates. These cycles are produced by a 28-day schedule of daily injections of changing mixtures of estradiol benzoate, progesterone, and testosterone propionate calculated to result in plasma hormone levels similar to those of intact females on each day of the cycle. The estradiol peak will then occur on day 12 of the cycle, and vaginal bleeding occurs every 28 days. There is a midcycle "ovulatory" (there are no ovaries) dip in basal body temperature and an LH peak on days 12 and 13. The artificial cycles are associated with changes in female proceptivity and copulatory frequency characteristic of natural menstrual cycles in intact females (Fig. 5-9). Similar behavioral changes, including the rapid decline in copulatory frequency during the "luteal" phase, occur when both progesterone and testosterone are omitted from the daily injections, showing that the rise and fall of estradiol levels is the most important factor for cyclic behavioral changes in the monkey (Chapter 16).

Activational Effects in the Male

Throughout the year in nonseasonally breeding species, overall androgen levels in male mammals generally vary little except for circadian rhythms, but testosterone levels increase in seasonally breeding species as the mating season approaches. One of the best examples of seasonal breeding is provided by the red deer stag of Scotland. The testes are small and quiescent between March and June, but rapid antler growth starts in August as the testes enlarge and plasma testosterone increases. The velvet of the antlers is cast and stags are said to come into rut for a 6–8 week period in September–October. They leave the all-male groups and begin to round up hinds (females). In response to greatly increased plasma testosterone levels, the stags start to bellow, masturbate, and thrash the undergrowth with their antlers. At this time, they engage in dominance fights

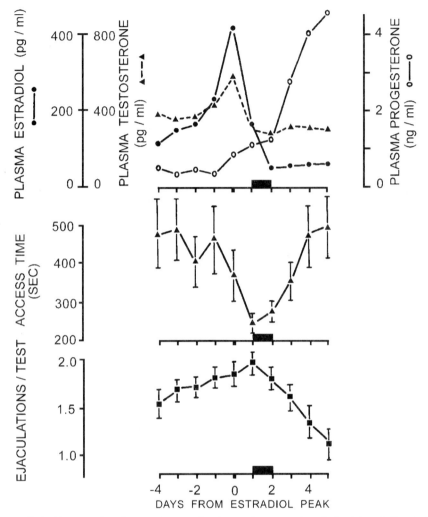

FIGURE 5-8. Mean periovulatory changes in plasma hormone levels (top) in relation to the female's sexual motivation measured by her rate of lever-pressing for a male (middle) and ejaculatory behavior (bottom). Solid bars give expected day of ovulation. (SOURCE: Reprinted with permission from Michael & Bonsall, 1977b, Peri-ovulatory synchronisation of behaviour in male and female rhesus monkeys. *Nature* **265**, 463–465. Copyright © 1977 by Macmillan Magazines Ltd.)

with other stags that are also attempting to round up females (male mate competition). Under the influence of testosterone, the mass of the neck muscles greatly increases so that stags can support and wield their large antlers. Mating activity becomes intense and they have little time for anything else, including feeding. By the end of October, stags have lost weight and are sometimes in poor condition, making it difficult for some to survive the harsh winter, and antlers are shed as testosterone levels fall abruptly. During

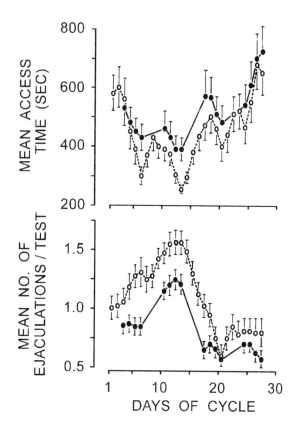

FIGURE 5-9. Changes in female sexual motivation, as assessed by the speed with which she attained access to the male (top), and copulatory behavior (bottom) were similar in artificial menstrual cycles (closed circles) and in natural cycles (open circles). (SOURCE: Reprinted with permission from Michael et al., 1982. Copyright © 1982 by the American Psychological Association)

the mating season, stags urinate a lot; the urine combines with secretions from preputial glands, and this material becomes embedded in the belly hair. It is thought to contain a powerful pheromone signal (Chapter 11, "Chemical Communication") attracting the female.

In male rhesus monkeys, another seasonally breeding species, there is a pulse of LH followed by a pulse of testosterone about every 90 minutes, so plasma testosterone levels remain high for several weeks at a time during the mating season. Even the seasonally breeding rhesus monkey will continue to mate and maintain high plasma testosterone levels when housed in a constant photoperiod and paired regularly with estrogen-treated females (Chapter 7, Fig. 7-12). The tonic secretion of androgens in the male contrasts with the large hormonal fluctuations occurring during the estrous or menstrual cycles of females and is regarded as the physiological basis for the fact that male mammals, unlike females, mate opportunistically and are constantly prepared to copulate with as many partners as possible (see also Chapter 13, Theoretical Considerations). It is also the reason why males rather than females are used more in pharmacological and other medical studies when more stable hormonal baselines are needed. Castration, which reduces plasma androgens to very low levels within hours, decreases sexual behavior and mating within days or weeks, although some *individuals* may continue ejaculating sporadically for a long time—several weeks in rats, and over a year in rhesus monkeys.

Androgens are also important for aggressive behavior, notably in male mate competition; it has been known for centuries that castrating domestic animals, such as stallions, bulls, rams, and dogs, makes them less aggressive and more tractable. The start of the mating season in many species, including some primates, is associated with a dramatic increase in aggression by males toward other males as they compete for mating territories, individual females, or groups of females. Aggression, although less severe, may even be directed toward females when they approach males, especially in the absence of other males. The positive correlation between changes in androgen levels and in the frequency of aggression toward females can be close, especially in species where the female poses no threat to the male because she is smaller and more vulnerable (Fig. 5-10). Both at puberty and repeatedly at the start of each breeding season, androgens, which are anabolic steroids, are also critical for the development of morphological characteristics that play a role in male mate competition. These include larger and stronger muscles than in females, the growth of horns or antlers in many ungulates, and thicker, protective pelage (hair) around the vulnerable neck region in lions, baboons,

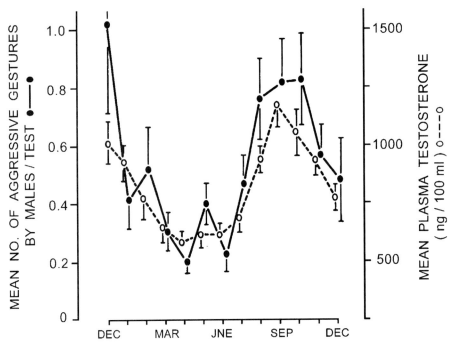

FIGURE 5-10. Annual changes in the frequency of aggressive gestures directed by male rhesus monkeys at their estrogen-treated female partners closely parallel the changes in the plasma testosterone levels of males. (SOURCE: Modified from Michael & Zumpe, 1978a, Annual cycles of aggression and plasma testosterone in captive male rhesus monkeys. *Psychoneuroendocrinology* 3, 217–220. Copyright © 1978, with permission from Elsevier Science.)

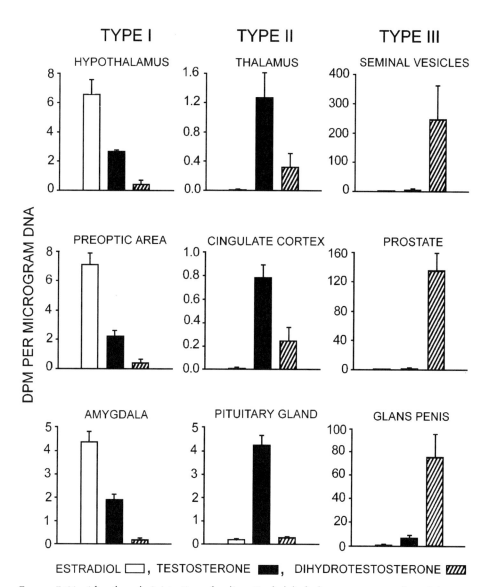

FIGURE 5-11. After the administration of radioactively labeled testosterone, radioactivity was primarily in the form of estradiol in the hypothalamus, preoptic area, and amygdala, brain regions important for sexual behavior and gonadotropin secretion (left panels), and in the form of DHT in genital tissues (right panels). (SOURCE: Reprinted from Bonsall et al., 1989, Identification of radioactivity in cell nuclei from brain, pituitary gland and genital tract of male rhesus monkeys after the administration of [³H]testosterone. J. Steroid Biochem. 32, 599–608. Copyright © 1989, with permission from Elsevier Science)

and sea lions. Some may have heard of "steroid rages," sudden episodes of uncontrolled anger and violence by athletes following the dangerous but fashionable practice (illegal in many sports) of taking high doses of anabolic steroids to enhance endurance, strength, speed, and muscular development.

Mechanisms of Action of Testosterone in the Male

It will be recalled that testosterone can be metabolized locally in the brain and elsewhere in the body both to DHT and to estradiol. The administration of estradiol, especially when given together with DHT, was almost as effective as testosterone itself in restoring sexual behavior in castrated male rats and rabbits, but not in castrated guinea pigs. This suggested that testosterone exerts some of its behavioral effects via its two metabolites. As methodology has improved, it became possible to examine which types of hormone receptors are present both in different brain areas and in peripheral tissues, and to use radioactive labeling and high-performance liquid chromatography (HPLC) to determine the chemical form in which hormones are taken up by receptors. There are numerous androgen, estrogen, and progestin receptors in the brains of both female and male mammals, particularly in the preoptic area, hypothalamus, and amygdala—areas implicated in the control of sexual behavior and gonadotropin secretion. In castrated male rhesus monkeys injected intravenously with tritiated testosterone (^{3}H-T), tissues fell into three types (Fig. 5-11). In Type I tissues (preoptic area, hypothalamus, amygdala), estradiol was the principal metabolite, and this was true for fetuses, neonatal males, and adults. In Type II tissues (thalamus, cingulate cortex, pituitary gland), unchanged testosterone was the major hormone, while in Type III tissues (seminal vesicles, prostate gland, glans penis), the major metabolite was DHT. Findings such as these, together with the results of behavioral studies in several species, have demonstrated that in both males and females, estradiol plays an important role in activating sexual motivation and behavior. Additionally, the findings in primate fetuses indicate that the estrogenic metabolite of testosterone, estradiol, plays a major organizational rose in masculinizing the male fetus during intrauterine life.

CHAPTER **6**

Behavioral Endocrinology
Stress and Adrenal Hormones

DEFINITION

As with sex, everyone believes they know about stress. But the concept of stress is an unsatisfactory one in several ways. It has always been difficult to define, and the problem of definition, even today, occupies a central position on the opening day of most conferences on the subject. A stress for one person, for example, getting married with its attendant anxieties, may be a pleasant, even joyful, experience for another. Consider the prospect of addressing a large audience: easy for some, anxiety-provoking and very stressful for others. Anything causing stress is termed a *stressor*, but whether or not a stressor turns out actually to be one depends entirely on the reaction created in the mind and body of the individual. So the term *stress* is often used sloppily, referring partly to stimulus situations and partly to the bodily responses caused by the stimulus.

TYPES OF STRESS

Physical stress is rather straightforward. When a car accident occurs and someone's bones are broken, there is agreement that the individual has been stressed and, apart from the direct effects of the injury itself, the responses of the body are called *stress responses*. But without any physical damage, for example, the mother of a child awaiting life-saving surgery may experience prolonged and profound emotional stress and exhibit all the stress responses of the body. So we immediately have two classes of stressors, the physical and the emotional: Physical stressors are largely external, and emotional stressors are often internal. Violence and abuse are pervasive in the community, the home, the school, the office, and the streets. Particularly for children and young adults living in cities, this exposure, whether actual or implied, is an almost daily occurrence and a major source of continuing stress of which the individual may be largely unaware. These social stressors are as potent and pervasive as national disasters such as earthquakes or even war. All the foregoing (there are many others) are external stressors that can generally be easily identified, but there is a good deal of stress that originates internally, particularly in the developing child, of which we may be unaware. These are the "ordinary" stresses of getting to school or office on time, of pleasing everyone—friends, family, and workmates included. All this would be regarded as a normal part of growing up in most Western communities, but there are societies in various parts of the world where such "ordinary" stress is virtually unknown, and they are portrayed as idyllic. Many social institutions are responsible for both creating and resolving internal stresses because society needs to direct and mold its norms. The family, the school, and the religious institution may inculcate standards and goals that the individual cannot always live up to, and this causes much guilt, fear, and shame, so that emotional breakdowns occur in susceptible individuals if the strain becomes too severe. These intrapsychic or internal causes of stress are important in persons with nervous or anxious dispositions. It is how the individual copes with and handles a stress that is critical for bringing about successful adaptations. Persons very worried about the outcome of a major game or major exam may become so apprehensive that, as well as being stressed, they have actual anxiety attacks complete with sweaty palms, dry mouth, palpitations, and hyperventilation, all of which make matters worse. We do not know for certain whether there is a cause-and-effect situation here or whether the anxiety and stress reactions are bundled into a single emotional and physical response.

THE STRESS RESPONSES OF THE BODY

We can now ask: What *is* the stress response, and is it helpful or not? Although stress has been vaguely known about for centuries, it was the Canadian physiologist, Hans Selye (1973), who began to address this question in the mid-1930s, particularly as it related to clinical and medical syndromes. He recognized that, despite the many different types of stressors, all seemed to give rise to much the same pattern of bodily responses, which he therefore regarded as being nonspecific. Thus, a stressor may comprise both pleasant (excessive lovemaking) and unpleasant (excessive cross-country running) experiences, but the stress responses are closely similar in the two situations and depend on the intensity of the demand being made on the capacity of the body to cope and adapt. These responses constitute Selye's general adaptation syndrome, the first stage of which in most animals and in humans is an immediate response to threat or injury, the so-called alarm reaction, by which the sympathetic nervous system mobilizes the body's defenses for "fight or flight," the famous phrase of American physiologist W. B. Cannon.

THE ADRENAL MEDULLA AND SYMPATHETIC AROUSAL

The alarm reaction is a rapid response and, as we have said, under the control of the sympathetic division of the autonomic nervous system (the other being the parasympathetic division) which, in the case of the adrenal medulla, is via preganglionic fibers from the spinal cord that pass through the coeliac ganglion in the abdomen to terminate on the chromaffin cells of the adrenal medulla. These cells produce the adrenal's principal neurotransmitters, adrenalin (or epinephrine) and noradrenalin (or norepinephrine) and, to a lesser extent, dopamine, all of which are catecholamines having adjacent hydroxyl groups on the 6-carbon ring. These transmitters are released from the medulla into the bloodstream. They increase blood pressure, heart rate, blood sugar, and free fatty acids; they contract the spleen (increase blood volume), cause the hair to stand up (piloerection), cause pupillary dilation, and help redirect the blood supply from the viscera to the muscles. All these physiological responses prepare the individual for action (flight or fight).

THE HYPOTHALAMUS AND THE ADRENAL CORTEX

There is a concomitant response to stress within the hypothalamic–pituitary–adrenal axis (HPA). To summarize, corticotropin-releasing factor (CRF) or hormone is liberated from neurons widely distributed in the brain but particularly concentrated in the paraventricular nuclei and the median eminence region of the hypothalamus, whose nerve terminals are situated immediately adjacent to the primary vascular plexus of the pituitary portal system (Figs. 6-1 and 6-2). The portal system transmits CRF as well as

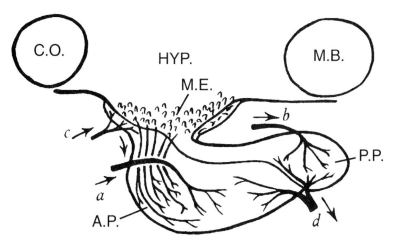

FIGURE 6-1. Diagram of a sagittal section through the pituitary gland of a rabbit, illustrating the hypophysial blood supply. A.P. = anterior pituitary; C.O. = optic chiasm; HYP. = hypothalamus; M.B. = mammillary bodies; M.E. = median eminence; P.P. = posterior pituitary; a = anterior hypophysial artery; b = posterior hypophysial artery; c = arterial twigs to the pars tuberalis plexus; d = venous drainage. (SOURCE: Harris, 1955, with permission)

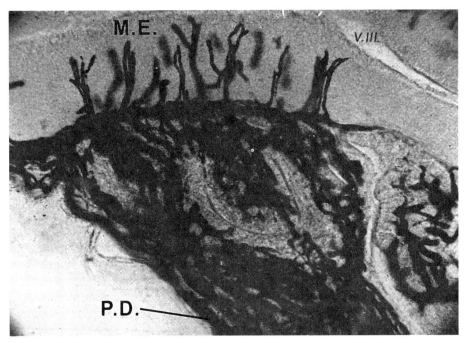

FIGURE 6-2. Photomicrograph through the median eminence of a dog. The blood vessels are injected with India ink. M.E. = median eminence containing vessels comprising the primary vascular plexus; P.D. = pars distalis; V.III = third ventricle. (SOURCE: Harris, 1955, with permission)

other hormones to the anterior pituitary gland via this vascular link, whose importance was first recognized by the English physiologist, G. W. Harris (1955). Cells in the anterior pituitary gland then liberate adrenocorticotropic hormone (ACTH) into the systemic blood supply to reach the adrenal cortex. There is also a high concentration of cortisol receptors in neurons of the hippocampus, which are involved in inhibiting the response of the HPA to stress.

CORTICOSTEROID PRODUCTION

The adrenal cortex, which surrounds the medulla, has at least three layers: the outer layer, or zona glomerulosa, secretes aldosterone (probably not under the control of ACTH), the main mineralocorticoid, which increases the reabsorption of sodium ions by the kidneys and is very important for electrolyte balance. The middle layer, or zona fascicularis, produces cortisol and cortisone, the main glucocorticoids, responsible for mobilizing protein and converting it to carbohydrates during stress, which is important for the control of inflammation. The innermost layer, or zona reticularis, next to the adrenal medulla, produces sex steroids, mainly androgens, but, to a lesser extent, also

BEHAVIORAL ENDOCRINOLOGY

progesterone and estradiol. The androgens of the adrenal cortex play a role in the initiation of puberty (adrenarche).

CORTICOSTEROID METABOLISM

The starting point for the biosynthesis of all steroids is cholesterol, which is derived both from the diet and by biosynthesis. Mineralocorticoids, glucocorticoids, and progestins are all C21 steroids and contain that number of carbon atoms. Androgens are C19 steroids that contain 19 carbon atoms, and estrogens are C18 steroids containing 18 carbon atoms (Fig. 6-3). The four rings making up the steroid nucleus are not in one flat plane, and the different substitutions (hydrogen, hydroxyl, methyl, and ethyl) project above (beta forms) or below (alpha forms) the plane of the steroid molecule. This is

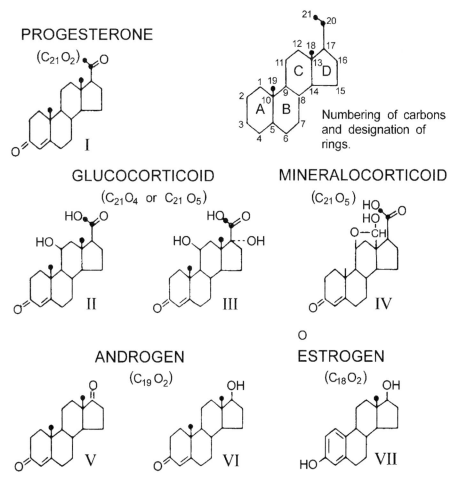

FIGURE 6-3. Molecular structures of C21, C19, and C18 steroids.

important because, for example, 17β-estradiol is biologically active, whereas 17α-estradiol is biologically inactive. The activity of a corticosteroid depends on hydroxyl-OH groups at the C11 and C21 positions, ketone groups at the C3 and C20 positions, and the double bond between C4 and C5. This gives one an idea of the structure–activity relations, which are also different for androgenic and estrogenic activity. Many different enzymes are responsible for the different oxidative and reductive reactions, and for the cleavages of side-chains that result in the different steroids. The pathways can be simplified and summarized as follows (but many important intermediaries are omitted): cholesterol to pregnenolone to progesterone to cortisone and cortisol to aldosterone. Also, we have progesterone to androstenedione to testosterone to estradiol (Fig. 6-4). Important clinical syndromes, such as congenital adrenal hyperplasia (CAH), occur when the genes coding for one or more enzymes are missing. A small enzymatic defect, which in 90% of cases is impaired 21-hydroxylase activity, has profound effects on the developing genetic female *in utero* by reducing cortisol production and therefore its feedback on the pituitary gland. As a consequence, adrenal androgen production increases and virilizes the fetus. The clitoris becomes enlarged into a phallus, labioscrotal folds fuse into a scrotum, and the baby may at delivery be mistaken for a boy (Fig. 6-5). If recognized early (and it may not be, in rural communities in which parents are embarrassed by genital abnormalities), CAH is easily treated by cortisone administra-

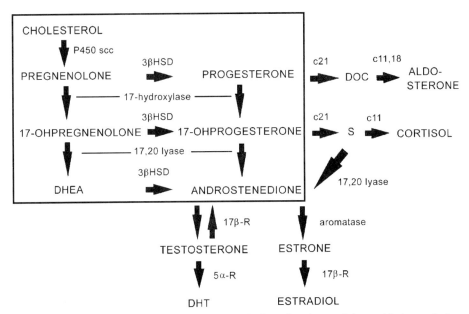

FIGURE 6-4. Metabolic pathways of synthesis of adrenal and gonadal steroids from cholesterol. DHEA = dehydroxyepiandrosterone; DHT = dihydrotestosterone; scc = side-chain cleavage; 3βHSD = 3β-hydroxysteroid dehydrogenase; 17β-R = 17β-reductase; 5α-R = 5α-reductase; DOC = deoxycorticosterone; S = 11-deoxycortisol. (SOURCE: Rosenfield & Lucky, 1993, with permission)

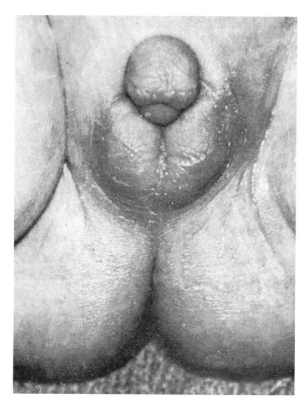

FIGURE 6-5. Genitalia of a genetic female infant virilized by congenital adrenal hyperplasia.

tion. If left untreated for months or even years, adult behavior is somewhat masculinized and comes to occupy a position on a masculinity–femininity scale that is intermediate between a normal female and a normal male.

The adrenocorticoids also have anti-inflammatory actions by inhibiting inflammatory cytokines, as well as both immunostimulatory and immunosuppressive actions.

HABITUATION TO STRESS

Sustained high levels of plasma corticoids are one of the hallmarks of the second stage of the general adaptation syndrome, and its length depends on both the chronicity of the stress and the coping mechanisms of the individual. Repeated, prolonged stresses can damage hippocampal CA3 region neurons by liberating excitatory amino acids such as glutamate. Many stressors lose their potency as the individual adapts and habituates; indeed, this phenomenon is utilized in several types of training. Parachutists, for example, have large catecholamine and corticoid responses after their first jump, but the magnitude of these responses rapidly declines with further training (fourth or fifth jump). Adaptation to stress is also seen in monkeys that learn to press a lever to avoid an electric shock. Adaptation is important for outcome. If habituation and attempts at coping fail, the third stage of the general adaptation syndrome follows, characterized by a state of overwhelming stress and exhaustion that can lead to death. Why some individuals succumb and some survive may be analogous to the body's responses to a severe infection with a specific organism. The outcome of the illness depends on both the virulence of the organism and the resistance of the host. Similarly, some stressors are more virulent for some individuals than for others, and resistance varies widely, so that different responses, different lesions, and different outcomes result, as observed in experimental work with laboratory rats, and in hospital wards with patients.

In severely stressed rats, the adrenal glands are enlarged (hypertrophic) but depleted of steroids; adrenal ascorbic acid is also depleted. Plasma corticoids increase, as do urinary 24-hour catecholamines; testosterone levels fall, and the thymus and lymphatic glands involute (diminish in size) and become atrophic; there is widespread ulceration of the gastric mucosa. In patients, acute gastric ulcerations and adrenal enlargement have long been known to accompany severe burns and major surgery, especially if associated with infection. We have seen, however, that much human stress occurs without any of these major somatic changes: Coping and adaptation occur rapidly and the stress is expressed only by mild symptoms such as fatigue, dyspepsia (indigestion), and insomnia. It appears, however, that in aging humans, the negative feedback effects of high plasma cortisol are reduced, so that the HPA axis does not shut down so rapidly after challenge and remains hyperactive for prolonged periods after the occurrence of a stress in the elderly.

FUNCTIONS OF STRESS

Stress, then, is an integral part of life, and the stress responses can be either normal or abnormal. What are their functions? Are they beneficial or are they injurious? Clearly, the alarm reaction helps the body to deal with acute threats and shocks; it improves the chances of survival and, in this respect, is beneficial. But if it occurs when the threat is trivial, it is no longer adaptive and does not help anything. These bodily responses are phylogenetically old, and as has been pointed out, stress-induced increases in plasma free fatty acids, which are associated with heart disease, would not be beneficial for modern, civilized humans and so could be regarded as maladaptive. These free fatty acids, which provide the energy for rapid and often violent physical activity, are not used in many modern, stressful situations when physical activity is not an option. For example, if stranded in a car in a major traffic jam, free fatty acids can be mobilized without the action leading to their depletion, and this type of stress situation can lead to heart disease. As you will have understood, there is no simple answer to the question originally posed.

PSYCHOSOMATIC MEDICINE

We now come to the field of psychosomatic medicine, and it is perhaps a little unusual to introduce this topic in an introductory text. But there is no harm in doing so. Psychosomatic medicine deals with the influence of bodily disease on the mind and the emotions. If someone has a heart attack, even if it is not life threatening, it changes the mood, attitudes, and emotional outlook of the victim. In the past, this aspect of the illness has not always been considered important. Psychosomatic medicine is a two-way process, that described above, and also the effect of the emotions on bodily reactions. People with severe emotional stress for one reason or another can experience bodily changes, and the emotional symptoms are then said to have become somaticized. The precise symptomatology expressed by the body cannot be described here because it is very complicated and would require extensive treatment. When the emotional component of a psychosomatic illness is treated, for example, by the short-term use of benzodiazepines, although the underlying pathology is not directly impacted, the symptomatology may improve dramatically and the patient feels better. The syndrome of post-traumatic stress disorder is recognized now as a clinical entity and appears as such in the various diagnostic manuals of mental disorder. There is usually an unmistakable stressor, and the symptoms involve reexperiencing past traumatic situations. The reexperiencing results unpredictably in a variety of autonomic and dysphoric symptoms (feeling ill) that are very distressing. The original stressors are mostly severe, such as a violent rape, torture, or a violent battle situation from which there is no respite and no escape. The individual relives the events both in recurring nightmares and when awake. During these episodes, there is intense emotional arousal and much stress. There may be decreased responsiveness to the outside world and, more rarely, a dissociative state that lasts hours or even days; this syndrome is most frequently reported in veterans who have been in combat. Reliving these highly traumatic experiences from the past produces sympathetic secretion and arousal, and high plasma corticoids, although basal HPA activity is lower than normal. There is a dread of sleep because of nightmares, which leads to exhaustion and debilitation. In some well-documented cases, the symptoms may only emerge several years after the traumatic experience, when the signs of stress are reactivated. Interestingly, children who have lost both parents, or those raised in dysfunctional families, show higher blood pressure responses and corticoid responses to stress as adults than do normals. These early traumata are thought to set the level of responsiveness of the HPA for later in life. Autoimmune disorders such as rheumatoid arthritis are associated with an inability of patients to show a good response to stress, so they lack the anti-inflammatory steroid response, and steroids must be provided for effective treatment.

TWO PSYCHIATRIC SYNDROMES

Major depressive disorder is a psychosis now defined by a series of fairly precise criteria. Some of the clinical signs of depression suggest a hypothalamic component—such as disturbances of mood or affect, sleep, sexual function, and appetite. A high percentage of patients with the subtype termed endogenous depression, which has a genetic component, have hypersecretion of cortisol, probably due to a disinhibition of ACTH secretion. The normal diurnal variation in plasma cortisol (Chapter 7, Fig. 7-8), with a fall in the evening, may be blunted or lost. Dexamethasone is a potent synthetic steroid that blocks pituitary function, and it has been used to help in the diagnosis of patients with depression. Perhaps 50% of patients with endogenous depression and with hypersecretion of cortisol do not show the normal suppression of plasma cortisol levels observed in controls after dexamethasone administration. Several factors can invalidate the test so that its usefulness as a diagnostic tool is somewhat restricted, but the current view is that the hypercortisolemia is central to the neuropathology of the illness rather than a response to the stress brought on by the illness. In the other major psychosis, schizophrenia, the situation is different. Plasma cortisol only increases in the acute phases, when the patient is in turmoil and highly disturbed, but levels become normal as the condition becomes more chronic and the patient is more withdrawn.

Mention should be made of suicide, which, as far as we can ascertain, is unique to the human. Some animal species, particularly domesticated ones, are reported to become depressed at times, to go off their food and die, but it is impossible to ascertain whether there is an underlying disease process. Suicide attempts and actual suicide appear to be different with different demographics. Attempted suicide is associated with hysterical personality disorder and poor impulse control. Attempts are often repeated and made conspicuously, when intervention is more likely and the means are less lethal. The age group is younger than with actual suicides. Self-poisoning occurs in 70% of all suicide attempts. Actual suicide accounts for between 0.4% and 0.9% of all deaths and is the eighth leading cause of death in the United States. Adolescent suicide is the third leading cause of death in this age group. There are about 1,000 suicides a day worldwide, and these data do not include suicide attempts. First-degree relatives of suicides have an eight times greater than normal incidence of suicide, and 25% of all suicides occur in males over 65 years of age. There is a close association between actual suicide and other psychiatric disorders, notably, major depression. Theories abound, but we really do not know why this form of aggression (directed against the self) is unique to *Homo sapiens*.

CHAPTER 7
Biological Rhythms

All living creatures show rhythmic changes in their physiology and behavior. Table 7-1 lists some examples of physiological rhythms in the human. Biological rhythms, like physical rhythms such as sound waves, can be described by four properties (Fig. 7-1):

1. *Phase* defines a specific time within the cycle (e.g., the maximum or the minimum).
2. *Period* defines the duration of the rhythm as measured between two identical phases (e.g., between successive maxima).
3. *Frequency* defines the rate of the rhythm as measured by the number of cycles per unit of time.
4. *Amplitude* defines the magnitude of the rhythm as measured by the difference between the maximum phase and the mean between the maximum and minimum phases (Fig. 7-1, horizontal line).

A terminology often used in the analysis of biological rhythms depends on fitting a cosine curve accurately to the data. If it can be done, this facilitates statistical comparisons. In this terminology, the mean of all values is called the mesor, and the time of the maximum phase is called the acrophase.

FUNCTIONS OF BIOLOGICAL RHYTHMS

Biological rhythms have at least two primary functions. The first is to facilitate adaptation to an impending change in environmental conditions that allows the organism

Table 7-1. Approximate Periodicity of Some Human Rhythms

Type of rhythm	Frequency	Period
EEG α-rhythm	10/sec	0.1 sec
Cardiac	70/min	1.0 sec
Respiration	10/min	6.0 sec
Sleep cycle	5/7.5 hr	90 min
Body temperature	1/day	24 hr
Menstrual cycle	1/month	28 days
Testosterone levels in men	1/year	365 days

to prepare physiologically and behaviorally for the change. It is no accident, therefore, that major biological rhythms take their periodicity from rhythmic, predictable, geophysical changes such as the 24-hour day–night rhythm, the 12.4-hour tidal rhythm produced by the effects of the moon on the oceans, the 14.8-day rhythm produced by the combined effects of the sun and moon on the oceans, and the 365-day rhythm produced by the rotation of the earth (seasonal changes). A species feeding on nocturnal moths would not benefit were it to hunt randomly throughout the 24 hours, because during daylight it would not obtain any food, and a nocturnal activity cycle prevents this. The other primary function of a biological rhythm, certainly of the 24-hour rhythm, is to be able to measure the passage of time in the same way we do by consulting a watch. This is important for compass orientation and migration (Chapter 8) because it allows the individual to compensate for the apparent movement of the sun through the sky as time passes. The 24-hour rhythm has been studied more than any other. Halberg (1959) coined the term *circadian* (Latin *circa* meaning "about" and *dies* meaning "a day"),

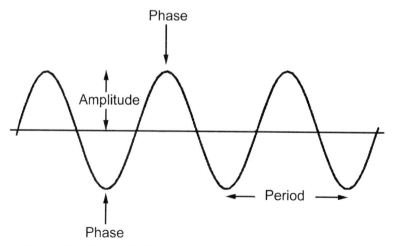

FIGURE 7-1. Schematic defining the properties of rhythms (see text).

Table 7-2. Types of Biological Rhythms

Aschoff terms	Alternate terms	Periodicity	Example
Circadian	Daily	24 hr	Human core body temperature
Circatidal	Lunar day	12.4 hr	Fiddler crabs
Circalunar	Semilunar	14.8 days	*Clunio* midges
—	Monthly	29.5 days	Ant lion pit size
Circannual	Annual	365 days	Monkeys, starlings

and Aschoff (1981) followed the nomenclature for other major rhythms, which are sometimes known by somewhat confusing English names (Table 7-2).

CIRCADIAN RHYTHMS (23–26 HOURS)

The circadian rhythm, thought to be determined by a self-sustaining oscillator called the "internal clock," has six main properties:

1. *Ubiquity.* The rhythms occur in all animals and plants, at least in all eukaryotes (cellular organisms).

2. *Periodicity.* The period is about 24 hours but, when not entrained (see below) by light or temperature, it varies between 22 and 28 hours in plants and between 23 and 26 hours in animals. The variance is due to differences among species and among individuals of the same species. In humans, the average period is 25 hours plus or minus 7 minutes. In an individual, the clock is very accurate, so that the periodicity can vary by as little as 1 minute per day.

3. *Entrainment.* Even when considerably longer or shorter than 24 hours, the rhythm can be entrained or "reset" every day by environmental cues called *Zeitgeber* (German for "time-giver"). The principal *Zeitgebers* for circadian rhythms are light and temperature. Entrainment is analogous to resetting one's watch to synchronize with a radio time signal, the *Zeitgeber*, every morning, because the watch gains or loses a little every 24 hours.

4. *Persistence.* Under constant environmental conditions such as constant light, constant dark, or constant temperature, in the absence of any *Zeitgeber*, the rhythm adopts the intrinsic periodicity of the internal clock and persists indefinitely, for example, for many generations in fruitflies (*Drosophila*) and mice. It is then said to be "freerunning."

5. *Phase shifts.* The phase of the rhythm can be shifted by changing the timing of a light or temperature *Zeitgeber*. This happens when we fly across several time zones, for example, across the Atlantic from Atlanta to London, where we land during daylight at 8:00 A.M. local time but at 3:00 A.M. Atlanta time, when we would normally be asleep. Since, in humans, the circadian rhythm can be advanced by only about 1.5 hours every 24 hours but delayed by about 2 hours without incurring jet-lag effects, it takes longer to recover after flying from Atlanta to London (west to east, which requires that the internal clock be advanced) than after flying from London to Atlanta (east to west, which

requires that the internal clock be delayed). There is, of course, no jet-lag when flying long distances from north to south and vice versa.

6. *Temperature-compensation.* A clock that speeds up when it is warm and slows down when it is cold would be useless to most living organisms, remembering that only birds and mammals are homoiothermic. It is no surprise, then, that the periodicity of the circadian rhythm remains constant under temperature extremes, but the temperature-compensating mechanism remains a mystery.

One might wonder why there is so much individual and species variation in the circadian rhythm's periodicity, and why it is so sensitive to entrainment. Would it not have been simpler to evolve an internal clock running at a period of precisely 24.0 hours? This alternative would be analogous to having a very precise clock without any means for resetting it. The reason why such a clock has never been produced is that even minute errors compound into very large ones over time. For example, an error of just 2 seconds per day would compound to over 14 hours during the lifetime of a long-lived species such as the human, and would have especially disastrous consequences for similarly long-lived (70 years) migrant species such as geese. Since the sun appears to move 15° each hour, a 3-hour deviation of an animal's internal clock produces a 45° navigational error. Entrainment of the clock is essential for species that migrate in directions other than due north and due south, and use sun compass navigation (Chapter 8). Without the ability to adapt to the change in the timing of dawn and dusk as they cross different longitudes, migratory species could not find their way. Given that the internal clock is entrainable, the periodicity of the circadian rhythm can vary between 1.5 and 2.0 hours from the 24-hour period without any adverse consequences.

Studying Circadian Rhythms

Circadian rhythms were first studied in the laboratory by measuring changes in locomotor activity. Nocturnal rodents, such as rats and hamsters, typically spend a good deal of their waking time wheel running, when a wheel is provided, and the rotations of the wheel can be monitored readily by a pen that makes a vertical mark on graduated paper mounted on a drum rotating at a fixed rate, once every 24 hours. For other species, the movements of a suspended cage floor can be monitored similarly. Each day's record is then copied and the two versions are "double-plotted" side by side, and the double-plots of successive days are mounted from top to bottom, aligned by clock time. This plotting method is used to facilitate easy visual detection of any changes in the start and end of activity periods where these cross over the midnight mark. In the schematic (Fig. 7-2), the top four traces illustrate activity entrained to the 24-hour light–dark cycle. The next four traces show a gradual delay in the onset and end of activity when the animal is maintained in constant light or constant dark, that is, while the circadian rhythm is freerunning. The last four traces show entrainment after restoration of the 24-hour light–dark cycle, whose onset is now 10 hours later than originally. Figure 7-3 shows an actual double-plot of an isolated human living without any external time cues.

The way in which light entrains a free-running rhythm has been studied in many species, mostly in nocturnal animals, including mammals such as hamsters and flying squirrels (*Glaucomys volans*). The schematic in Figure 7-4 is based on several studies in

BIOLOGICAL RHYTHMS 103

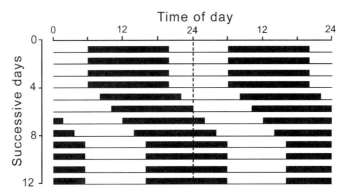

FIGURE 7-2. Schematic of double-plots of a circadian activity rhythm shown by black bars (see text).

which nocturnal mammals showing a free-running activity rhythm, with a period of 25 hours in constant darkness, are exposed to a 1-hour pulse of light at different times throughout the 24-hour day. If the light pulse is given during the subjective day of the animal, when it is inactive, there is no change in its rhythm (Fig. 7-4, A). If the light pulse is given early, during the first half of the animal's subjective night, when it is active, the rhythm is delayed (Fig. 7-4, B and C), but if it is given late during the animal's subjective night, the rhythm is advanced (Fig. 7-4, D and E). In diurnal species, as in nocturnal animals, responsiveness to light is largely restricted to subjective night, although data for diurnal mammals are sparse. Thus, in both diurnal and nocturnal animals, early morning light advances, while evening light delays the rhythm. In Figure 7-4, the dawn light pulse (E) produces a 2-hour phase advance, while the dusk pulse (B)

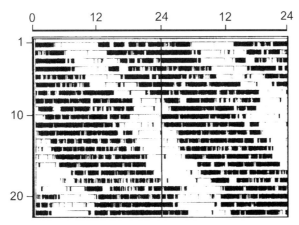

FIGURE 7-3. Free-running activity rhythm of a human living without external time cues. Numbers along top give hours of the day, and numbers along vertical axis give successive days. (SOURCE: Moore-Ede et al., 1982, with permission)

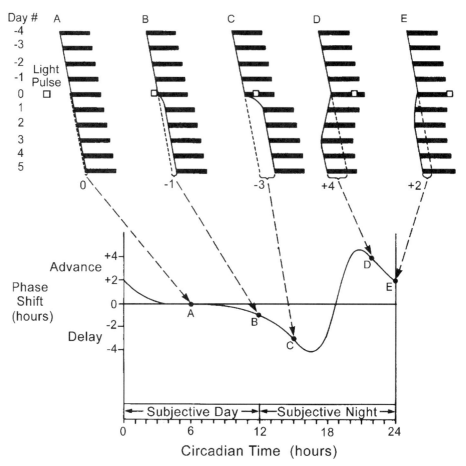

FIGURE 7-4. Schematic illustrating how light entrains the circadian rhythm. The horizontal bars represent activity (nocturnal animal) or rest (diurnal animal) during a free-running activity rhythm with a 25-hour period. The bottom part shows the phase response curve resulting from 1-hour light pulses given at different times during the animal's subjective day and night. See text for other explanations. Day number 0 = day of light pulse. (SOURCE: Moore-Ede et al., 1982, with permission)

produces a 1-hour phase delay; this results in an overall 1-hour phase advance that resets the intrinsic 25-hour rhythm to 24 hours each day. In simple organisms, the mechanisms underlying entrainment differ in different taxa but seem to depend either on the activation of a gene or the destruction of a gene protein product.

Because phase advances usually require several more circadian cycles ("transients") to become fully established than do phase delays (compare C and D in the upper part of Fig. 7-4), a circadian rhythm can be delayed more rapidly than it can be advanced. This is important for workers rotating through day and night shifts. Impairment of well-being and efficiency will be minimized if (1) the interval between the daily starting times of the old and the new shifts is as short as possible; (2) the starting time of the new shift is later than that of the old shift, so that the rhythm is delayed; and (3) the worker is given a sufficient number of days off between old and new shifts to adjust to the new activity rhythm, namely, one day of rest for every 2 hours of delay or every 1.5 hours of advance. It seems ironic that medical administrators make no such allowances for emergency room physicians.

Location of the Internal Clock

Even simple organisms such as bread molds have circadian rhythms, and one might expect that complex organisms would have several, for example, in different organs or tissues. Furthermore, they might be entrained or synchronized by some central master clock. This may be so in humans, as evidenced by data obtained from volunteers who lived alone for weeks in environments such as sound-proofed cellars and mine shafts without any *Zeitgebers*, namely, in temporal and social isolation. Under such conditions, changes in autonomic functions, such as body temperature and excretion of urinary constituents, all show free-running rhythms, with a period of about 25 hours. Typically, free-running rhythms of the sleep–wake cycle do so too, remaining internally synchronized with rhythms of autonomic functions. In some cases, however, subjects have experienced desynchronization of their circadian activity rhythms, whose periods increased to 35 hours or more. Interestingly, subjects were completely unaware that their "days" had lengthened by about 10 hours, and they did not eat more during meals or lose as much weight as expected. In a recent study, researchers used controlled exposure to the light–dark cycle on a forced desynchronization protocol to determine the period of the human circadian pacemaker, as measured by core body temperature, and melatonin and cortisol levels. To prevent feedback effects of light that might lengthen the apparent circadian period, younger (21–30 years) and older (64–74 years) volunteers were subjected to 28-hour "days" by scheduling bedtime 4 hours later each day over about 3–4 weeks (Czeisler *et al.*, 1999). Under these conditions, the pacemaker's period deviated very little from 24 hours, averaging 24.18 hours in both younger and older volunteers (Fig. 7-5). Internal desynchronization of some circadian rhythms under constant conditions also occurs in nonhuman mammals. One could therefore conceptualize a hierarchy of different internal clocks.

The master clock appears to be located in the brain, which is essential not only for circadian rhythms in behavior but also for most other rhythms. Removal of the cerebral ganglia (equivalent to the brain) of two species of moth, one active in the morning and the other active in the afternoon, abolished their circadian activity rhythms. When ganglia were transplanted between the two species, they resumed circadian activity whose timing corresponded to that of the donor species. In birds, the pineal gland, which is sensitive to light penetrating the skull, secretes melatonin during the night and is essential for circadian rhythms. Excision of the pineal gland eliminates circadian rhythms, and pineal transplantation between individuals exposed to light either early or late in the day produced in recipients, as in moths, activity rhythms characteristic of donors.

In mammals, light also penetrates the skull but does not affect the pineal gland directly, although, as in birds, melatonin is secreted only during the night in both diurnal and nocturnal species. The role of melatonin has been well established in rats. If

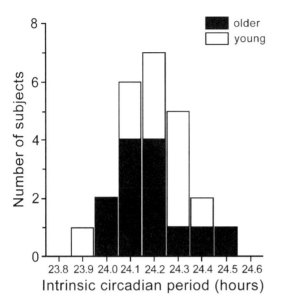

FIGURE 7-5. In humans subjected to a forced desynchronization protocol (see text), the intrinsic circadian period (average of body temperature, and melatonin and cortisol periods) was very close to 24 hours and did not change with age. (SOURCE: Reprinted with permission from Czeisler et al., 1999, Stability, precision, and near-24-hour period of the human circadian pacemaker. Science **284**, 2177–2181. Copyright © 1999 by the American Association for the Advancement of Science)

precisely timed, its injection entrains circadian rhythms of wheel-running activity, locomotor activity, drinking, body temperature, and so on. The photic input necessary for entrainment is received by the retina, which is connected by monosynaptic retinohypothalamic pathways to the suprachiasmatic nuclei (SCN) of the anterior hypothalamus. The suprachiasmatic nuclei, each consisting of some 10,000 neurons, are the brain structures essential for circadian rhythms in mammals, and are the site of the master clock (Fig. 7-6). The SCN, via a polysynaptic pathway, projects to the paraventricular nuclei (PVN) of the hypothalamus, which then project via the medial forebrain bundle (MFB) to the upper thoracic spinal cord, and thence to the superior cervical ganglion, which eventually innervates the pineal gland. Some pinealocytes appear to translate photic information into a neuroendocrine response: a concept first advanced by von Frisch (1911). Destruction of the SCN, either experimentally in animals or by tumors or lesions in humans, destroys all circadian rhythms. In blind persons, or those with a neurological lesion that stops the light signal reaching the SCN, circadian rhythms become quite variable. But when these people are shielded from all environmental cues that betray the time of day, such as changes in traffic noises, birdsong, temperature, and so on, their free-running rhythms are just as precise as those of seeing people. This supports the view that the light–dark cycle is the primary *Zeitgeber* in humans. Some people suffer from a sleep disorder called delayed sleep phase insomnia, which is caused by an inability to advance the clock. Some individuals tend to stay in bed longer in the mornings and retire later in the evenings during weekends and vacations than when working. The ensuing delay of the clock must therefore be reversed as soon as the weekend or vacation is over. People with delayed sleep phase insomnia are incapable of making this adjustment; they go to bed earlier and get up earlier but fail to fall asleep for the first few hours. Consequently, they become debilitated by chronic sleep deprivation.

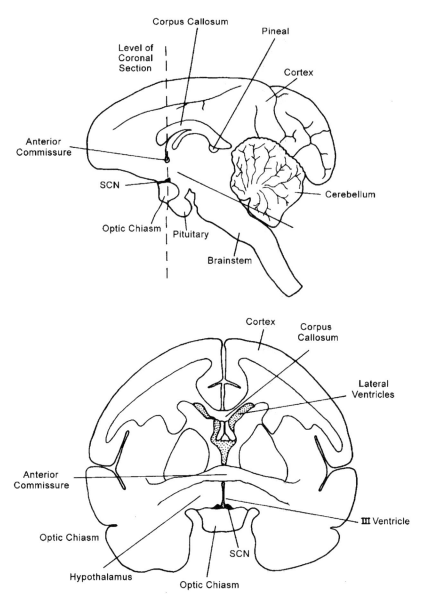

FIGURE 7-6. The location of the suprachiasmatic nuclei (SCN) in midline sagittal (top) and coronal (bottom) sections of the squirrel monkey brain. (SOURCE: Moore-Ede et al., 1982, with permission)

The remedy is to take them into a sleep laboratory and, over the course of several weeks, incrementally delay their rhythm all the way round the 24-hour clock, until it is entrained to the timing required for their working lives. Fortunately, this condition is rare.

It has not been established conclusively that the internal clock is indeed entirely internal and not, as thought by some, driven by rhythmic changes known to occur in several geophysical variables such as background radiation and geomagnetic and geoelectrical forces. Proponents of the latter view hypothesize the existence of an internal "autophaser" that synchronizes ("autophases") the organism's responses with the changes in geophysical forces. The following observations point to an internal mechanism and, while they cannot conclusively rule out either clocks or autophasers, are generally taken as evidence for an internal clock:

1. Circadian rhythms have a free-running periodicity that varies not only between species but also between individuals of the same species.
2. Within individuals, free-running circadian rhythms can be very precise, and accurate to within seconds or minutes per 24 hours.
3. Free-running circadian rhythms persist under constant conditions (e.g., constant light) for generations in species as diverse as fruitflies and mice.
4. There is a genetic basis for circadian rhythms, as shown by hybrid studies with plants and with mutant *Drosophila* flies having short, long, or no circadian rhythms (Chapter 4).
5. Rapid translocations of animals over several time zones do not affect the phase of the free-running rhythm, as demonstrated, for example, in bees that were trained to expect food between the hours 20:15 and 22:15 French summer time in Paris. After being flown overnight to New York, they looked for food in their controlled experimental chambers at the appropriate time in Paris, namely, 5 hours earlier than the corresponding time of day in New York (Fig. 7-7).
6. Attempts to counteract geophysical forces, for example, by maintaining plants in the Antarctic on a turntable rotating in a direction opposite to the earth's rotation, or to eliminate them by taking bread mold into space did not reliably disrupt circadian rhythmicity.

Circadian Rhythms in Humans

In addition to sleep–wake cycles, humans have circadian rhythms in many different physiological variables as well as in cognitive and physical abilities. Figure 7-8 shows changes in six hormones: Cortisol, prolactin, and testosterone all peak during the latter half of the sleep period and decline after wakening, while growth hormone is high during the early part of sleep and declines before awakening. Luteinizing hormone (LH) peaks at night, and this is important in endocrinology because the earliest sign of approaching puberty is the onset of rises in nocturnal LH in boys that, in turn, stimulate the gonads. In male rhesus monkeys, and possibly men, the arcuate nucleus of the hypothalamus contains the oscillator, which controls the approximately 1-hour (circhoral) rhythm of LH pulses, each of which elicits a testosterone pulse. It is the slight shortening of this periodicity at night and lengthening during the day that results in the

BIOLOGICAL RHYTHMS

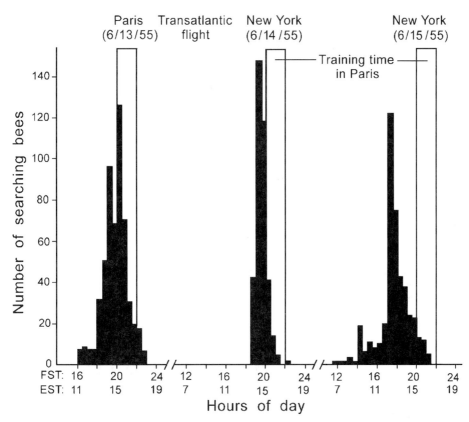

FIGURE 7-7. Honeybees, under constant conditions in an enclosed chamber, were trained to expect food between 20 and 22 hours in Paris (open columns). When tested just before and immediately after a flight to New York, and again during the second and third (not shown) days in New York, they continued searching for food at or before the appropriate time of day in Paris, not New York. They clearly showed a free-running rhythm of less than 24 hours. FST = French Summer Time; EST = Eastern Standard Time. (SOURCE: Modified from Figures 14 and 15 in Renner, 1957. Copyright © 1957 by Springer-Verlag)

circadian plasma testosterone rhythm. Figure 7-9 illustrates the familiar circadian rhythm in basal body temperature, together with rhythms in cognitive and physical task performance, of 6 volunteers subjected to 4 days of an artificial light–dark cycle (LD) and then to 4 days of continuous darkness (DD). Throughout this study, however, the subjects were kept on a rigid 24-hour schedule and wakened several times a night for the performance measures; in the presence of these *Zeitgebers*, none of the rhythms became free-running during the 4 days of continuous darkness. More recently, it has been found that circadian rhythms also play a role in the efficacy of many over-the-counter and prescription drugs; the same dose administered at different times of day can have different potencies.

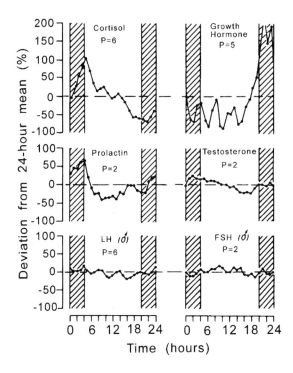

FIGURE 7-8. Circadian rhythms of six hormones in humans. Shaded areas represent sleep. P = number of publications from which data were drawn. (SOURCE: Figure 11 in Aschoff, 1978. Copyright © 1978 by Springer-Verlag)

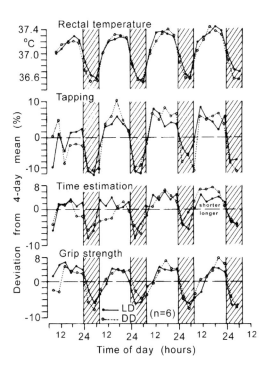

FIGURE 7-9. Circadian rhythms in body temperature and performance measures in humans. Shaded areas represent sleep. n = number of subjects. (SOURCE: Aschoff et al., 1972)

BIOLOGICAL RHYTHMS

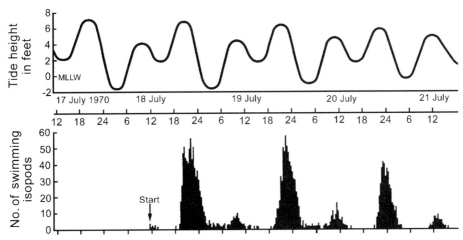

FIGURE 7-10. The activity of the isopod *Excirolana chiltoni* in the laboratory varies with the timing of high tides in the collection area. (SOURCE: Modified from Figure 2 in Klapow, 1972. Copyright © 1972 by Springer-Verlag)

CIRCATIDAL RHYTHMS (12.4 HOURS)

The gravitational field of the moon produces bulges in the ocean surface that move with the earth's rotation, producing two high tides every lunar day, namely, once every 12.4 hours. It is not surprising that organisms adapted to an intertidal habitat have activity rhythms adapted to this periodicity. Fiddler crabs, for example, come out on the beach each low tide to feed and seek mates, and maintain this activity rhythm for many days under constant conditions. A small isopod crustacean that lives buried in sand high in the intertidal zone comes out to feed when reached by the incoming surf. Its activity pattern in a sandy-bottomed beaker of seawater precisely mirrors the rhythm and height of tides at its home beach (Fig. 7-10).

CIRCALUNAR RHYTHMS (14.8 DAYS)

Spring tides (highest high tides and lowest low tides) occur when the gravitational field of the sun summates with that of the moon at the time of new moon and full moon, whereas neap tides (lowest high tides and highest low tides) occur when the sun's gravitational field is at right angles to, and therefore antagonizes, the gravitational field of the moon during its first and last quarters. Thus, spring tides (and neap tides) occur every 14.8 days, twice every lunar month, and some species possess a circalunar rhythm that permits them to synchronize certain activities with the timing of these tidal variations. For example, the pupae of a small midge (*Clunio marinus*) live at the lowest extreme of the intertidal zone, which is exposed to air only for some 2 hours during the spring low tide, the only time during which the adult insect can hatch and survive. In the laboratory, it was found that to synchronize the 14.8-day hatching rhythm in a population of pupae maintained on a 12:12 hour photoperiod, a small lamp simulating a full moon was necessary for 4 nights. This suggested that the bright light from a full moon, rather than day-length, is the *Zeitgeber* that entrains the free-running hatching rhythm in these midges. Grunions are unusual marine fish in that they spawn on the beach. During the spring and summer seasons, adults gather in the millions just offshore for 3–4 nights after the highest spring tides and, at high tide, swim to shore to spawn in the sand (Fig. 7-11). This timing allows the eggs to develop for about 10 days in undisturbed sand before the surf of the next high spring tide stimulates a massive hatching.

MONTHLY RHYTHMS (29.5 DAYS)

Many species exhibit monthly rhythms that have a periodicity of about 29.5 days, the interval required for the moon to rotate once around the earth and, therefore, the interval between two successive full moons. Monthly rhythms include the size of the pit within which ant lions (*Myrmeleon obscurus*) ambush ants, the activity of rats with thalamic lesions, the initial bearings taken by homing pigeons, the color sensitivity of humans, and the macaque and human menstrual cycles. In many cases, the functions of monthly rhythms are not understood.

FIGURE 7-11. Grunions (*Leuresthes tenuis*) gathering on a California beach to spawn in the sand. (Photo courtesy of Michelle Riehle)

CIRCANNUAL RHYTHMS (365 DAYS)

In temperate zones with marked seasonal changes in photoperiod, temperature, rainfall, and, consequently, vegetation (biomass) and food availability, many species show circannual rhythms with a periodicity of about 365 days. Probably the greatest change in biomass on the planet is a circannual one that depends upon the spring warming of the polar ice caps. Within pools of water from melting ice there occurs an abrupt and massive growth of vegetation (including algae and mosses) and of the tiny animals and plants that live on it. In turn, there is a massive increase in the shrimp, krill, and small fish around the polar ice that brings the birds and mammals that feed on them. In tropical species, seasonal rhythmicity may be attenuated or absent.

The function of circannual rhythms is generally to synchronize mating, birth of young, migration, or hibernation with environmental conditions that are optimal for these activities. Some animals respond to increasing day-length (photoperiod) and mate in spring (birds, rodents, pig, horse); others respond to decreasing day-length and mate in the fall (sheep, cattle, bison, red deer, rhesus monkeys) (Chapter 14). The fact that these rhythms can be phase-shifted 6 months by moving animals across the equator, for example, by transporting Merino sheep from Europe to Australia, demonstrates the importance of the photoperiod and other environmental variables as the *Zeitgeber.* It has been shown for some species that circannual rhythms persist under constant conditions. Free-running circannual rhythms in the testis size and molt of starlings (*Sturnus vulgaris*) can be entrained to a periodicity as short as 1.5 months by contracting the normal annual photoperiod changes (Gwinner, 1981). In rhesus monkeys, circannual rhythms in plasma testosterone levels persisted for 3 years in a constant photoperiod and, probably because they were free-running, lengthened to 13 and 14 months in the second and third years, respectively (Fig. 7-12). Annual changes in plasma testosterone levels and in sperm counts of a large sample of male U.S. veterans (Chapter 17, Fig. 17-22) may also be true circannual rhythms, but researchers cannot establish that they are free-running under constant conditions in a human population.

RHYTHMS IN HUMAN DISEASE

In many human pathological states, there are rhythms whose periodicities vary widely and do not conform directly to any of the five primary biological rhythms listed in Table 7-2. It has not been possible to determine if they would be free-running under constant environmental conditions, and the mechanisms producing them are unknown. They are remarkably striking and regular, and can be useful diagnostic tools. Richter (1976) recognized three types of biological clocks: homoeostatic clocks that involve feedback mechanisms typified by the ovarian ovulatory cycle; central clocks that depend on intrinsic rhythmicities of hypothalamic–pituitary mechanisms; and peripheral clocks as exemplified by intermittent hydrarthrosis. This condition is characterized by recurrent episodes of swelling of the knee joints, often at 7- to 10-day intervals. Figure 7-13 shows data on a 47-year old man whose left and right knees swelled alternately, each maintaining a 7- to 8-day rhythm, apparently independently of the other. Cyclic agranulocytosis is characterized by intermittent periods of very low white cell counts in

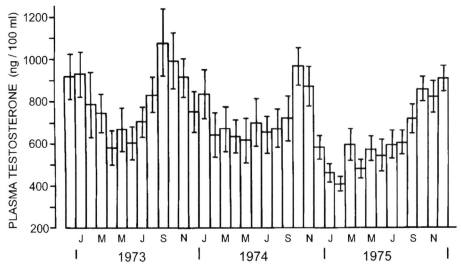

FIGURE 7-12. Changes in the plasma testosterone levels of male rhesus monkeys maintained in a constant photoperiod and paired daily with ovariectomized, estrogen-treated females throughout 3 years. The 13- and 14-month intervals between successive peaks suggest a free-running annual rhythm. (SOURCE: Michael & Bonsall, 1977a, with permission)

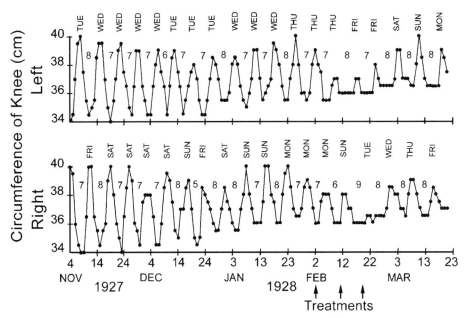

FIGURE 7-13. Intermittent hydrarthrosis. The numbers above the horizontal axes give the lengths of the cycles in days. (SOURCE: Richter, 1976, with permission)

BIOLOGICAL RHYTHMS

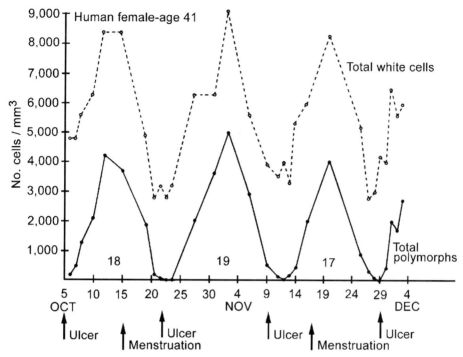

FIGURE 7-14. Cyclic agranulocytosis. The numbers above the horizontal axis give the lengths of the cycles in days. (SOURCE: Richter, 1976, with permission)

blood. Figure 7-14 shows 17- to 19-day cycles in total white cells and total polymorphonuclear cells per cubic mm of blood in a 41-year-old woman. Figure 7-15 shows the well-recognized Pel–Ebstein 24- to 26-day cyclic temperature rhythm in a 19-year-old male with Hodgkin's disease (lymphoma). There are also cyclic disturbances of mood and sleep in persons afflicted with affective disorders. A 5-day sleep cycle occurred in a 30-year-old woman who experienced 1–2 days of excitement, with as little as 2.5 hours of

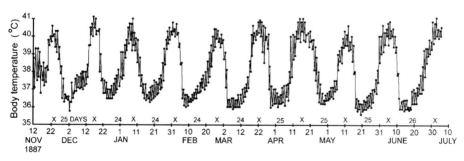

FIGURE 7-15. Hodgkin's disease. The numbers above the horizontal axis give the lengths of the cycles in days. (SOURCE: Richter, 1976, with permission)

sleep, followed by 2–3 days of sluggish behavior, with up to 10 hours of sleep. Over a period of 110 days, while kept in bed and on a constant diet, a 22-year-old male catatonic schizophrenic experienced a 19- to 21-day rhythm of wild excitement alternating with normal or slightly depressed mood.

In summary, as far as we can tell, all cellular organisms have circadian rhythms that may underlie rhythms with longer periodicities. In habitats subject to marked seasonal changes, many animals have circannual rhythms, and those living in or near tidal zones may also possess circatidal, circalunar, or monthly rhythms that allow them to anticipate predictable environmental changes. As described in Chapter 8, circadian rhythms are also essential for animals that use the sun to navigate, and both circadian and circannual rhythms are necessary for species that migrate or hibernate to escape periods of unsuitable environment conditions.

CHAPTER **8**

Orientation and Navigation

Animals need to orient themselves to various environmental stimuli to find shelter (nests, lairs, burrows), food, and mates, to nurture their young, to escape predators, and, the primary topic of this chapter, to stay within or return to favorable environments. *Orientation* as used here comprises turning the body in a particular direction, together with movement over short distances, while *navigation* is the term reserved for long-distance travel, including migration. Orienting stimuli can involve several different sensory modalities. Visual stimuli include the sight of mates, predators, or prey in many birds. Acoustic stimuli are used in echolocation by bats and porpoises to avoid obstacles and find prey. Chemical stimuli include insect and mammalian pheromones that are used to locate mates. Tactile signals enable male clawed frogs to detect an appropriate mate. Sharks use electrical fields to detect prey buried in sand, while certain snakes use thermal cues to detect warm-blooded prey. In addition to signals produced by animals, a variety of other exteroceptive cues, such as gravity, light, temperature, barometric pressure, humidity, magnetic fields, and air and water currents, can serve as orienting stimuli within an animal's immediate environment.

ORIENTING RESPONSES

An early classification of orienting responses was that of Kühn (1919), who developed the terminology and intended it to reflect the physiological mechanisms involved. The two primary categories, kinesis and taxis, are distinguished by whether or not the organism orients itself with respect to the *source* of the orienting stimulus.

Kinesis

Kinesis, in which the organism does *not* orient itself at an angle to a stimulus source (light, sound, odor), can nevertheless lead a simple organism into a more favorable environment and maintain it there by means of either orthokinesis or klinokinesis.

Orthokinesis

The animal changes its *velocity* on entering favorable or unfavorable conditions. For example, woodlice require a humid environment and will move quickly in areas of low humidity and slow down as humidity increases, thereby congregating under damp stones and in rotting vegetation.

Klinokinesis

The animal changes its *rate of turning* on entering favorable or unfavorable conditions. For example, the human body louse is affected by humidity, temperature, and odors. It moves in a straight line in an unfavorable environment and even when conditions improve. It begins turning in increasingly tight circles in a favorable environment and, consequently, remains where conditions are favorable.

Taxis

Most organisms orient themselves by the source of a stimulus such as light and gravity, maintaining some constant angle to it. The term used for this in animals is *taxis*, which implies movement of the entire organism. Another term, *tropism*, implies growth or movement of part of the organism and is typically used for plants (leaves, stems, roots). In practice, taxis is used in an animal even when only part of the body moves. There are five categories of taxis.

Klinotaxis

Klinotaxis involves the stimulation of a single receptor or several receptors that are not equally accessible to stimulation from all directions. For example, the fly maggot must move into a dark environment to pupate and has only a single anterior photoreceptor that is not stimulated by a light source located immediately behind it. The maggot swings its anterior end from side to side; each time the photoreceptor is stimulated, the animal twists its anterior end away from the light source, so that it finally comes to move in a straight line away from the light.

Tropotaxis

Tropotaxis involves the simultaneous comparison of stimulation received from bilaterally symmetrical receptors. For example, like many other insects, the fly tends to move directly toward a light source by orienting so that both eyes are stimulated equally.

A tropotaxis can be demonstrated experimentally by eliminating one of the two bilateral receptors, which causes the animal to move in circles.

Telotaxis

During telotaxis, the animal orients toward or away from a configurational stimulus, for example, toward the odor of a mate or away from the image of a predator. Eliminating one of the pair of bilateral receptors does not produce circling.

Menotaxis

Menotaxis involves orientation by maintaining a constant angle to a stimulus, for example, by keeping the sun at 90° to the right. The menotaxis of a solitary desert ant finding its way home after a hunting foray can be demonstrated by deflecting the sun by 180° with a mirror: the ant turns around and heads off in the opposite direction.

Mnemotaxis

Mnemotaxis involves orientation based on the memory of landmarks. By placing a ring of pine cones around the nest of a digger wasp and then relocating the cones several meters to one side after the wasp had left the nest to hunt, Tinbergen found that the wasp located its nest by memory, because it searched for it within the ring of cones upon its return (Fig. 8-1).

There are other classifications. One, unlike that of Kühn, differentiates whether the animal orients toward or away from the stimulus and between types of orientation stimulus, for example, light, gravity, water currents, sound, or odor. According to this classification, an animal is said to show a positive phototaxis if it moves toward a light source, and a negative phototaxis if it moves away from a light source. Similarly, animals may show a positive or negative geotaxis in response to gravity, rheotaxis in response to water currents, phonotaxis in response to sound, or chemotaxis in response to chemicals, including odors.

Tinbergen (1951) separated the taxis component from the behavioral component of orientation mechanisms, and pointed out that a fixed action pattern and its associated taxis may mature at different times during ontogeny. He illustrated this with the food-begging, gaping response of European blackbird nestlings, which are blind for the first 8 days after hatching. During this time, the gaping response is elicited when the nest is jarred; the nestlings simply stretch their necks and point their beaks upwards. After their eyes open, the same behavior is elicited by visual stimuli, but it takes several additional days before the nestlings orient toward the stimulus (Chapter 3, Fig. 3-3). An animal can possess several orienting mechanisms for different purposes: sound or odor to find a mate, and vision to find prey. Orienting processes may be affected by an organism's drive state. Many aquatic animals, including fish, possess a "dorsal light reaction" that enables them to maintain an upright position, or a ventral light reaction if they usually stay upside-down, as in the case of some aquatic insects found directly beneath the water

FIGURE 8-1. Mnemotaxis in a digger wasp (*Philanthus triangulum*): When leaving its nest, it quickly memorized a ring of cones placed round the nest (top). When it returned, it looked for the nest within the ring that was moved away from the nest in its absence (bottom). (SOURCE: *The Study of Instinct* by N. Tinbergen (1951) with permission of Oxford University Press)

surface. By changing gravitational pull using centrifuges and altering the position of light sources, von Holst (1950) found that angelfish (*Pterophyllum*) use both light and gravity to maintain a normal swimming posture, with the dorsum toward the light source above and the ventrum toward the pull of gravity below. Changing the positions of the two orienting stimuli relative to each other produced reliable deviations from the normal upright posture, but when prey was introduced into the tank of a hungry fish, the latter's

ORIENTATION AND NAVIGATION

posture became more closely aligned to the light source, probably because vision is important for hunting. An animal's internal state may also change the direction of its movement with regard to a stimulus—toward it at one time and away at another. Water beetle larvae orient away from light to hunt for prey in deep waters, but when in need of air, they swim toward light, first forwards and finally backwards, to obtain a bubble of air by contacting the water surface with the abdomen (Fig. 8-2).

NAVIGATION

Navigational Mechanisms

As noted earlier, navigation comprises orientation during travel over large distances, when an animal returns to its nest after long foraging or hunting trips, or when it

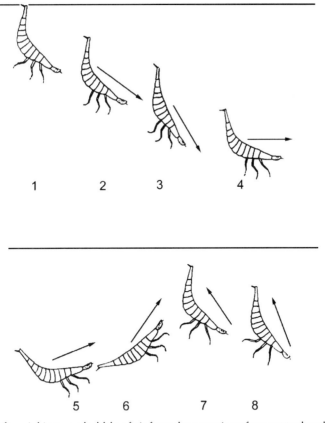

FIGURE 8-2. After picking up a bubble of air from the water's surface, water beetle larvae will swim away from light to hunt in deeper waters (top). When in need of air again, they swim toward light, first forward and then backwards (bottom). (SOURCE: Figure 2 in Schöne, 1962. Copyright © 1962 by Springer-Verlag)

migrates from its primary feeding to its breeding sites and perhaps back again. Four different navigational methods can be distinguished.

Piloting

Piloting is the simplest method of navigation because it relies on the use of memorized landmarks, such as those we use when going for walks. A good example of animal piloting is the mnemotaxis of Tinbergen's digger wasp, which found its nest by means of the ring of pine cones. Landmarks are perceived by different sensory modalities, including visual, auditory, and olfactory cues. Piloting can only be used for finding a precise location in a limited area. It cannot account for a migration over hundreds or perhaps thousands of miles, although it may be involved in the final leg of long journeys, just as we find our way home from the local airport after a transcontinental flight. This is the only type of navigation that does not require compass orientation.

Compass or Directional Navigation

Compass navigation, which involves travel in the direction of a specific compass bearing, is used in the migration of many birds. They travel in a given direction for a certain period of time; it is thought that they have a mechanism by which they can time events, together with the internal clocks to achieve this. It is known that the direction of travel is largely innate in some species. This has been shown by cross-breeding two populations of blackcap warblers with different migration directions (Chapter 4). More commonly, it has been shown by translocation studies. There are two distinct populations of European storks, a western group in West Germany and an eastern group in the Baltic. In fall, the western group flies southwesterly via Gibraltar to Africa, while the eastern group flies southeasterly over the Bosporus, also to Africa. If experimentally

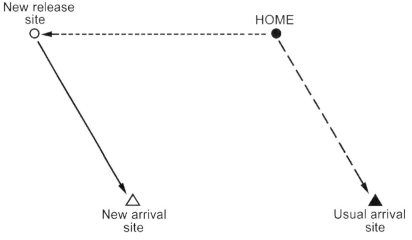

FIGURE 8-3. Schematic illustrating compass navigation (see text).

ORIENTATION AND NAVIGATION 123

translocated to the west, Baltic storks still take off to the southeast when released (Fig. 8-3), and consequently fail to reach their wintering grounds. However, if naive young Baltic storks meet up with western storks, they join the latter and follow their southwesterly course via Gibraltar.

Dead Reckoning

As implied by this nautical term, *dead reckoning* is a type of navigation similar to that employed by sailors who must compensate for deviations from their set compass course caused by wind and tide to find their home port. It requires precise records of the direction and distance traveled on each leg of the journey. The information is then integrated to determine the present position relative to home (Fig. 8-4). The desert ant, for example, travels long distances along irregular paths while on hunting forays but travels back to its nest in a straight line once it has captured prey. Since it cannot find its way home if it is transported by an experimenter, it seems that it remains aware of its position relative to its nest during the outward-bound trip by registering each turn and the distance traveled in the new direction.

Homing or "True" Navigation

The term *homing* is used when an animal can maintain its reference to a goal, regardless of its location and independent of any landmarks. It can find its destination even when it is transported inside a dark box to an unfamiliar location hundreds of miles away (Fig. 8-5). This involves compass navigation and also requires some kind of internal map to tell the animal its precise location at all times. At present, we know of very few species capable of this prodigious feat; foremost among them is the homing pigeon.

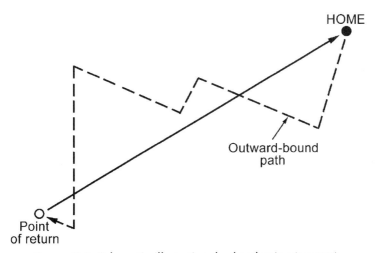

FIGURE 8-4. Schematic illustrating dead reckoning (see text).

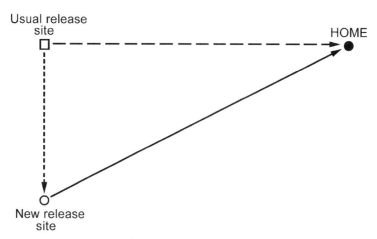

FIGURE 8-5. Schematic illustrating homing (see text).

The Compass Mechanism

All types of navigation, piloting excepted, involve the perception of a compass bearing. Although some migratory birds, such as European storks, have innate knowledge about which heading to take, it still leaves open the question of how they orient themselves during their travels. Many diurnal species use the sun for orientation (menotaxis), but the sun appears to move from east to west by 15° per hour (360° divided by 24 hours). Moreover, its compass bearing (azimuth) varies both with seasons and with latitude.

The use of the sun as an orienting cue, together with a timing mechanism that compensates for changes in position, was demonstrated in experiments with migratory birds such as European starlings (Kramer, 1952). Individual birds were placed outdoors in a closed, circular chamber provided with six narrow windows equally spaced around its circumference. So that researchers could work throughout the year and not just before the annual migration, the birds were trained to expect food in dishes that were always set at the same compass bearing. When tested at a given time of day without food, they oriented in the direction of the food dish at an angle to the sun entering the chamber. At the same time of day, the birds' angle to the sunlight was maintained when light was deflected by mirrors mounted to the right or left of windows (Fig. 8-6), demonstrating that the birds used the direction of sunlight as the orienting cue. As time progressed between 12:00 and 13:00 hours, however, the angle changed slightly (about 15°), suggesting that they were compensating for the change in the sun's position.

Involvement of the circadian rhythm in this compensation was shown by experiments in which the circadian rhythms of pigeons and starlings were phase-shifted by changing the light–dark cycle. The schematic (Fig. 8-7) illustrates how a phase shift of 3 hours either delaying (middle panel) or advancing (right panel) the circadian rhythm would cause a navigational error of 45° either to the right or to the left of the target

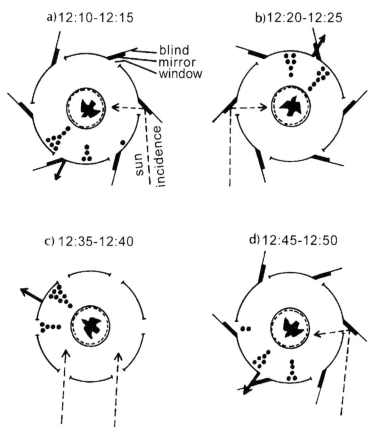

FIGURE 8-6. Compass orientation of starlings in test chamber varied with experimental condition [a) and d) mirrors at 135° left of windows; b) mirrors at 135° right of windows; c) no mirrors] but remained relatively constant with respect to the angle of sunlight (dashed lines), demonstrating orientation by the sun. However, the bird's angle to sunlight also changed about 15° with time [compare a) with d)], showing that the bird compensated for changes in the sun's azimuth. (SOURCE: Kramer, 1952, with permission)

direction. In fact, when released outdoors on a sunny day, pigeons whose circadian rhythms had been advanced by 6 hours set off 90° to the left of the correct direction of the home loft taken by control birds (Fig. 8-8, left). Interestingly, on heavily overcast days, the phase-shifted birds set off in the true direction like the controls (Fig. 8-8, right), indication that they have a backup system when they are unable to determine the sun's position (Keeton, 1969).

Navigational Cues

Animals rarely rely on a single navigational cue. Many sensory channels provide access to information that could be used in navigation:

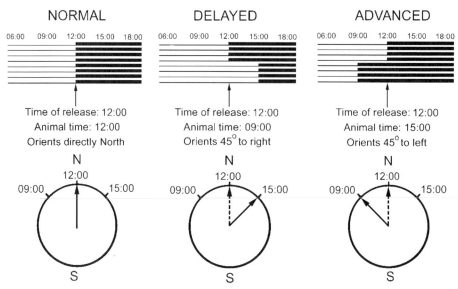

FIGURE 8-7. Schematic illustrating the relationship between the circadian rhythm of activity (top) and the predicted orienting response (bottom) of a hypothetical animal navigating due north (left) when its circadian rhythm is either delayed (middle) or advanced (right) by 3 hours.

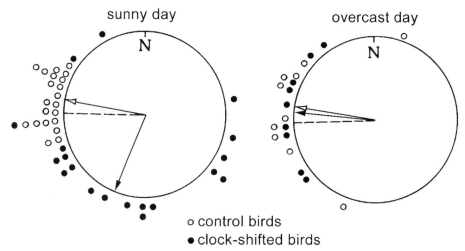

FIGURE 8-8. Under clear skies (left), clock-shifted pigeons (6-hour advance) navigated by the sun and set off 90° to the left of the true direction home, but under overcast skies (right), they set off in the correct direction, indicating a navigational backup system. Broken line = direction home; each dot = "vanishing bearing" of one bird; arrows = mean bearings for both groups. (SOURCE: Reprinted with permission from Keeton, 1969, Orientation by pigeons: Is the sun necessary? *Science* **165**, 922–928. Copyright © 1969 by the American Association for the Advancement of Science)

Visual: (1) landmarks, (2) sun,* (3) stars,* (4) moon,* (5) polarized light.*
Magnetic: (1) polarity,* (2) inclination,* (3) intensity.*
Electric: electrolocation
Sound: echolocation
Olfactory: odor

*The asterisks describe variables presently known to be used in compass, dead-reckoning, and true navigation.

It is beyond the scope of this chapter to review them, but a fascinating variety of navigational mechanisms have been uncovered by ingenious experiments with invertebrate and vertebrate animals. Many insects use polarized light. Certain aquatic insects use polarized light reflected from water to find suitable habitats; others, such as honeybees and desert ants, use the pattern of polarization to determine the position of the sun when it is blocked from view. Polarized light is used by birds that migrate at night but set their course at dusk, after the sun has set. Thereafter, these birds navigate by the stars. Studies on warblers and indigo buntings, for example, were conducted in planetaria, in which the positions of the stars can be changed. It appears that these migratory birds use the constellations that rotate around the stationary Pole Star (Northern Hemisphere); they do not need to see the Pole Star itself. Orientation by the moon, which changes position and shape during successive nights and is often invisible, is known in only a few species, notably, some shore-living crustacea such as beach hoppers. The earth's magnetic field can provide three relatively reliable navigational cues: its polarity, currently positive at the north pole; its inclination or dip, namely, the angle of magnetic force relative to the angle of gravity, which is smallest at the poles and about 90° at the equator; and its intensity or strength, which is greatest at the poles. The inclination of the magnetic field appears to be used as a navigational cue by more species than is polarity or intensity, for example, by loggerhead turtles, European robins, and homing pigeons. In some species sensitive to magnetic cues, for example, certain bacteria, honeybees, and many vertebrates, particles of magnetite, a magnetic form of iron oxide, have been found in clumps or chains that, in vertebrates, are usually located in the head. Just how these particles are involved in magnetic orientation remains speculative. Magnetic fields may be detected and used for orientation by induction of electric potentials. Certain cartilaginous fishes, including sharks and skates, which possess electrosensors used to locate hidden prey, generate electric potentials on their skin by swimming through the magnetic field in seawater, and voltage differences between different skin regions (e.g., on the back and belly) are thought to encode compass direction.

A species can use different navigational mechanisms (piloting, compass navigation) for different distances from the target, and different orienting cues (sun, magnetic fields) for different environmental conditions. The homing pigeon uses *compass navigation* by the earth's magnetic field and by the sun, and *piloting* by olfaction near its target and by visual landmarking when within about 10 miles of its target. There is some evidence that the "map" it requires for *homing* may be provided by regional anomalies in the earth's magnetic field, so-called geomagnetic anomalies.

One might ask how a compass, whether it depends on the sun, polarized light, stars, or other cues, is calibrated in the inexperienced animal. This has been investigated in a few vertebrates and imprinting-like learning processes appear to be involved. In pigeons, initial reliance is on the geomagnetic compass and several months' experience of both the 24-hour solar rhythm and the sun's apparent movement through the sky is required before animals use the sun as their primary compass. Indigo buntings migrate at night using a star compass, and fledglings learn to recognize Polaris as the star about which the constellations rotate. Those raised in a planetarium in which the constellations were set to rotate around Betelgeuse, a bright star in the constellation Orion, subsequently oriented with reference to Betelgeuse rather than Polaris. Salmon hatch in freshwater streams and then migrate to the ocean, where they spend several years before returning to breed in the streams in which they were born. While in the ocean, navigational cues appear to include the sun, polarized light, and magnetism, but olfactory cues are used to locate the particular river mouth and then the precise river tributary of origin. At least some of these olfactory cues are imprinted before salmon leave for the ocean, as shown by experiments in which salmon smolts were exposed to water tainted with either morpholine or phenethyl alcohol and, 18 months later, allowed to select between 19 streams, two of which had either morpholine or phenethyl alcohol added to them (Scholz et al., 1976).

MIGRATION

Migration is the term generally reserved for movements by an animal from one region to a distant region and its subsequent return. Many insects, birds, and mammals migrate from feeding to breeding areas in spring and back again in fall. Great distances may be traveled; for example, Arctic terns migrate from Greenland and other Arctic islands to Antarctica. Not content with this, they first cross the Atlantic to Europe, then turn south along the west coast of Africa and cross the southern Atlantic to reach Antarctica, a 23,000–25,000 mile annual round-trip. Some sandpipers journey 24,000 miles, so the Arctic tern is not unique, and banding studies have established the routes taken. The advantage of north–south movement is that the birds enjoy two summers. Barn swallows migrate from Alaska to central Argentina, a distance of 7,000 miles, and all these migrations are under the control of the photoperiod, but are sometimes triggered by a sharp temperature change. Apparently, in late autumn, after breeding is over and after the autumn molt, the pituitary and adrenal glands secrete increased amounts of prolactin and corticosterone that result in considerable fat accumulations for the energy needs of the journey. This does not happen in nonmigrating species. Migrating blackpoll warblers increase their body weights from about 11 grams to 22 grams before migration to fly nonstop over the Atlantic for 90 hours from the United States to South America. Most passerine birds migrate below 5,000 feet, but radar has shown that some species reach 20,000 feet and may travel 1,000 miles a day. Bar-head geese travel in spring from the northern Indian plains directly over the Himalayas to nesting sites in Mongolia. Certain species, including woodcocks, cuckoos, and thrushes, migrate mainly at night, and rest and feed during the day; perhaps this helps to avoid predation by hawks and gulls. Gray whales migrate some 6,000 miles from the Arctic to the Sea of Cortez, Mexico. Satellite transmitters attached to leatherback turtles have shown migrations southwest, from nesting sites on the west coast of Costa Rica to some 1,700 miles beyond the Galapagos Islands, using narrow and apparently unvarying pathways across the open Pacific.

Migrations are very costly in energy and mortality. Their functions are, however, thought to include a net energy profit when species leave cold latitudes for warm feeding grounds at the approach of winter, reduced competition by returning to less densely populated temperate zones, increased reproductive success due to longer days for parents to forage food for young in summer at higher latitudes, and perhaps escape from nonmigrating predators. The phylogenetic origins of migration remain unknown but may have involved responses to major climatic changes (glaciation) and continental drift that dissociated optimal feeding from breeding grounds.

CHAPTER 9
Feeding, Foraging, and Predation

A critical task for all living organisms is to obtain from their environment the nutrients required for energy expenditures, growth, and reproduction, and, in the short term, to maintain by homeostatic processes the constancy of the internal milieu. These various needs are mediated by behavioral mechanisms such as appetite and hunger, which lead to food-seeking behavior. Particularly in the human, these mechanisms can go awry and this results in disturbances in the control of body weight, leading to obesity or starvation.

Animals vary greatly in the foods they consume and how they acquire them. Many aquatic invertebrates, including mollusks such as clams and oysters, and some vertebrates, including various fishes and, among mammals, baleen whales, are filter-feeders; they have specialized structures that filter plankton and krill, which are small shrimp-like crustaceans, from the water. Terrestrial species are herbivores that feed on plants, insectivores that eat insects, carnivores that eat the flesh of other animals, and omnivores that eat both plants and animals. The type of food consumed is often associated with morphological and physiological specializations that have evolved to allow the animal to access, process, and digest its food. Familiar examples of such specializations include the grinding teeth and four-chambered stomach of ruminating, or cud-chewing, ungulates, including deer, camels, yaks, cattle, and sheep, with which they can grind up and digest grass and other vegetation. Anteaters have long thin snouts and protrusible, sticky tongues that allow them to extract ants from their nests. Many carnivores have claws and shearing teeth that help them to pull down prey, and to cut and tear flesh from the carcass. The adaptive radiation (divergent evolution) of Darwin's finches on the Galapagos Islands is one of the most famous examples of the relationship between morphological specialization and diet (Chapter 12). Dentition and diet are closely aligned. The omnivorous human possesses grinding molar teeth, remnants of canines, and sharp incisors. Examination of the dentition of ancient skulls has helped to identify the diet and mode of life of our extinct predecessors.

FEEDING BEHAVIOR

The association between morphological specialization and diet indicates that the type of food eaten, the sensory cues with which it is detected, and the behavior used to obtain and ingest it, are all at least in part genetically programmed. The degree of flexibility in the types of food eaten varies greatly between species. Among mammalian herbivores, for example, the giant pandas of China are famous for their almost complete dependence on bamboo shoots, which is one of several traits contributing to their impending extinction. Similarly, the koalas of Australia are dependent upon the leaves of eucalyptus trees, and are now carefully protected. In contrast, goats, also herbivorous, thrive in vastly different habitats throughout the world because they feed on a wide range of different plants, including lichen, moss, and bark, when grass and leaves are unavailable.

There is similar species variability in the extent to which experience and learning are involved in the selection of food items and in the behavior needed to acquire them. The leopard frog and European toad, for example, respond virtually automatically to very simple visual cues provided by their prey. These cues are sign stimuli for flicking out the tongue to capture fast-flying insects in the case of the frog or slow-moving earthworms in the case of the toad (Chapter 11, Visual Communication). In other carnivorous species, especially those with more complex brains, learning and experience are required to select appropriate prey and to catch it, and this is also true for noncarnivorous animals in choosing and obtaining access to appropriate foods. Learning from conspecifics, called socially facilitated learning, is most prevalent during development but may continue into adulthood.

Social Learning and Facilitation of Feeding
During Development

Kittens and puppies chase, pounce on, and bite any moving object almost from the time their eyes are open. However, learning from adults, as well as maturation and experience, are necessary to turn this basic tendency to hunt appropriate prey into the first independent kill. Cats are a solitary species and hunt alone. Most well-fed cats never learn to kill their prey effectively with a lethal deep bite, unless reared with a cat that does so, usually the mother. The role of social learning is greater in carnivorous mammals that hunt large prey cooperatively, for example, African hunting dogs, wolves, hyenas, and lions.

Before being weaned, the young of many birds and mammals, whether solitary or social, herbivorous, carnivorous, or omnivorous, are in close proximity to a parent when the latter is feeding, and are exposed to the food and behavior of the parent; moreover, they often sample scraps from the parent's meal. One of the most striking examples of the importance of this experience for the young comes from the omnivorous black rat (*Rattus rattus*), whose flexible, opportunistic feeding behavior contributes to its worldwide distribution. This species has found a niche, occupied elsewhere in the world by squirrels, wood mice, and other animals, in the relatively young forests of Jerusalem pine in Israel, which grow in acid soil that does not support any other vegetation or animals. The rats live and nest in the pine trees, rarely coming to the ground, and the energy-rich seeds of pine cones are almost their sole source of food. The seeds are located at the base of scales that are tightly packed in a spiral arrangement. The only way that a rat can obtain sufficient numbers of seeds to survive is to strip the scales individually, starting at the base of the cone, proceeding systematically upward in a spiral fashion, until only the central shaft remains. A series of laboratory experiments with black "stripper" rats from the pine forests, and "naive" rats from other habitats in Israel, have shown that the efficient pine cone-stripping technique is socially rather than genetically acquired (Aisner & Terkel, 1992). Naive adults and those paired with strippers gnawed randomly at the cones but were unable to develop the stripping technique, as were pups born to either stripper or naive mothers and reared by naive mothers. In contrast, pups born to either stripper or naive mothers but raised by stripper mothers acquired the stripping technique (Fig. 9-1); they observed the mother opening cones and eventually stole and continued stripping cones partially opened by her. Naive adults also developed the technique when supplied with cones that had already been partially stripped, but such cones are virtually never found on the ground in the pine forests. Studies on Norway rats (*Rattus norvegicus*) showed that the diet preferences of pups is also influenced by flavor cues in the milk of the mother that reflect her diet during lactation (Galef & Henderson, 1972). To control for this possibility in the black rats of the experiments, the naive mothers were fed exclusively on manually extracted pine seeds throughout lactation.

In Adults

Within limits, what is consumed and whether the animal eats at all at a given point in time can be influenced by the feeding behavior of other individuals nearby. Many of us have been reluctantly introduced to novel food items from other cultures and have come to enjoy them, and are familiar with the urge to eat when seeing others consuming something delectable. Some species of carnivorous butterfly fishes (*Chaetodontidae*) living in coral reefs are very difficult to maintain in captivity, partly because they very slowly starve to death in the absence of their normal prey, since they only occasionally sample substitute foods such as small shrimp, chopped mussels, and earthworms. If in the presence of an individual, not necessarily of the same species, that has fully accepted the substitute foods and eats them hungrily, however, they may be stimulated to eat more themselves and may eventually accept the foods fully.

In social species, the effect of socially facilitated learning is particularly strong; the highly social, omnivorous rat is again a good example. Rat colony members frequently investigate the head and snout regions of companions returning to the nest and are believed to obtain information about what has been eaten from odor and perhaps taste

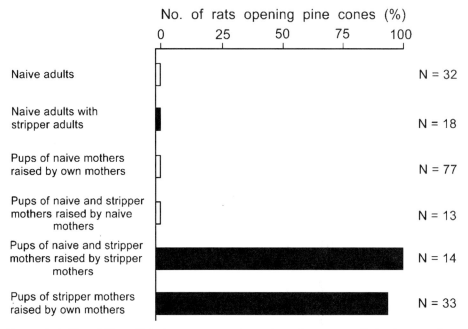

FIGURE 9-1. The ability of black rats to open Jerusalem pine cones efficiently enough to prevent starvation is not genetically determined. Learning the skill required that the naive individual be housed with a skilled "stripper" adult (black bars) and be a pup with access to partially stripped cones stolen from the female raising it. N = number of rats. (SOURCE: Aisner & Terkel, 1992, with permission)

cues clinging to the returnees' muzzles. If the food is a familiar one, colony members are informed by their inspections that the food is available nearby and they can search for it or follow the companion on its next foray. If, on the other hand, a rat finds a novel food, it eats only a small portion and will not return to it for some hours. If the food is poisonous the rat will become ill but, due to the small amount ingested, is unlikely to die. Associating the illness with the novel food, the rat never touches it again, and other members of the colony, presumably by associating its odor with the malaise of their companion, also avoid that particular food for several generations. This is why it is difficult to eradicate rats by baiting food with poison. Unlike rats, humans can not only expel toxic food by vomiting, but like rats, we also have a powerful avoidance response, perhaps lasting many years, to anything eaten before becoming nauseated and sick, even when the illness is known not to be caused by food poisoning. Some individuals will totally avoid a vacation spot if it has been associated with eating bad food.

FORAGING

Whether an animal feeds on plants or animals, it must locate and ingest the food efficiently enough to ensure that the energy per unit time expended doing so is greatly exceeded by the energy per unit time gained from caloric intake. For convenience, in this section "prey" will refer to both animal and plant food items.

Optimal Foraging

All species have to secure food, avoid predation, and reproduce. It has been suggested (Wilson, 1975) that time–energy budgets have evolved so as to match the times of greatest stringency, accounting for the observation that animals are often idle when food is plentiful. It is also reasonable to propose that the time and energy allocated to each activity are proportional to the payoff in fitness. Time and energy are usually allocated, in decreasing order, to securing food, avoiding predation, and reproduction. Periods of rest and sleep are also important. Species differ greatly in the proportion of their total time–energy budgets devoted to feeding because of differences in their size, mobility, and food requirements and in how their food is distributed. For example, to obtain adequate amounts of food, ungulates spend a large portion of the day grazing on evenly distributed vegetation, during which energy expenditure per hour is quite low. Cheetahs, on the other hand, typically engage in high-speed, mostly unsuccessful chases to catch large prey, during which energy expenditure is so high that they must rest for prolonged periods to recover. Optimality theory (Chapter 1) has been used extensively to investigate the influence of intrinsic and environmental variables in determining the foraging and feeding behavior of different species in terms of (1) the "decisions" made by the animal, such as which type of prey to consume and when to leave a depleted feeding site for a new one, (2) the "currency" being maximized, for example, net energy intake, and (3) the limitations imposed by the animal's physical attributes, which determine the nature and size of prey it can utilize (Stephens & Krebs, 1986). In addition, constraints on optimal foraging are imposed by environmental variables such as competition from conspecifics and threat from predation.

Food Item and Site Selection

Optimality theory predicts that animals behave so as to maximize their net energy intake, in other words, to obtain as much energy from their food, with as little energy expenditure as possible in obtaining it. Net energy intake is the sum of energy expended in locating the food, capturing it, handling it (e.g., the pine cone stripping by black rats to reach the seeds), and consuming it. Ancillary energy expenditure may accrue from defending the food source from competitors, either retrospectively, as when lions defend a kill against hyenas and vultures, or prospectively, by establishing and maintaining a feeding territory to keep away competing conspecifics, a common phenomenon in many fishes, birds, and mammals.

The prediction is supported by two studies on coastal birds preying on mollusks. Northwestern crows (*Corvus caurinus*) feed on whelks on the western coast of Canada. At low tides, these crows search along the waterline and, after picking up a whelk, fly away from the water and then almost vertically upward to drop the whelk on the rocks below. If the shell does not break, the whelk is dropped again until it does, allowing the bird to eat its contents. Crows generally select the largest, rarely the medium-size, and virtually never the smallest whelks. Experiments showed that all sizes were equally palatable, but that larger whelks broke more easily than smaller ones when dropped from the same height. By selecting large whelks and dropping them from an average height of about 5 meters, the crows maximized the amount of energy obtained and minimized the amount of energy expended in flying upward to break them; calculations showed that a net energy loss would result if crows selected medium or small whelks (Zach, 1979). Another shorebird, the oystercatcher (*Haemoptopus ostralegus*), preying in winter on mussels (a bivalved mollusk) in estuaries of northwest Europe, maximizes net energy intake in a somewhat different way. Oystercatchers either hammer a hole into one shell to extract the prey or stab between the two shells. Birds that hammer into the shell were found preferentially to select mussels around 40 mm long, not necessarily the largest (Fig. 9-2, left). However, when the numbers of available mussels were corrected for the fact that the birds generally avoid mussels that are difficult to open because the shells are either encrusted with barnacles or too thick, data showed that oystercatchers select the largest and most profitable prey (Fig. 9-2, right), as predicted by optimality theory (Meire & Ervynck, 1986).

Where different types of prey are available and equally easy or difficult to handle,

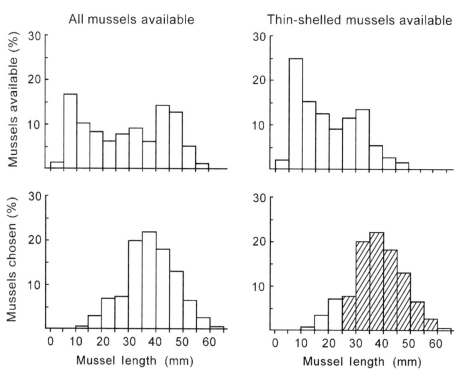

FIGURE 9-2. Prey selection by oystercatchers that extract mussels by hammering a hole in the shell. The birds did not preferentially select the largest mussels available in a location (left) but were found to select the largest mussels available when those with thick or barnacle-encrusted shells were omitted from the data (right). The hatched bars show the predicted optimal diet. (SOURCE: Modified from Meire & Ervynck, 1986, with permission)

optimality theory predicts that animals choose the most profitable prey, even if it is less abundant than the others, and that the less profitable items are eliminated from the diet as the most profitable item becomes more abundant. This prediction has been supported in experiments on several species and also seems to hold in humans. When hiking in the backwoods, we may be content to subsist on a variety of snacks and, if necessary, berries and insects—items that are abandoned as soon as there is a prospect of palatable cooked meals. In some cases, however, animals have nutritional needs that are not fully met by the energetically most profitable prey. For example, some large herbivores do not obtain sufficient salt from their usual diet and therefore periodically expend energy finding it. Elephants will travel hundreds of miles to find a cave containing rock salt to lick, and moose periodically eat aquatic plants that contain less energy but more sodium than terrestrial plants.

The food of many species is not distributed evenly but occurs in patches, being more abundant in some areas than in others at a given time; moreover, food abundance within a patch may be stable or fluctuate either predictably or unpredictably. Several models have been proposed to predict how animals would behave to assure optimal foraging and to avoid starvation under different environmental conditions. Considering the many variables involved, including the size and mobility of a species, and the nature and distribution characteristics of its prey, it is not surprising that individual models fit some species better than others. In general, however, it appears that animals do, in fact, maximize their food intake while minimizing energy expenditure by staying within or moving between patches. Chipmunks (*Tamias striatus*), for example, feed on seeds that have dropped from deciduous trees, and since trees drop seeds at different times, the animals depend on several patches with fluctuating food supplies. As predicted by one model, chipmunks began to spend more time and energy visiting other locations as the food density decreased at the exploited location.

Search Techniques

Even assuming that animals know the characteristics (visual, auditory, chemical, or tactile properties) of their appropriate prey, the latter are often intermingled with many other objects and species. Moreover, many prey are camouflaged to blend into their surroundings (see section on antipredator defense). This raises the question of how animals search for cryptic prey. Two techniques have been proposed and tested in a few bird species. The first technique is the formation of a search image, once a prey item has been found, whereby attention is focused on stimuli that readily distinguish that item from its surroundings. This search technique predicts an increase in the rate at which that particular prey type, but not others, is found. The second technique is a decrease in search rate, once a food item is found, whereby the animal slows down its scanning of the environment. This technique predicts that all types of cryptic prey are found equally easily, but each type is found more slowly than if a search image were employed. The following analogy illustrates the two concepts. You are asked to find the plays of Shakespeare and the novels of Alexandre Dumas on your friend's disorganized bookshelves, and know that each play and novel is separately bound. You happen to find two plays, both of whose spines have blue with gold lettering, after which you scan all the shelves and quickly discover all of Shakespeare's plays because you used the search image "blue spine with gold lettering." You begin scanning the shelves again for the novels, and find the first. Unfortunately, nothing distinguishes it from the surrounding books, so in the absence of a search image, you decrease your search rate to read the spine of every book in order to find the remaining Dumas novels. To distinguish between the two search strategies, ingenious experimenters have used operant techniques to measure the pecking responses of birds shown slides of several real (moths on bark) or artificial prey (letters of the alphabet) made more or less cryptic, or have tested the feeding responses of birds to real prey (beans and wheat) made more or less cryptic by changing the substrate. The results suggest that animals may use both techniques, although the evidence seems stronger for search image formation that for search rate adjustment.

Constraints on Optimal Foraging

Under natural conditions, optimal foraging is often constrained by competition with conspecifics and predator avoidance.

Competition

Where food is not widely and evenly distributed, optimal foraging can be constrained because competitors, generally conspecifics, reduce access to it. If a patch contains a renewable source of food adequate to sustain one individual or group over time, and if it is small enough for the animal or group to patrol and defend, a feeding territory may be established (Chapter 12, Territorial Aggression). This minimizes extensive travel in the search for food but incurs the costs of repelling intruders and prevents the territory owners, and especially any intruders, from foraging optimally. Immature butterfly fishes, for example, are constantly darting about on the coral reef defending their territories, so that interspersed feeding bouts can be rather brief. Moreover, the strongest individuals monopolize the richest areas, so that others must make do with poorer ones. In social groups, whether or not they defend a feeding territory, group members often have a dominance hierarchy in which subordinate individuals give way to higher-ranking individuals rather than risk a fight (Chapter 12, Dominance Hierarchy). When lions make a kill, for example, large, powerful, adult males are generally the first to feed, then the adult females, and finally the immature individuals. As described in Chapter 12 for herbivorous Japanese macaques (*Macaca fuscata*) that were supplemented with wheat dispensed daily in a small area, the nutrition and consequently reproductive success of subordinate individuals can be negatively affected by such seemingly benign competition. Where the food source is both patchy and ephemeral, so that it cannot be permanently defended, overt fighting is common. Vultures keep a close watch on each other as they circle the skies looking for a carcass; as soon as one bird spots a kill and drops toward it, others quickly follow until it is entirely covered by vultures pecking as often at each other as at the meat. Almost alone among vultures, the turkey vulture amazingly locates prey by smell; it needs to do so because it hunts over the forest canopy in Central America, while the carrion is mostly located on the forest floor, where it is invisible.

Predation

With the exception of a few species, including modern humans, most animals are preyed upon by others, so that optimal foraging is constrained by the constant need to be vigilant to the approach of predators and to take evasive action if necessary. There is now considerable evidence from fishes, birds, and mammals that animals are very sensitive to predation risk and alter their foraging behavior accordingly. In the Tsaobis Leopard Park, located in a semidesert region of Namibia, four social groups of chacma baboons (*Papio cynocephalus ursinus*) live in a location around an ephemeral river that comprises four habitats that differ with respect to food availability and relative risk of predation by leopards and formerly by lions. In this location, trees provide the sole source of food for baboons. The four habitats are, in order of decreasing food availability, woodland along the edge of the riverbed, the bed itself that also contained some trees, plains beyond the woodland, and the adjoining rocky foothills; the latter two habitats provide virtually no food at all. Leopards ambush and attack their prey from a distance of up to about 10 meters, so baboons are at greatest risk of predation in the woodland, where their visibility is restricted, and at least risk in the foothills. It was found that the baboons spent less time feeding in woodlands, and more in the river bed, than predicted by the energy-based ideal free distribution (foraging effort directly proportional to the amount of available food, which is one parameter of optimal foraging). However, they preferentially rested and groomed in the safety of the hills (Cowlishaw, 1997b). Moreover, the baboons spent much of their time on or near a safe refuge, a tree or cliff face (Fig. 9-3), and spent more time being vigilant when more distant rather than close to a refuge (Cowlishaw, 1997a). Similarly, highly vulnerable smaller species, such as various rodents, tend to forage near the safety of their burrows, where the food supply is already depleted, rather than in more plentiful but distant areas. Their feeding is frequently interrupted as they scan the surroundings, and the flight distance is greater when the burrow lies between the animals and their predator, so that they must run toward the predator to reach safety, than when the animals are between their burrow and the predator. As might be expected, the degree of risk taken increases with increasing hunger.

Coping with Changes in Food Supply

All animals are affected by changes in their food supply. These changes may be unpredictable, for example, caused by unusual weather patterns resulting in droughts, floods, or unusual temperatures, or predictable because they are a consequence of regularly recurring circadian and annual rhythms. As described in Chapter 1, r-selected species are adapted to coping with unpredictable environmental changes in a number of ways, which include having a short life span and a single, huge reproductive effort, so that at least some offspring survive to reproduce. In contrast, K-selected species are adapted to coping with predictable habitats; they typically have a long life span, reproduce repeatedly, care for their young, and compete with conspecifics for resources. Being longer-lived, they are exposed to more predictable seasonal changes in their food supply in most regions of the world, and several different mechanisms for coping with

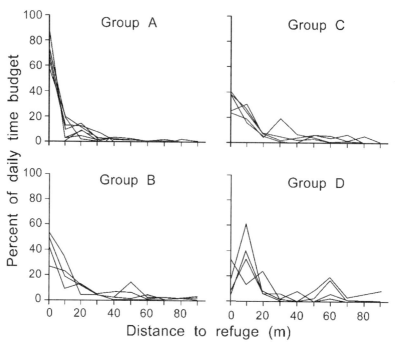

FIGURE 9-3. In four social groups of chacma baboons, adults spent a large proportion of their time on or close to a refuge (tree or cliff face) providing safety from predation by leopards. Each line represents one monitored adult. (SOURCE: Modified from Cowlishaw, 1997a, with permission)

this problem have evolved. One of these is migration, whereby animals follow their food source. Many large herbivorous species, such as wildebeest and other ungulates in Africa, regularly migrate hundreds of miles north and south each year as grasses and other vegetation sprout in one location shortly after the rains and die off in another, and carnivores that prey on these species follow them. Another way to provide against predictable food shortages is hoarding. This is primarily restricted to seed-feeding species, since most other plant parts and meat do not survive long-term storage. Squirrels and ground squirrels begin to bury nuts and seeds in the ground during the fall and retrieve them during the winter. In very cold climates, such as on high mountains or near the Poles, dwindling food supplies during winter do not compensate for the increased energy expenditure due to increased effort in searching for food and increased loss of body heat. Nevertheless, some species survive by hibernating during the winter. Alpine marmots and omnivorous bears of North America, for example, accumulate extensive fat deposits during the summer and fall, and then retreat into burrows or caves, where they fall into a deep sleep. Their metabolic rate remains at low levels until it gradually increases again, as the animals arouse in the spring.

K-selected species are, of course, also exposed to unpredictable food shortages due to events such as extreme droughts, floods, or the abrupt loss of the species on which they prey. This usually results in the starvation and death of many individuals in the population, or at least in reproductive failure. Barn owls lay fewer eggs or none at all if the rodent population in their home range declines drastically: The adults conserve their energy and postpone reproduction until the next year, when circumstances may be more favorable. Unpredictable, short-term fluctuations in food availability, even when not resulting in starvation, can have profound effects in highly K-selected species, as demonstrated by a laboratory study on lactating female bonnet monkeys (*Macaca radiata*) (Rosenblum & Paully, 1984). Bonnet monkeys are highly social macaques that have a multimale social organization (Chapter 16). Three groups of bonnet monkeys, each consisting of 5 mother–infant dyads, were housed in separate pens. All infants were 4–17 weeks old at the beginning of the study, too young to utilize solid food. Each pen was equipped with several food containers that consisted of many individual compartments. The monkeys could not see which compartments contained food and which did not, so they had to forage inside each compartment to obtain food. In one pen, all compartments were supplied with food, so that the mothers had a "low foraging demand" (LFD) to feed themselves. In another pen, only a few compartments in each food container were supplied with food, so that mothers had a "high foraging demand" (HFD) to meet their daily nutritional needs. In the third pen, the adult monkeys were subjected to a "variable foraging demand" (VFD), namely, LFD for 2 weeks, alternating every 2 weeks with HFD. This schedule continued throughout the 14-week study period, during which all mothers maintained their body weight and all infants exhibited

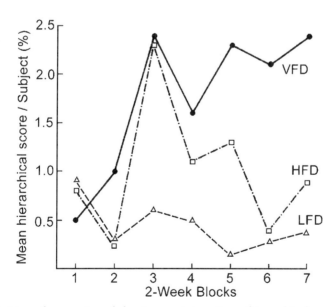

FIGURE 9-4. Mean frequencies of dominance interactions (hierarchical score) increased temporarily between bonnet monkey mothers subjected to high foraging demand (HFD) conditions but permanently between those experiencing variable foraging demands (VFD). (SOURCE: Rosenblum & Paully, 1984. Copyright © 1984 by the Society for Research in Child Development, Inc.)

the weight gain typical for their age, suggesting that the foraging demands did not result in any nutritional deficits.

Nevertheless, the different foraging demands greatly affected the feeding, social, and maternal behavior of the monkeys. The LFD mothers wasted about 50% of the food and spent little time foraging, whereas the HFD wasted virtually no food and foraged for fivefold the amount of time spent foraging by LFD females. The VFD mothers showed similar patterns, wasting food and spending little time foraging during LFD periods and not wasting food but foraging longer during HFD periods. However, it took them several days to adapt to each condition. HFD mothers exhibited a temporary increase in agonistic interactions, combined with a decrease in social and affiliative interactions, but a similar increase in social disruption was permanent in VFD mothers (Fig. 9-4). Infants at this age normally spend increasingly more time off their mothers playing with peers, and are responsible for initiating such bouts. The infants of both the HFD and the LFD mothers showed these developmental changes, but the infants of the VFD mothers spent *less* time off their mothers (Fig. 9-5), and tried to maintain contact with their mothers, which themselves forced infants away (Chapter 15, The Conflict Model). Evidently VFD mothers could not adjust to short-term changes in foraging

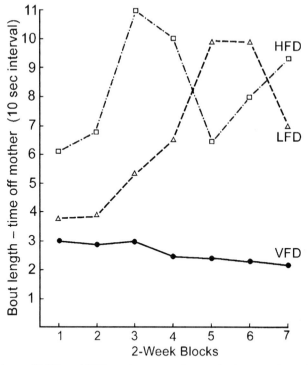

FIGURE 9-5. Infants of HFD and LFD mothers spent increasingly *more* time off their mothers as they got older, whereas the infants of VFD mothers tended to spend *less* time off their mothers. (SOURCE: Rosenblum & Paully, 1984. Copyright © 1984 by the Society for Research in Child Development, Inc.)

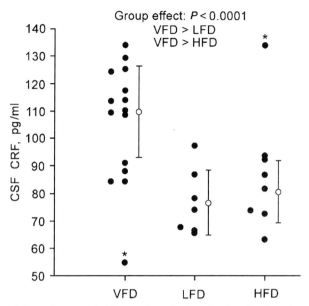

FIGURE 9-6. As adults, offspring of VFD monkeys had higher levels of corticotropin releasing factor (CRF) in their cerebrospinal fluid (CSF) than did other offspring—a phenomenon associated with major depression and anxiety disorders in humans. (SOURCE: Modified from Coplan et al., 1996. Copyright © 1996 by the National Academy of Sciences, USA)

demands and consequently they attempted to wean their infants prematurely, which would enable them to survive this period of unpredictable food availability and to reproduce again under more favorable circumstances. Under natural conditions, such short-term fluctuations in food availability would be rare because food sources typically do not die off and replenish every 2 weeks.

Rejection by their mothers had profound and permanent effects on the VFD infants, which developed emotional disturbances similar to those of Harlow's socially isolated rhesus monkey infants (Chapter 16). Three of the five infants were severely depressed, sitting curled up and clasping their own bodies or that of a peer. This prompted the termination of the study. Subsequent studies on the infants from all three groups as adults showed that the physiology relating to stress and coping responses was permanently affected in the VFD group. Their cerebrospinal fluid (CSF) contained elevated concentrations of corticotropin releasing factor (CRF) (Fig. 9-6) and lower cortisol levels than the offspring of LFD and HFD mothers. Elevated CRF levels are found in patients with major depression and anxiety disorders and, together with depressed cortisol levels, in patients with posttraumatic stress disorder. It is not unreasonable to suppose that the results of these studies on bonnet monkeys may have relevance for humans, not least in developed and developing countries, for whom social and financial perturbations beyond individuals' control can easily produce short-term fluctuations in access to food and other essential resources.

Feeding in Humans

Diet

Most primate species are either insectivorous or herbivorous. Among Old World primates, some species supplement their herbivorous diets with insects, crustacea, and vertebrates, including mammals. For example, cynomolgus monkeys, also known as "crab-eating" macaques, may eat crabs, chimpanzees eat termites, and baboons and chimpanzees kill and eat small mammals. Chimpanzees are also known to hunt, kill, and eat colobus monkeys. Arguably, humans are the only true omnivores among primates, and this omnivory was facilitated by the use of fire to boil and thereby break down cellulose and deactivate toxins in plants. Meat is also tenderized and "cured" for preservation by boiling. Cultural differences in the food items considered edible and desirable, and in how they should be prepared, are profound, but all humans require certain nutrients in their diet that reflect evolutionary selection pressures that presumably shaped our omnivorous diet thousands of years ago. Humans need a stable supply of carbohydrates, typically obtained from rice, wheat, corn, and other grains, or from starchy tuberous roots such as manioc and potatoes, vegetables rich in vitamins and minerals, and proteins that are present in some plants and seeds but found in much greater amounts in meat and animal products, including eggs and milk.

Humans also require small amounts of sugars, salt, and fats. During much of human history, these three nutrients would have been scarce in most habitats. The primary sources of sugars are fruits and honey. Fruit is an ephemeral sugar source because most trees and shrubs fruit only briefly once a year, and wild honey is difficult and dangerous to procure because nests are usually located high in trees and bees take violent exception to disturbance of their nests. Salt, for eating and for preserving fish and meat, is also a rare and precious commodity in most habitats, being found principally near coasts or in certain rock formations. The importance and value of salt is reflected by the fact that part of the regular pay of Roman legionnaires was money specifically for buying salt, called *salarium* (from the Latin *sal*), from which the word *salary* is derived. Fats are also important for humans, especially during the first few years of life. Fats, together with proteins, other nutrients, and antibodies, are provided by the mother's breast milk; in most preindustrial cultures, infants were nursed for about 3–4 years. During most of human evolution, before animals were herded and domesticated, fats would have been available only occasionally and in small amounts, as anyone will know who has eaten game, namely, meat from wild birds and animals. Game is so lean that it needs to be cooked with additional oil, butter, or lard.

Taste and Odor

It is probably not a coincidence that humans, like some other mammals, have a well-developed taste for both sugar and salt, which can be detected at very low concentrations and are considered pleasant. Moreover, we find foods containing either salt or sugar combined with fats more palatable and satisfying than those without these ingredients. There is reason to suppose that the tastes for sugar and salt evolved to ensure that humans made the effort to find these essential nutrients when they were scarce and difficult to obtain. Where they are plentiful and cheap, and included in large amounts in virtually all processed foods, even the most sedentary among us continue to eat them in quantities that are far beyond our nutritional needs. Unless under doctor's orders because of obesity, diabetes, or elevated blood pressure, few can resist adding some kind of dressing to the salad, putting salt and either sour cream or butter on a baked potato, and finishing the meal with a piece of sweet pie or ice cream. The fact that sugar, salt, and fat are consumed to excess when readily available suggests that they may be supranormal stimuli (Chapter 3). Although overindulgence is associated with severe health problems and premature death, symptoms usually do not occur until late middle age, namely, after people have reproduced. Consequently, there is no selection pressure to counteract our predilections for excessive salt, sugar, and fat consumption.

Humans are also very sensitive to sour and bitter tastes. Within limits, sour taste, as in the tartness of slightly unripe apples or the citric acid in ripe citrus fruits, is often considered pleasant and may have evolved because it signals the presence of vitamin C. Bitter taste, on the other hand, is usually considered unpleasant and avoided, perhaps because some foods taste bitter when they spoil due to bacterial action (e.g., milk). The odor of rotting vegetation, meat, and eggs is considered very unpleasant and remains a very useful signal to protect us from many forms of food poisoning.

Some Physiological Aspects of Feeding

Bilateral destructive lesions of the ventromedial nucleus of the hypothalamus (VMH) result in hyperphagia (Brobeck *et al.*, 1943). In lesioned rats, the circadian feeding pattern is lost, so eating continues throughout the 24 hours. Meals are both larger and more frequent, the weight gain can be enormous, and it appears that animals are unable to attain satiety. Lesions in the lateral hypothalamic area (LHA) (Fig. 9-7), on the other hand, result in hypophagia and weight loss (Anand & Brobeck, 1951), suggesting that this area is responsible in some way for appetite and hunger. There is an abnormally high level of insulin in obese animals, and it has been hypothesized that there are different receptors in different hypothalamic neurons with glucostatic, aminostatic, and lipostatic functions. But we lack hard evidence both for the existence of these receptor neurons and for the precise neurotransmitters involved in their regulation. It has been proposed more recently (Panksepp, 1974) that there are both short-term and long-term satiety mechanisms involving the LHA and VMH, respectively. The short-term mechanisms are responsible both for the initiation of feeding and its short-term inhibition by gastric distension and other signals from the gut and liver via the vagus nerve to the LHA (lesions of which result in hypophagia). On the other hand, the VMH is responsible for the longer-term regulation of energy balance (VMH lesions result in hyperphagia). Our understanding of the regulation of food intake and the energy balance took a step forward with the sequencing of the *obese* gene that expresses a protein product (Zhang *et al.*, 1994) subsequently termed leptin. This hormone is secreted by adipose tissue, and its circulating levels reflect the mass of fat in the body. It is now regarded as a signal regulating long-term energy balance, and injections of leptin reduce food intake and reverse experimental obesity. Several appetite-enhancing neuropeptides are known, including neuropeptide Y, opioids, and melanin-concentrating hormone, but recently endogenous peptide ligands have been identified, leading to the discovery of orexins A and B, together with their receptors, which are localized in the LHA (Sakurai *et al.*,

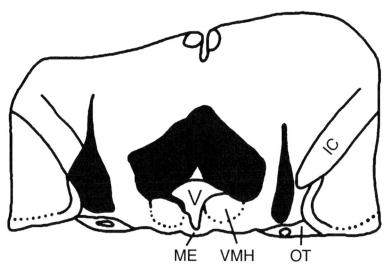

FIGURE 9-7. In a rat, lesions (black areas) in the VMH induced hyperphagia and obesity, and subsequent lesions in the LHA (lateral hypothalamic area) abolished feeding. IC = internal capsule; ME = median eminence; OT = optic tract; V = ventricle; VMH = ventromedial nucleus of the hypothalamus. (SOURCE: Modified from Anand & Brobeck, 1951, with permission)

1998). When administered to rats, orexins stimulate a short-latency increase in food consumption. The clinical implications for controlling obesity are promising.

It should be mentioned that the small intestine does not digest nonstarch polysaccharides efficiently. These are the main component (cellulose) of plant foods consumed by protein-producing, herbivorous cattle and sheep, our main source of meat. Herbivores depend on bacteria for digestion. In ruminants, the fermentation chamber is a multichambered stomach. Animals with this are known as foregut fermenters. Other herbivores, such as rodents, lagomorphs (rabbits and hares), and apes, do not have multichambered stomachs but have very enlarged bowels and caecum, and are known as hindgut fermenters. In some of these species, feces eating, or coprophagy, is employed to allow the products of fermentation in the large bowel to be absorbed via the small intestine during the second passage.

PREDATORY TECHNIQUES AND ANTIPREDATOR DEFENSE

Somatic Adaptations

Energetically the most cost-effective method for capturing prey and avoiding capture is to remain undetected. Being largely visually oriented ourselves, humans are aware of adaptations in the appearance of animals that make them difficult to detect, but it is possible that there may be olfactory and auditory mechanisms similar to the two visual ones described below: crypsis and mimicry.

Crypsis

Many animals are difficult to detect in the wild because they blend into their surroundings, and this may be achieved in several ways.

Matching Color to Background. Some invertebrates (e.g., cephalopods such as the octopus, insects), fishes, amphibians, and reptiles are capable of changing their appearance quite rapidly to match their current backgrounds by expanding and contracting pigmented cells within or just below the integument. The proverbial example is the chameleon. Birds and mammals, of course, cannot do this, but some in montane or polar habitats shed their feathers (ptarmigan) or pelage (Arctic fox) twice each year so that they are brown or gray during the summer and white during the winter. Other animals do not change their appearance at all, but tend to remain or come to rest on the substrate that best matches their appearance; many moth species look exactly like the bark of the species of tree on which they feed and deposit their eggs. Among snakes, the venomous copperhead is practically invisible in dry leaves.

Countershading. Many invertebrate and vertebrate species are darker on their upper than their lower surfaces. This phenomenon is known as countershading. The backs of deer, for example, are brown; the color becomes increasingly pale down the sides, and the belly is virtually white. Since light comes from above, the back is usually brightly illuminated while the belly is in deep shadow. Consequently, at a distance, the visual effect is that of a two-dimensional brownish patch. Countershading is also effective in water when animals are seen from the side; moreover, predators and prey coming from below see white against the bright surface light, while from above, they see dark against dark.

Breaking Up the Outline. In addition to countershading, many animals have colors similar to those of their normal background, together with spots or stripes that effectively break up their outline. In woodland habitats where light shining through trees produces irregular patches of light and shadow, animals often have dark spots on a paler background or vice versa, for example, leopards and some deer. In habitats with tall, slender vegetation (e.g., bamboo), irregular vertical stripes, as on tigers, are more effective. In prey living in dense herds on open grasslands, such as zebra, stripes may also break up the outline of animals sufficiently to make it difficult for the predator to target a specific individual.

Mimicry

Mimicry is found in many behavioral contexts and can also be an advantage for both predators and prey. In some cases, the entire body takes on the general appearance of inanimate objects such as leaves or twigs, for example, the praying mantis and stick insect; this appearance is associated with minimal and slow locomotion. In other cases, the entire body, or part of it, mimics another species, and this involves movements similar to those of the mimicked species.

Among predators, the mimicked animal is one that poses no threat to the prey. The alligator snapping turtle (*Macroclemys temminckii*), resting with its jaws wide open on the bottom of muddy lakes and rivers, attracts fish into its mouth by wiggling two red, worm-like protrusions at the tip of its tongue. On coral reefs, many fish have their gills and bodies cleaned of ectoparasites by "cleaner" fishes, notably, the Pacific cleaner wrasse *Labroides dimidiatus*, which is strikingly colored, with a longitudinal black band surrounded by neon blue fading to a white belly. It performs a dance involving vertical undulation that attracts other fish, which spread their gill covers and fins to be cleaned. Both the appearance and the behavior of the cleaner wrasse is mimicked by a predator, *Aspidontus taeniatus*, which then darts in and takes a quick bite out of the unsuspecting fish waiting to be gently cleaned.

Among potential prey, the animal being mimicked is usually a species that advertises with warning coloration that it is inedible because it stings or is poisonous, this is known as Batesian mimicry. Several wasp species, for example, are easily recognizable because they have very striking black bands alternating with yellow ones. They sting when captured, and insectivores rarely try to catch one after a painful experience. Their appearance or buzzing is mimicked by perfectly edible flies.

Predatory Techniques

Most of the many specialized methods for catching mobile prey found throughout the animal kingdom can be classified into one of the following three main categories.

Trapping

Trapping generally involves the construction of pits, webs, or other devices and staying nearby to capture prey that stray into them; the orb webs of garden spiders are a familiar example. Humpback whales use a somewhat different form of trapping, whereby they surround a school of herring or other fish with a net of air bubbles that prevent the fish from scattering. The whales dive below the fish and then spiral upward around them while blowing air from their blow holes, after which they quickly gulp in huge numbers of fish.

Ambushing

Ambushing involves moving stealthily to a location with many prey and lying in wait until one of the prey is sufficiently close to be captured; it is the ambushing predators that are frequently well camouflaged. Examples are the praying mantis among insects, and leopards, tigers, and sometimes lions among mammals. Olfactory signals are important for prey detection in mammals, so ambushing mammalian predators typically attempt to lie in wait downwind of prey.

Chasing

Many predators rely on superior speed to chase down their prey. The cheetah, for example, typically stalks its prey until it is within relatively close range and then gives chase at about 40 miles per hour and, in short spurts, in excess of 60 miles per hour, which is energetically too expensive to maintain for very long. Other familiar mammalian predators, including lions, hyenas, and wolves, are social species that chase in groups, enabling them to kill prey that are larger and stronger than themselves (Chapter 10).

Antipredator Defense
Repelling Predators

Predators are repelled with a variety of morphological and physiological adaptations that, in combination with appropriate behavior patterns, allow prey to fight off or deter their attackers. Many ungulates, for example, can effectively fight and damage predators with weapons such as sharp hooves or horns. Some species, usually ones that cannot move very rapidly, have protective armor. When threatened, woodlice and armadillos curl up into a ball that is completely covered by their hard dorsal plates; hedgehogs do likewise while raising their sharp spines, and porcupines also erect their spines. Noxious substances can also repel or deter predators. In real or perceived acute danger, many birds and mammals, including nonhuman primates and humans, explosively defecate and urinate on taking flight. Other animals, such as bombardier beetles and skunks, expel noxious chemicals from specialized glands when startled. As noted earlier, some species, including certain insects and snakes, advertise with warning coloration that they contain poisons or unpalatable substances, or administer them with stings or bites; there is evidence that predators instinctively avoid preying on animals, including Batesian mimics, with particularly dramatic color patterns.

Diversion

Some species have bilateral eyespots, usually black surrounded by a pale area, located on an area distant from the head, for example, on the wingtips of some butterflies and moths, near the tail of many fishes, or on the posterior of some toads. Sometimes the real eyes are camouflaged—by a vertical black stripe in the case of butterfly fish. It is thought that small eyespots confuse the predator, which expects the prey to move eyes-first, while large eyespots may startle the predator into hesitating long enough to allow the prey to escape. When threatened, several small lizards and the slowworm, a legless salamander, detach their tails, which continue to wriggle and convulse; this diverts the predator's attention while the animal escapes unharmed. Other types of diversion rely on behavioral mechanisms. Several species of ground-nesting birds, including killdeer and meadowlarks, feign injury when a predator approaches the nest, luring it away by fluttering slowly and apparently helplessly away from the nest. Some animals exploit the fact that many predators kill only live prey by feigning death until the predator departs; the famous mammalian example is the opossum, which gave rise to the phrase "playing possum."

Group Defense

As described in Chapter 10, group living provides several advantages in predator defense and avoidance. Among the latter is increased vigilance, which in some species

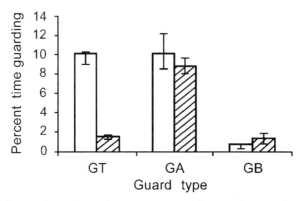

FIGURE 9-8. Vigilance of social meerkats in a nature park (open bars) and on ranchland with fewer predators (hatched bars). Guarding by a sentinel from a raised vantage point (GT) was more prevalent where predator pressure was higher, but vigilance behavior during foraging (GA) and at a burrow entrance (GB) did not differ with predator pressure. Error bars give interquartile range. (SOURCE: Reprinted with permission from Clutton-Brock et al., 1999, Selfish sentinels in cooperative mammals. Science **284**, 1640–1644. Copyright © 1999 by the American Association for the Advancement of Science)

involves individuals in the group taking turns to act as sentinels for the entire group. Sentinels often take up position on raised ground or trees to scan a large area, and utter alarm calls when they detect an approaching predator. One such animal is a species of African mongoose (*Suricata suricatta*) known as meerkats, whose behavior was studied in the Kalahari Gemsbok Park, where predator pressure is high, and in neighboring ranchland, where predator pressure is lower (Clutton-Brock et al., 1999). Meerkats dig into soil to find invertebrate and small vertebrate prey, and look up occasionally while foraging ("guarding away," GA). Group members take turns acting as sentinels on a mound or dead tree near a burrow entrance ("raised guarding," GT). They utter repeated calls to communicate that a guard is on duty and alert others to predators with graded alarm calls. Figure 9-8 shows that sentinel behavior, unlike other vigilance behavior, is more frequent in locations with high rather than low predator pressure. It is interesting to note that guarding behavior is generally regarded as hazardous, since the sentinel is more exposed than others to predation, and therefore is believed to have evolved through either direct ("kin") selection or reciprocal altruism (Chapter 10). Meerkats sentinels, however, were less likely to be killed than other group members, probably because they stayed close to a burrow entrance and were the first to detect a predator. This, together with the finding that food provisioning increased GT, suggests that this form of guarding may have evolved through direct selection, because it seems to be the optimal activity when the individual is not hungry and there is no other individual on guard.

CHAPTER **10**

Social Behavior

SOCIAL SYSTEMS

Different types of social systems have evolved independently under the influence of selection pressures that affect the inclusive fitness of the individual. Major selection pressures are imposed by the need to obtain food, to avoid predation, to reproduce, and to ensure the survival of offspring and other kin. These selection pressures are not always compatible, resulting not only in the benefits but also the costs of sociality (see below). Assemblages of conspecifics can be classified into three basic types of social group:

1. *Aggregations* form when many individuals congregate in an area because it provides environmental or climatic features that attract and suit them. There is no attraction between the animals themselves, which remain anonymous. For example, woodlice typically congregate in warm and humid areas under rotting logs (Chapter 8), thousands of monarch butterflies congregate in certain sheltered sites in Mexico and California for the winter, and snakes that hibernate during winter amass by the hundreds in protected underground sites.

2. *Anonymous groups* consist of individuals that are attracted to groups of conspecifics but, as in aggregations, individuals are anonymous. Anonymous groups may be *open* and joined by other conspecifics, for example, flocks of starlings and schools of herring, or they may be *closed* when acceptance is restricted to individuals identified as group members, for example, by odor, as in many social insects (ants, termites, honeybees).

3. *Individualized groups* comprise animals that recognize each other individually, for example, elephants, lions, wolves, rats, nonhuman primates, and humans.

Social systems are, of course, largely a consequence of how conspecifics interact with each other, and social behavior is defined as *behavior that influences and in turn is influenced by the behavior of other conspecifics*. Even animals that live solitary lives other than for brief mating encounters influence and are influenced by conspecifics, and are then said to have a solitary social organization. Immature butterfly fish, adult hamsters, tigers, and some bears, for example, avoid direct contact with conspecifics and, if their territory is invaded by one, they will actively repel the intruder (Chapter 12, Territorial Aggression). Social behavior also occurs in open and closed anonymous groups and in individualized groups, but not in aggregations, in which individuals mainly respond to environmental cues rather than to conspecifics.

Before examining the benefits and costs of sociality, it is worthwhile to consider a few unusual examples of social organization that occur primarily among invertebrates, and to summarize briefly the basic types of social organization in vertebrates.

Coelenterate Colonies

Animals of the phylum *Coelenterata* are aquatic and include single individuals, such as the hydra, but most species are colonial, such as corals. The basic structure of hydra, and of the polyp or zooid of colonial species, is a tube consisting of two cell layers, with an opening or mouth surrounded by a ring of tentacles. In most colonial species, different polyps are specialized to perform different functions: feeding, reproduction, and defense (their tentacles contain stinging cells called nematocysts). Colonies of sea anemones, such as *Anthopleura elegantissima*, defend themselves against each other with smaller "warrior" polyps around the colony boundary that have more nematocyst-laden fighting tentacles than do polyps in the middle of the colony. Hydrozoans reproduce sexually but can also do so asexually by budding off new polyps, and some colonial species consist of unrelated zooids, others of mixtures of unrelated and related zooids, and still others exclusively of related zooids. In the latter case, the entire colony originates from a single zygote and may be considered a single individual with different organs performing different functions. A striking example of this is the Portuguese man-of-war (*Physalia*), which is feared by coastal vacationers in Europe. It consists of four types of zooids: One type forms a large float that is visible above the water, a second type specializes in feeding, a third in reproduction, and the fourth in defense, producing long tentacles that inflict very painful stings on unwary swimmers.

Eusociality in Insects

Many insects are solitary and others are group living, but some species of ants, wasps, termites, and bees have an extreme form of sociality called *eusociality*. Eusociality is characterized by a social "caste" system in which some individuals reproduce while others are sterile workers whose activity increases the reproductive output of the entire colony. This type of social system is associated with the overlap of different generations and a high degree of relatedness between members of different castes.

SOCIAL BEHAVIOR

The honeybee is perhaps the best known example. The young queen bee is fertilized by a male "drone" once in her life, during her "nuptial flight," and stores the sperm in her spermatheca. She returns to the home hive to become an egg-laying machine, while nonreproducing workers feed the queen, tend the young, build the nest, guard the nest entrance, and forage for food. The queen lays fertilized eggs throughout the summer that develop into females (workers and potential queens), but when the sperm store is depleted, she lays unfertilized eggs that develop into drones. From the pharyngeal gland on their heads, workers secrete nutritionally rich "royal jelly" that is briefly fed to all the youngest larvae, most of which become female workers, but those destined to become queens (which later fight each other to replace the mother) are fed royal jelly throughout larval life. The queen secretes a pheromone, oxodecenoic acid, known as the queen substance, which is taken up by the workers tending her and passed throughout the colony by food sharing. This suppresses the ovaries of workers and prevents them from rearing new queens. If the queen dies or becomes ill, so that the workers are no longer under the influence of the queen substance, they begin to rear queens and may even lay eggs. In the course of adult female workers' 6 weeks of life, their roles in the colony change. During the first week, they clean brood cells and then feed older larvae a mixture of pollen and honey. During the second week, they feed royal jelly, after which their pharyngeal glands regress and wax-secreting glands develop on the abdomen, and they transition to cell construction. A few days later, they guard the hive entrance and undertake orientation flights. From the start of the third week until death, they forage for food. This timetable is somewhat flexible and may change according to the needs of the colony as a whole.

In honeybees, then, all females except the queen rotate through the different tasks that ensure the well-being of the colony, but in other species, various castes may be specialized for different functions, for example, defense or for food provision. Until recently, it was believed that eusociality was restricted to insects, but there is at least one eusocial mammal, the naked mole-rat (*Heterocephalus glaber*), which has a social system not unlike that of the termite. In both species, males as well as females are diploid. The single, very large, reproductive female "queen" of the colony, together with a few male "kings" in the case of the naked mole-rat, inhabits the central chamber within a nest consisting of an extensive network of tunnels that are built, maintained, and defended by large numbers of smaller, nonreproductive individuals.

As described in Chapter 1, the evolution of eusociality was a great puzzle until Hamilton (1964) introduced the concept of inclusive fitness. In the honeybee, for example, a worker can transmit 50% more of her genes if she raises a sister rather than a daughter, for the following reasons. A queen mates with only one male, which is produced by an unfertilized egg and is therefore haploid. Consequently, all (diploid) workers produced by the queen share 100% of the half of their genes derived from the father and, on average, 50% of the other half of their genes, derived from their mother. So workers have 75% of their genes in common, and are more closely related to each other than to either their mother or any daughter they might produce that shares only 50% of a worker's genes. In species such as termites and naked mole-rats, eusociality may have evolved in the absence of haplodiploidy due to extensive inbreeding within the colony, resulting in a high coefficient of relatedness between colony members. DNA fingerprinting of naked mole-rats has demonstrated very close genetic similarities between colony members but not between members of different colonies (Reeve *et al.*, 1990).

Vertebrates

Social systems can be classified very broadly into five main categories: solitary, one male with one female, one male with several females, several males with several females, and, rare among birds and extremely rare in mammals, one female with several males. The effects of different compromises between various selection pressures for sociality have independently, by convergent evolution, given rise to similar social systems in many different taxa. For example, tigers and orangutans are solitary; geese and swans, foxes, and certain voles all form lifelong pair-bonds with a single individual of the opposite sex; groups of deer, elephant seals, and gorillas contain only one fully mature male; and hyenas, lions, wolves, and chimpanzees live in groups containing several adult males and females. However, within each of these broad categories the social systems can differ considerably. For instance, although wolf and chimpanzee groups contain several adult males and females, in wolves only the dominant male and female reproduce, whereas in chimpanzees, all individuals do so. Moreover, quite closely related species can have different social organizations. For example, both olive baboons (*Papio anubis*) and hamadryas baboons (*P. hamadryas*) live in social groups containing adult males and females, and their young, and the species readily interbreed where populations overlap. A male olive baboon sequesters one female at a time, only temporarily, when she is near the time of ovulation. In contrast, a male hamadryas baboon sequesters a small number of females into a permanent "harem," often before the females become sexually mature. The females are constantly guarded and prevented from straying and interacting with other males. The difference in male behavior results in differences in both social structure and the size of groups, which are much larger in hamadryas (comprising many males with their harems) than in olive baboons.

These examples illustrate the point that social organizations are strongly influenced by mating systems, which themselves are determined in part by the amount of care each parent provides for the offspring (Chapter 15). In animals that breed seasonally, the social system may change annually. For instance, many songbirds form large flocks when hunting for food during winter but break up into male–female pairs during the spring and summer breeding season. In some species, the social system changes during development: Many butterfly fish are solitary in feeding territories during early development and change to male–female pairs after sexual maturity. Moreover, social systems can be flexible, and populations of a single species may have somewhat different social organizations in different environments as a result of different environmental conditions, such as more or less predator pressure and different food distributions. For example, in some locations, groups of langur monkeys (*Presbytis entellus*) in North India consist of several adult males and females, and their young, but in other locations they contain only a single adult male, while the remaining males travel singly or in small, all-male bands. This kind of flexibility tends to occur more often in birds and mammals than in fish, amphibians, and reptiles, and contrasts with the rigidity of even the most complex social systems among invertebrates.

BENEFITS OF SOCIALITY

Sociality has four main benefits.

Reduction in Predator Pressure

The first advantage of sociality is one that was exploited by fighter and bomber squadrons and merchant ship convoys during World War II, namely, that predators are more likely to be detected and repulsed when more individuals are available to act as lookouts and participate in defense. Another advantage is that an individual must spend less of the daily time budget avoiding predation when living in a social group than if living solitarily. In pigeons, increasing flock size results in earlier predator detection and, consequently, decreased hunting success by the predator (Fig. 10-1). Even when predator detection fails and defense is not an option, animals in groups can reduce their likelihood of becoming prey by means of the selfish herd effect (Hamilton, 1971). If all animals in a group coordinate their activities, the individual can selfishly avoid predation by using others as a shield. This results in clumping, which makes it more difficult for the predator to focus on a single prey to catch and kill. For example, despite the presence of leopard seals waiting for them, Adélie penguins must jump off their breeding grounds

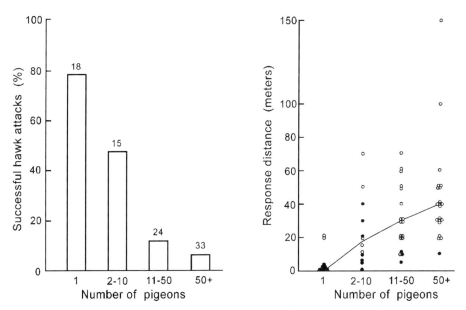

FIGURE 10-1. Group size and vigilance in pigeons. Left: the success of goshawk (*Accipiter gentilis*) attacks on pigeons decreases with increasing numbers of pigeons in the flock. Numbers of attacks are given above bars. Right: the distance at which pigeons respond to an approaching goshawk increases with increasing flock size. Closed circles = successful goshawk attacks; open circles = unsuccessful attacks; line gives median response distance. (SOURCE: Kenward, 1978, with permission)

into the sea in order to feed. The penguins enter the sea in large groups, not in small groups or singly. Likewise, guillemot juveniles still incapable of flight must leave their nests high on cliff ledges to get into the sea to feed; they jump off during the same 4-hour period every day. There is also the dilution effect: If the group contains more individuals than a predator can consume at any one time, the odds of one animal being killed decrease with increasing group size. This might not be effective for the prey of European foxes, which wantonly kill most or all of the chickens and ducks in an enclosure when they manage to enter it. Finally, protection is also afforded by numbers when the size of the prey species is sufficient to make defense possible. Some large ungulates, such as wildebeest, can fend off lions or hyenas as a group but not individually. Others, for example, musk oxen, form a circle, horns outwards, around more vulnerable group members (calves), which makes it virtually impossible for predators (wolves) to make a kill.

Improved Foraging and Hunting Efficiency

Another advantage of group living is that it improves the efficiency of foraging for food, especially where food sources are unevenly distributed or when they are ephemeral, such as ripe fruit available for only short periods each year in dispersed trees. We have seen that many individuals are better than one at spotting a predator. They are also better at discovering food sources. Vultures, for example, roost together but spread out over a wide area searching for carrion. When a vulture spots a food source and approaches it, others quickly converge, resulting in the familiar sight of a crowd of vultures, each trying to feed on the carrion. Groups of cooperatively hunting predators are better able to detect prey and to capture and kill animals that are considerably larger than themselves, giving each individual a better chance of a full meal while reducing the risk of injury (e.g., hyenas, wolves, Cape hunting dogs, and lions). A less familiar example is a bird, Harris's hawk (*Parabuteo unicinctus*), which preys on rabbits that are two to three times the weight of the hawks. These birds hunt in small family groups of up to about six birds that split into smaller parties and take turns scouting for rabbits. Once a rabbit is detected, several hawks converge on it from different directions. If the rabbit reaches cover, the hawks will flush it out into the talons of another family member positioned nearby. Another strategy is to chase a rabbit continuously and to exhaustion; as the first bird pounces and misses, it is immediately replaced by another. All family members share in the kill. It has been shown that both the kill rate of a group and the energetic intake of its members increase with increasing group size.

Improved Defense of Limited Resources

Cooperation of individuals in a social group facilitates defense of limited resources (space, food, mates) against other groups of conspecifics. Prides of lions consist of several females and either one or several males that are often closely related, for example, brothers. Each pride maintains a territory thought to be important for exclusive access to prey, for protecting cubs from infanticidal nomadic males, and for preventing neighboring prides from intruding on the living area. Territory boundaries are regularly patrolled by the males of the pride, which are better able to defend the territory than are single males.

Improved Care of Offspring

In many species of social birds and mammals, there is communal defense of the young, and in some species, there is also communal care and feeding of the offspring. This often occurs when older, adult offspring remain in the family group for a while instead of dispersing to begin reproducing themselves, perhaps because there are shortages of food, breeding areas, or mates. Among birds, breeding pairs of scrub jays and acorn woodpeckers, for example, often have several helpers that participate in rearing their younger siblings, and their presence has been shown to increase the odds that the eggs hatch and the fledglings mature to adulthood. Similar observations have been made in mammals such as the monogamous blackbacked jackal, *Canis mesomelas*. Typically, a few older offspring remain with the parents and help them rear their next litter; these helpers participate by feeding regurgitated food to the pups as well as to the mother and defending them from predators. The number of surviving pups increases with increasing numbers of helpers.

COSTS OF SOCIALITY

The four main benefits of sociality are offset by the following four major costs.

Increased Competition between Conspecifics

Group living brings with it competition between group members for food, mates, nest sites, and other limited resources. The intensity of the aggression stimulated by this competition is often reduced by the formation of dominance hierarchies (Chapter 12) among males, females, or both. While intense and injurious aggression may occur during the formation of a dominance hierarchy, once established, it is maintained by mild threat because higher-ranking individuals are given priority of access to disputed resources by lower-ranking individuals. The potentially serious effects on those of lower rank are seen in the reduced caloric intake and reduced reproductive success of low-ranking female Japanese macaques (*Macaca fuscata*) deprived of access to a nutritious food supplement by higher-ranking females (Chapter 12).

Increased Risk of Infection

As we know from our own experiences with ailments such as head lice and influenza, close proximity among group members greatly increases the chances of infection with parasites and diseases. Cliff swallows (*Hirundo pyrrhonota*) build their nests of mud in dense colonies on cliffs, church towers, or under bridges and in the eaves of tall buildings. The odds of nestlings becoming infested with swallow bugs (related to human bedbugs) increase with the size of the colony, the density of nests in the colony

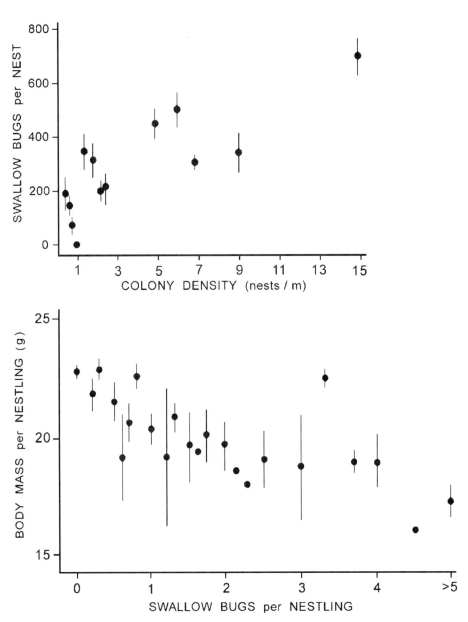

FIGURE 10-2. Swallow bug infestation of cliff swallow nests increases with increasing colony density (top). As the number of swallow bugs per nestling increases, the mean body mass of cliff swallow nestlings decreases (bottom). Vertical lines give standard error of means. (SOURCE: Modified from Brown & Brown, 1986, with permission)

(Fig. 10-2, top), and this infestation has severe costs in terms of delayed development and stunted growth (Fig. 10-2, bottom).

Increased Risk of Mating Interference and Parental Exploitation by Conspecifics

Harassment of the mating pair and actual interference with mating by group members is not uncommon in several social species, including macaques. In addition, an individual can reduce its own parental efforts by imposing on a social companion. Cliff swallows, for example, will enter the nest of a neighbor and quickly lay an egg. The owner of the nest generally takes care of the egg, and appears to lay fewer eggs as a consequence of the brood parasitism, so that her fitness is reduced (Brown & Brown, 1989). This is, of course, a benefit for the first individual but a cost to the second.

Increased Risk That Offspring Are Killed by Conspecifics

In a variety of species, it is not unusual for an animal to kill the offspring of another group member. Gulls readily cannibalize the newly hatched fledglings of other gulls on the communal nesting site, and female acorn woodpeckers may toss out an egg of another female in the communal nest. Among mammals, infanticide by males has been recognized in many species, including lions, baboons, and the hanuman langur (*Presbytis entellus*) of India, in which the phenomenon was first described (Sugiyama, 1965; Hrdy, 1974). In many social mammals, females remain in the group in which they were born, whereas males at adolescence emigrate to join other social groups (see below). Infanticide occurs when males that have emigrated from their natal groups succeed in deposing the resident males of other groups. The immigrants systematically hunt down and kill the youngest offspring of the resident females. Several theories for the evolution of this behavior have been proposed. It has been suggested that the behavior increases the reproductive success of the infanticidal male in at least two ways. First, he kills the offspring of other males, his reproductive competitors. Second, females that lose their infants soon became sexually receptive again and available for mating; this occurs more rapidly than would be the case if they continued to lactate until the infant was weaned.

PHILOPATRY AND DISPERSAL

In monogamous species, adolescent offspring of both sexes leave their family group to find new mates and raise young. In social species, young adults also leave the social group into which they were born to join another social group or to form a new one but, depending on the species, either the males or the females emigrate, while the other sex remains in the natal group (see below). In mammals it is usually the females that are philopatric, remaining in the natal group and home range, while the males disperse from the natal group, a process called natal dispersal. In many species, males continue to change social groups periodically throughout their lives, a phenomenon known simply as dispersal. It should be noted, however, that in other mammalian species, females instead of males, or both sexes, may disperse. Natal and subsequent dispersal is very

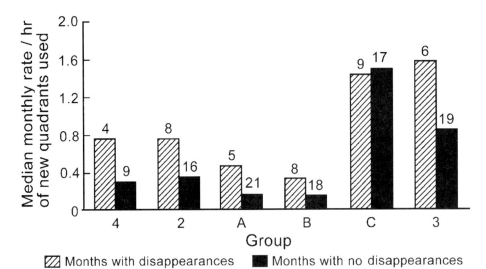

FIGURE 10-3. Cost of dispersal in vervet monkeys (*Cercopithecus aethiops*) living in leopard country. Except in Group C, more individuals disappeared (possibly predated) in months when groups used unfamiliar habitats more often than when they used them less often. Numbers of months are given above bars. (SOURCE: From Figure 1 in Isbell et al., 1990. Copyright © 1990 by Springer-Verlag)

costly and can even be fatal for the individual. Social species are much more vulnerable to predation when individuals travel alone or with just one or two companions. Moreover, animals have to deal with unfamiliar habitats and do not know where food and shelter are to be found, or the degree and type of threat posed by predators (Fig. 10-3). Acceptance into the new group is also fraught with danger.

Dispersal Hypotheses

Despite the high costs of dispersal, it must be adaptive for the many species that practice it. One of the major adaptive functions of natal and subsequent dispersal is believed to be inbreeding avoidance. Inbreeding is sufficiently reduced if only one sex leaves the social group. Evidence is now accumulating that avoidance of infection by parasites and sexually transmitted diseases might also be important. Another adaptive function of dispersal is the prevention of excessive intraspecific competition for mates, as well as for food and other resources.

Male dispersal is prevalent among mammals such as lions (Fig. 10-4), and female dispersal is prevalent among birds, although there are exceptions. Three main hypotheses have been proposed for the evolution of sex differences in natal dispersal. Greenwood (1980) proposed that the different sex-biased dispersal patterns of birds and mammals can be better understood by the interaction of two hypotheses. The first suggests that dispersal, while costly, is important for inbreeding avoidance, so a good compromise would be for one sex to have the advantages of philopatry while the other bears the costs of emigration. The second hypothesis is that the dispersing sex is the one

SOCIAL BEHAVIOR

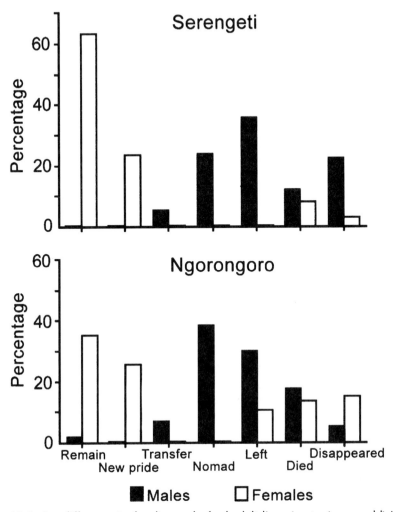

FIGURE 10-4. Sex difference in the dispersal of subadult lions (up to 4 years old) in two locations. Females tended to remain with the natal pride or form new ones, whereas males transferred into other prides, became nomadic, or left the area under observation. (SOURCE: Modified from Pusey & Packer, 1987, with permission)

for which resource competition is fiercest. Mammals typically have polygynous, "mate-defense" mating systems in which one male defends a group of females from other males, whereas birds typically have monogamous, "resource-defense" mating systems in which males compete with each other for territories that attract females (Chapter 15). In mammals, then, young males may have little chance of competing for mates with established older, stronger males, and disperse to take their chances elsewhere. In birds, however, familiarity with the area may be more important for males trying to establish a territory, while females might be more likely to find a better territory by dispersing. Another hypothesis, that of Dobson (1982), is restricted to mammals and based on competition for mates. Polygyny predominates in mammals, so mate competition is more fierce between males than between females; consequently, male-biased dispersal should predominate in mammals. Among monogamous mammals, mate competition should be less male-biased, so that both males and females would be expected to disperse. Dobson compiled data for 69 species and found that male dispersal predominated in 81% of 57 promiscuous or polygynous species, while both sexes dispersed in 92% of 12 monogamous species. The hypothesis of Liberg and von Schantz (1985), unlike the others, suggests that the difference in sex-biased dispersal in birds and mammals is a consequence of reproductive competition between parents and offspring; it assumes that the parents force offspring to leave, for which there is, however, little evidence in many species (see below). The hypothesis depends on interactions between two variables, namely, the reproductive physiology of females and the mating system of the species. Egg laying in birds enables daughters to "cheat" their parents by laying an egg into the parental nest and thus imposing parental care on the parents; this is impossible in mammals because of gestation and lactation. In monogamous mating systems, it is difficult for sons to "steal" matings from their fathers, but in polygynous or promiscuous mating systems, they may do so by fertilizing some of the potential mates of their father. The predicted consequences of interactions between these variables (Table 10-1) correlate quite well with observed behavior.

Many factors probably determine dispersal, and these vary in different species. In several primates (Chapter 16), for example, there is strong observational and experimental evidence that young males are not evicted but leave the natal group voluntarily, and that the proximate mechanism for this is the decline in sexual activity between males and females in the group resulting from increasing familiarity. This also appears to be a factor in the spotted hyena (*Crocuta crocuta*), in which natal males have higher rank but lower testosterone levels than immigrant males, yet exhibit much less sexual activity (Fig. 10-5).

EVOLUTION OF COOPERATIVE BEHAVIOR

Cooperative behaviors can be classified into three main categories according to the following two variables: (1) the type and timing of the benefits to both the donor and the

Table 10-1. Hypothesis of Sex Biased Dispersal in Birds and Mammals

Mating System	Birds (Daughter can cheat parents by laying eggs in family nest)	Mammals (Daughter cannot trick parents into caring for her young)
Monogamous (Son cannot steal matings from father)	Daughters disperse	Both sexes disperse
Polygynous/Promiscuous (Son can steal matings from father)	Both sexes disperse	Sons disperse

SOURCE: Modified from Liberg & von Schantz, 1985. Copyright © 1985 by the University of Chicago Press.

SOCIAL BEHAVIOR

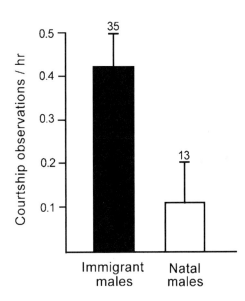

FIGURE 10-5. Immigrant male hyenas, although they are low-ranking and invariably lose fights with natal males, direct more sexual behavior toward resident females than do natal males. Numbers = numbers of males. (SOURCE: Holekamp & Smale, 1998, with permission)

recipient of the behavior, and (2) the evolutionary mechanisms thought responsible for them. As described in Chapter 1, natural selection is now viewed by some as acting exclusively on genes packaged in the "survival machine," the organism itself, in such a way that genes are replicated to the maximum possible extent through the generations. This is achieved if the genes ensure the survival and inclusive fitness of the individual bearing them; it follows from this that helpful behavior toward others would not have evolved unless the donor received a net gain from its actions.

Cooperation (Mutualism)

Cooperation, or mutualism, is a term used when two or more animals show helpful behavior to each other and all benefit immediately by their actions, for example, when carnivores such as wolves, lions, or Harris's hawks hunt cooperatively and share the kill. It is not difficult to understand how such behavior can evolve by individual selection if it increases the reproductive success and thereby the fitness of every individual compared with that of each individual hunting alone.

Reciprocity (Reciprocal Altruism)

Other behaviors are potentially costly to the individuals performing them but of benefit to others, for example, birds mobbing a predator, or a male baboon acting alone as a sentinel while the rest of the group is feeding, or someone treating a friend that is down on his luck to a free lunch. This fits the general definition of altruism as used for human behavior. Such behavior could evolve by individual selection if the individual performing it gets it returned later in kind and with interest by the recipient benefiting from it. This special case of cooperation is called reciprocal altruism (Trivers, 1971) or reciprocity. Since it requires that the participants interact again after the initial act, and must therefore be able to recognize each other, such behavior can only evolve in species living in individualized groups. Mutant genes in animals that "cheat" by not repaying the debt can spread to a limited extent within the population. Without going into detail, game theory, specifically the Prisoner's Dilemma, has shown that reciprocal altruism can indeed be an evolutionarily stable strategy (ESS) (Chapter 1).

In some cases, however, apparently altruistic behavior turns out to have greater costs than benefits for the recipient. Adoption of unrelated infants has been recorded in over 120 species of mammals and 150 species of birds. There have been several hypotheses as to why parents should waste their time and energy rearing unrelated young; it seems totally unadaptive. One hypothesis (Eadie & Lumsden, 1985) postulates that adults may decrease predation on their young by (1) earlier detection of predators due to larger brood size, and (2) dilution of predation on their own young, based on Hamilton's selfish herd effect. This has recently been examined in Canada geese (*Branta canadensis*), which are monogamous and often form mixed broods of adopted and natural offspring. In this study, mixed groups were about twice the size (about 6–7 goslings) of unmixed groups of either all adopted or all natural goslings. The survival effects were significant. The natural offspring in mixed broods lived longer than did either natural offspring in unmixed groups or adopted offspring (Fig. 10-6). Moreover, disproportionately more natural offspring in unmixed groups, and adopted young in mixed and unmixed groups, died within the first 6 months of life, when goslings are most vulnerable to predation. So both the parents and their offspring have fitness gains from the presence of nonkin in the brood, whereas there are fitness costs for the parents and offspring of adopted goslings. Similar observations have been made in a few other species, including mouth brooding cichlid fishes.

Altruism (Kin Selection)

Altruism, when not preceded by "reciprocal," is a term in evolutionary biology that is restricted to cases where the animal in fact does have decreased reproductive success

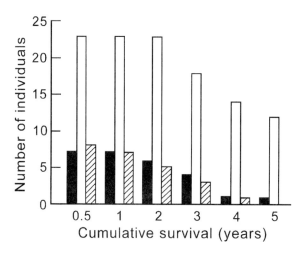

FIGURE 10-6. Natural offspring of Canada geese in mixed broods (open bars) had higher cumulative survival (years) than did natural offspring in unmixed broods (hatched bars) or adopted offspring in mixed and unmixed broods (black bars). (SOURCE: Nastase & Sherry, 1997, with permission)

and thereby reduced fitness as a consequence of helping others, whereas the recipient's fitness increases. A mutant gene causing its bearer to behave altruistically can spread through a population if the individual sacrifices its own reproductive success in order to care for other kin, as long the altruistic act increases its inclusive fitness, measured by the copies of its alleles passed on via its offspring and the nonoffspring kin that have been aided. An extreme case of this is eusociality, as in the honeybee (see above). But in other eusocial insects there are several queens in the colony, or a queen mates with several males, or both sexes are diploid, as in the case of termites; this considerably reduces the degree of relatedness of the workers. As noted earlier, increased rates of inbreeding between siblings and between parent and offspring may account for the evolution of eusociality in such species.

The selective force acting in the evolution of altruism is on helping kin other than offspring, namely, altruistic behavior that increases "indirect" fitness, whereas in the previous two categories of helping behavior, the selective force acts on helping offspring, namely, parental behavior that increases "direct" fitness. For this reason, the more prevalent term for altruism, *kin selection*, is not strictly accurate because, as originally defined, it refers to the evolutionary effects of aid given both to offspring and to relatives other than offspring (direct fitness + indirect fitness = inclusive fitness). For this reason, it has been argued that a distinction should be made between (1) direct selection as the evolutionary mechanism responsible for parental care and, by extension, for cooperation and reciprocal altruism, and (2) indirect selection instead of "kin selection" as the mechanism responsible for altruistic behavior.

MECHANISMS OF KIN RECOGNITION

Indirect selection, as well as inbreeding avoidance, requires that the animal behave differently toward kin and nonkin, and this means that it must be able to differentiate between related and unrelated conspecifics, and even between different degrees of relatedness among its relatives. Four possible mechanisms for kin recognition have been proposed.

Location

The newly hatched duckling learns to identify its mother because she is the individual closest to it that moves about uttering the correct sounds, while the mother goat learns to identify its newly born kid mainly by odor in similar fashion (Chapter 3, Imprinting). In a more general way, animals that do not disperse far from where they were born, but remain in or near their lairs, territories, or home ranges, are more likely to be related to each other than to individuals in other locations. This mechanism does not require the ability to recognize conspecifics individually; in the absence of individual recognition, any individual in the location would be treated as kin.

Familiarity

Animals that spend much time together, especially when one of them is young, are more likely to be kin than are strangers. This mechanism requires individual recognition.

The young in the nest or burrow learn to recognize each other and their parent or parents and later on treat familiar individuals differently from unfamiliar ones. This learning process is also possible across generations; for example, in many primate species a single infant is carried by the mother, and older siblings and their offspring also spend more time with the mother (and thus the infant) than with other group members. As noted earlier, familiarity appears to be an important kin-recognition mechanism for inbreeding avoidance in several mammals, including hyenas, some nonhuman primates (Chapter 16), and humans (Chapter 17).

Phenotype Matching

Another mechanism has been proposed for kin recognition, namely, that animals may be able to match their own phenotype, or the phenotype of their kin, against that of a stranger. Animals that share a larger proportion of morphological, physiological, and behavioral traits are more likely to be related than those that do not. Phenotype matching appears to be involved in kin recognition by some tadpoles and by some insects. In bees, the phenotypic label is an odor shared by all nestmates, derived in part from the shared food. Like honeybees, sweat bee (*Lasioglossum zephyrum*) workers guard the nest entrance and allow only nestmates, generally sisters, to enter. Greenberg (1979) showed that the proportion of intruders accepted into nests is directly related to the genetic similarity between intruder and guard (Fig. 10-7). In a subsequent study in which bees were raised either with sisters or nonsisters, it was found that guards admitted only those strangers that were sisters of the bees with which they had been reared, demonstrating that the phenotype of their nestmates had been learned by association and was used later as a referent to discriminate between strangers that were or were not allowed into the nest.

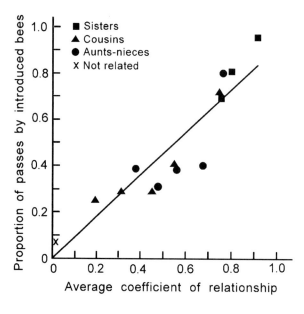

FIGURE 10-7. The proportion of strangers admitted to the nest by sweat bee guards increased as a function of the degree of relatedness between them. (SOURCE: Modified with permission from Greenberg, 1979, Genetic component of bee odor in kin recognition. *Science* **206**, 1095–1097. Copyright © 1979 by the American Association for the Advancement of Science)

Allele Recognition

There is some evidence that animals might inherit the ability to recognize kin without having to learn the phenotypic cues, namely, by inheriting a recognition allele, or group of alleles, that allow the animal to recognize bearers of the same allele. This mechanism has been dubbed the "green beard effect" to emphasize that the trait could be any conspicuous cue, provided that the same allele induces its bearer to behave appropriately toward other animals bearing it; alleles of the major histocompatibility complex (MHC) may provide such cues. The MHC, which is involved in the recognition of "self" and "other" by the immune system, contains much genetic diversity (estimated at about 30%) of often very ancient lineages of alleles, and any one of these could be responsible for signals that communicate relatedness. In house mice (*Mus musculus*), a gene in the MHC influenced mating preference in such a way that males from most strains preferred females whose alleles differed from their own, and the relevant cue in mice, and perhaps in other mammals, may be a pheromone in the urine.

ENVIRONMENTAL AND CULTURAL INFLUENCES IN PRIMATES

Where certain behavior patterns differ in populations of the same species that do not overlap geographically, it is possible that the behavior is more strongly influenced by direct environmental and cultural factors than by genetic effects. This is most likely for species with complex brains and developed cognitive skills, such as the social primates. By cultural effects, we mean learned responses that diffuse through the population and, in some cases, through subsequent generations within that population.

Nonhuman Primates

An example of a cultural influence on behavior comes from feeding in a group of Japanese macaques (*Macaca fuscata*). In some areas in Japan, groups are provisioned with supplemental food such as sweet potatoes. In one such group, it was observed that a juvenile washed its sweet potatoes in water before consuming them, and this behavior rapidly spread to other youngsters and eventually to adults in the group. In other groups, animals receiving wheat on a sandy shore began to drop handfuls of wheat, unavoidably including sand, into the water, where the sand grains sank but the wheat remained floating on the surface.

There are also environmental and cultural effects on social organization and behavior. The difference in the social organizations of langurs in different habitats in India was described earlier in this chapter. It is common among rhesus monkeys and baboons for social groups to be far larger when they have ready access to a great deal of food from garbage dumps or crops near human habitation than when they must forage for themselves. An interesting example of culturally determined sociosexual behavior comes from field observations on hamadryas and olive baboons where the distributions of these two species are sympatric (overlap). As noted earlier in this chapter, male hamadryas baboons collect females into their harems. A male prevents his harem females from straying by threatening, chasing, and biting them if they wander too far, and females respond to this by running *toward* him. Olive baboons live in multimale, multifemale groups. Males do not collect harems, and if a male threatens or attacks a female, she runs *away* from him. Hamadryas males have been observed to sequester olive baboon females into their harems. When these females strayed too far, the male would threaten and chase them, and females initially responded by running away, which intensified aggression by the male. However, these females soon learned to run toward the male when threatened, changing their responses to those of hamadryas females, which diminished male aggression.

Humans

We are all familiar with the cultural differences in food preferences and prohibitions, and with the many different ways in which food can be prepared and eaten. Cultural differences also account for variations in religious practices, clothing, and adornment. There are major differences in the laws of different societies dealing with birth, marriage, death, and the distribution of property. Socially acceptable behavior in one society may be regarded as criminal in another: In some societies, upon the death of a married man, his brother may be expected to marry the widow (see Shakespeare's *Hamlet*), which is regarded as incestuous in other societies. Humans transmit the knowledge of one generation to subsequent ones by oral and written traditions and are also more capable of changing their own environment than are any other species. These factors, among others, have brought about the rich differences in human social organizations. Cultural anthropologists have classified these organizations into four major types, namely, bands, tribes, chiefdoms, and states, whose increasing size and complexity can be related, at least in part, to the availability, production, and distribution of food and water.

Bands comprise kinship-based groups of less than 100 people that engage in foraging, often called hunting and gathering. Examples would be the !Kung-san bushmen of Africa and, until very recently, the Inuit of northern Alaska and Canada. The society is egalitarian, and there is no legal code or enforcement. In times of low food supply, the band may temporarily break up into nuclear families that forage on their own. In this social setting, and in that of the middle class in industrialized urban centers, the nuclear family is the basic unit and plays an important role in framing social structure. The common factor in both cases is probably the lack of ties to the land and consequent increased mobility. Tribes are larger and more settled descent groups, often living in villages. Descent groups are either matrilineal or, three times more commonly, patrilineal. Tribes are associated with a mixture of foraging and horticulture consisting of slash-and-burn cultivation in which land-use is neither intensive nor permanent. The Yanomami, some 20,000 natives of southern Venezuela and adjacent Brazil, are a modern example. There may be male heads of villages, male village councils, and men able to sway several villages into concerted action, but there is no enforcement of decisions, no centralized government or differentiation into social classes other than gender stratification, usually with women ranking below men. Chiefdoms and states are associated with increased population densities facilitated by increased food supplies due to (1) agriculture, that is, permanent, labor-intensive land use together with the use of domestic animals for food, clothing, transport and manure; (2) trade; and (3) industrialism. Chiefdoms are still kin-based, but there exists a permanent political structure and differential access to resources, which is typically determined by directness of descent. When wealth is handed on, there is a need to develop a priest caste to enshrine the rules governing its distribution, and this reinforces the authority of the chief. States or nation-states, which first arose about 5,500 years ago, are characterized by the presence and acceptance of socioeconomic stratification, for example, nobility, burghers, serfs, and slaves. There are kinship ties within but not between the different social strata. A central government is supported by a central military cadre, managerial authority, and enforced legal and ethical codes.

CHAPTER **11**

Communication

DEFINITION

Communication can be defined in a number of ways; some are problematical because they invoke unprovable concepts such as the "intent" to communicate. E. O. Wilson (1975) avoided such problems by defining communication as "an action of one organism (or cell) that alters the probability pattern of behavior of another organism (or cell) in a fashion that is adaptive to either one or both participants." In this wide definition, communication requires both a sender and a receiver, a response by the receiver, and adaptive value for either or both parties. For example, a scream of pain when you are hurt can be considered communication if you are overheard and there is a response with adaptive value (1) to you, if a stranger comes to your aid, (2) to the other person, if he or she uses the opportunity to rob you, or (3) to both of you, if your son comes to help, minimizing his chances of losing the parent who pays his tuition. So Wilson's definition includes signals that allow predators to detect prey and prey to detect predators, and also includes deceit both between and, more rarely, within species (see below). In its narrower sense, communicatory behavior is more commonly understood to occur between conspecifics and to have adaptive advantages for both participants.

FUNCTIONS OF COMMUNICATION

As shown in Table 11 1, communicatory signals may be broadly categorized by their function in permitting the recognition of other individuals as members of the same

Table 11-1. Functions of Communication

1. Recognition
    Species recognition
    Class recognition
    Deme* recognition
    Kin recognition
    Individual recognition
2. Reproduction
    Courtship
    Mating
    Parent–offspring interactions
    Synchronization of hatching
3. Agonism and social status
    Territoriality
    Dominance
4. Alarm
5. Hunting for food
6. Giving and soliciting care
7. Soliciting play

*Deme = the smallest local set of organisms within which interbreeding occurs frequently.

species or family; in facilitating reproduction, from courtship and mating to rearing the offspring; in differentiating between "friends" and "enemies"; in communicating social status; in warning of approaching danger from predators; in detecting and capturing prey; in giving or receiving care by feeding and grooming; and in play solicitation. These categories are neither exhaustive nor mutually exclusive but provide a glimpse of the importance of communication throughout the life of an animal. In addition to the communicatory mechanisms described below, other examples can be found elsewhere in this book.

"HONESTY" AND "DECEPTION" IN COMMUNICATION

We know from our own use of written and spoken language that communication can range from the very honest to the very deceitful, with many intervening gradations. Tactful comments about someone's appearance verge on lies, as does misinformation in military communiqués. Aesop's fable about the shepherd boy who cried "wolf" exemplifies the fact that advantages may accrue from deceit, but they decline as its frequency increases relative to the frequency of honesty. Thus, the boy was initially believed but those who ran to his aid and found he had lied subsequently ignored him, so when he was truthful and cried "wolf" in earnest, nobody came and he was devoured.

Deceit *between* species is common, especially in the form of mimicry whereby one species gains an advantage by imitating the appearance and behavior of another. Wasps are protected from predation by advertising their unpalatability and venomous sting with warning coloration of black and yellow rings. Various species of hover flies, which do not possess stings, have black and yellow rings, mimicking wasps, and are protected from bird predation. Some birds, including cowbirds and cuckoos, are nest parasites, laying their eggs in the nests of other bird species, where they are incubated; the other hatchlings are ejected from the nest by the interloper, which is fed by the host parents. The important point is that the parasite displays throat coloration and patterning identical to that of the host fledgling when begging for food, and this releases the feeding behavior of the host parents. Many other cases of such deception in communication between species are known throughout the animal kingdom. Less common are examples of deceit that are not as clearly based on phylogenetic adaptations. White-winged tanager-shrikes (*Lanio versicolor*) live in the canopy of the Amazon forest in Peru. They lead permanent mixed-species flocks of birds, use loud calls to maintain flock cohesion, and also act as sentinels by being almost always the first to give loud alarm calls when they detect approaching bird-eating hawks. The shrikes obtain most of their food by catching insects that are flushed out of cover or dropped by other species in the flock. They are significantly more likely to utter false alarm calls (that is, in the absence of any hawks) when other birds are pursuing the same prey than when they have no competition from flock members (Fig. 11-1). Since aerial competition for insects is extremely brief, lasting a second or so, momentary hesitation by the competing bird in response to the alarm call is sufficient to ensure the success of the shrike. In this case, only the sender of the signal profits in the short term; the cost to the receiver of the signal is probably outweighed by the benefit of repeatedly receiving "honest" early warnings of approaching predators.

Examples of apparent deception *within* species are rarer and interesting. Intraspecific deceit involves communication in which advantages accrue to the sender of the signal, but there may be some indirect advantages for the receiver too. This seems to be illustrated by another example of vocal communication in birds. Red-winged blackbirds (*Agelaius phoeniceus*) are territorial and each male has an extensive song repertoire:

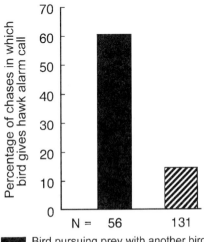

FIGURE 11-1. White-winged tanager-shrikes crying "hawk" to scare off other birds in the mixed-species flock that compete for the same insect. N = number of chases. (SOURCE: Reprinted with permission from Munn, 1986, Birds that "cry wolf." from *Nature* **319**, 143–145. Copyright © 1986 by Macmillan Magazines Ltd.)

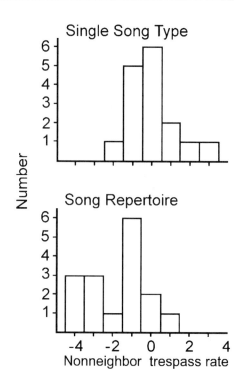

FIGURE 11-2. Playback of male red-winged blackbird song repertoires, but not of single songs, decreased trespass rates of intruders into territories. (SOURCE: Yasukawa, 1981, with permission)

songs can be sung either singly (conveying the impression of a single bird) or together (conveying the impression of several birds). Loudspeaker playback experiments showed that the territory of a bird was less likely to be invaded by unfamiliar individuals if several songs are played than if a single song was played (Fig. 11-2). This would clearly be advantageous to the singer, but it might also conserve an intruder's energy by preventing conflict to establish what might be an inadequate territory within the singer's territory, thus persuading the intruder to seek elsewhere. However, since the song repertoire increases with age and experience in this species, we cannot exclude the possibility that would-be trespassers might assess the territory owner's fighting ability by his song repertoire. Subsequent studies found no correlation between trespass rates per territory and male density in the area (Yasukawa & Searcy, 1985).

Partly because of such findings, it has been proposed that the functions of signals are not so much to maximize information transfer, the classical view, as to persuade the receiver to do something that is of advantage either to the sender or to both the sender and the receiver.

COMMUNICATORY SIGNALS

It is difficult to compare the numbers of basic communicatory signals in different species, but it has been estimated that the range is rather narrow, from 10–20 signals per species in social bees and ants to 30–40 signals per species in highly social vertebrates. At first glance, it might seem that these numbers must be an underestimate, but the number of *messages* that can be conveyed greatly exceeds the number of signals available; this is analogous to an alphabet of 26 letters (signals) from which thousands of unique words (messages) can be produced. Signals can be classified by properties that, with one exception, increase the number of messages conveyed:

1. Only one message can be conveyed by a discrete signal that gives binary "yes" or 'no" information, for example, when the zebra's ears are either upright, a neutral signal, or laid back, a threat (Fig. 11-3).
2. Several messages about the current state of the sender can be conveyed by a graded signal that varies from low to high intensity, for example, in the gradations of mouth-opening during greeting in the zebra.
3. A new message can be produced by composite signals, one of which alters the meaning of others; for example, mouth-opening by the zebra becomes a threat when the ears are laid back.
4. A new message can be produced by metacommunication, whereby a signal alters the meaning of the signals that follow. For example, by dropping its forequarters and wagging its tail, a dog signals that its subsequent fighting behavior is actually play-fighting.
5. A new message can be produced when a signal occurs in a different context, for example, when the food-begging behavior of infant birds and mammals is performed by an adult female toward the male during courtship.

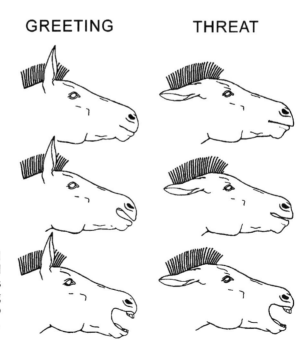

FIGURE 11-3. Composite signal in the zebra: Ears upward or laid back (discrete signal) changes the meaning of mouth opening (graded signal) from greeting to threat. (SOURCE: Trumler, 1959, with permission)

6. Messages about the current state of the sender can be conveyed by the degree to which the signal is exaggerated by specialized morphological structures, such as the spread of the peacock's tail during courtship (Chapter 2, Ritualization and Displays). Similarly, physiological changes, such as vasodilatation or vasoconstriction, urination, and defecation, may be sufficiently overt to be detected by others and acquire communicatory significance: A flushed face, for example, may signal emotional arousal (e.g., anger, sexual excitement, or embarrassment).

The examples given are from intraspecific *visual* communication, but the categories also apply to other sensory modalities and composites of them. In general, intraspecific communication is more complex than interspecific communication.

SENSORY CHANNELS OF COMMUNICATION

Communicatory signals can be received via one or more of five sensory pathways and may be classified accordingly:

1. *Visual communication*, usually by reflected light but also by bioluminescence.
2. *Acoustic communication*, including auditory communication.
3. *Chemical communication*, including olfaction, taste, and communication via the vomeronasal organ.
4. *Tactile communication*.
5. *Electrical communication*, confined to aquatic species and predominates in seeking prey.

There are wide species differences in the use of these various communicatory channels. Differences are related to habitat and way of life, and to the complexity of receptor organs. Wilson (1975) has estimated the relative importance of the three primary sensory channels (visual, acoustical, and chemical) in the displays of a few selected taxa. Microorganisms, moths, and social insects depend largely on chemical communication; arboreal lizards and butterflies depend primarily on visual communication, while crickets and seals principally use acoustical communication. For birds and diurnal primates, visual and acoustical channels are equally important, but in humans the use of language emphasizes acoustical communication. Finally, many mammals, including rodents and wolves, depend heavily on all three primary sensory channels.

Each sensory channel has a unique set of properties that make it more suitable for a given habitat or function than another. Table 11-2 lists nine key properties and approximately how they differ for visual, auditory, olfactory and tactile communication, but many exceptions come to mind. First, range refers to the distance at which the signal can be perceived. Second, transmission rate refers to the speed at which the signal is transmitted from the sender to the receiver. The third property is whether or not the signal can pass around environmental barriers separating the sender from the receiver. The fourth property is whether or not the signal can be perceived at night or in dark environments. Fifth, fadeout time refers to the speed with which the signal ceases after the sender stops transmitting; this determines how quickly the message can be changed. The sixth property is whether or not the receiver of the signal can use it to detect the

Table 11-2. Properties of Sensory Channels of Communication

Properties	Visual	Auditory	Olfactory	Tactile
Range	Medium	Long	Long	Short
Transmission rate	Fast	Fast	Slow	Fast
Effective around barriers	No	Yes	Yes	No
Effective in the dark	No*	Yes	Yes	Yes
Fadeout time	Fast	Fast	Slow	Fast
Localization of sender	Easy	Fairly easy	Variable	Easy
Indexical capacity	Yes	No	No	Yes
Representational capacity	Yes	No	No	No
Energetic expense	Medium	High	Low	Low

*Except bioluminescence.

location of the sender. Indexical capacity, the seventh property, refers to whether or not the signal allows the sender to direct it at only one of several potential receivers and also to direct the attention of the receiver at a third individual. The eighth property, representational capacity, refers to whether or not the signal can convey a representation of an object to the receiver. Indexical and representational components are often combined by humans, for example, when one man points at (indexes) the only man with a beard in a group and then, to ensure that only that man is referred to, strokes his chin to indicate the beard. Finally, energetic expense refers to how much energy the sender uses to transmit the signal.

These sets of properties give each sensory channel certain advantages and disadvantages. Consequently, pieces of the same message are often sent via two or more channels. Among birds and mammals, visual and auditory displays are frequently combined because they complement each other. For example, a dog threatening to attack bares its teeth and wrinkles its nose (visual), and this is accompanied by growling (auditory), which maximizes the chance that the message is received if the dog is momentarily not in view. Sometimes there is what might seem to be complete redundancy. Honeybee scouts encode the distance to a food source by the duration of the "straight run" of the waggle dance, which is transmitted by tactile communication, since it is dark in the hive (see Tactile Communication), and also by the duration of a buzzing sound that accompanies it. The two signals appear to be essentially independent because the sound lags behind the dance by several milliseconds (Griffin & Taft, 1992).

Visual Communication by Reflected Light

Humans' bias toward visual signals makes it difficult for us to recognize the importance of other communicatory channels. Among many species of bony (teleost) fish, for example, the individual often threatens conspecifics by orienting itself parallel to, and beating its tail toward, the opponent. We perceive this as a visual display, but tail beating actually creates a pressure wave in the pressure-sensitive lateral line organ of the opponent, which cannot be pleasant. In fact, a visual display may be derived from movements that are used to deliver a signal via a different sensory modality. Male dogs deposit communicatory scent marks on objects by trickling a few drops of urine on them.

To do so, they lift one hind leg in a characteristic stance that itself attracts other dogs, even when no urine is produced because the bladder is empty.

Signal Properties

The variables of visual signals with communicatory importance are light intensity, color, shape, size, and movement. The importance of each variable depends on the properties of a species' visual receptor system. Color is irrelevant for a species such as the dog, which does not possess color vision, but is important for those that do, such as butterfly fish and humans. Visual communication has the advantages of (1) fast transmission and fadeout times that facilitate rapid information flow, (2) the capacity to signal changes in motivational state, (3) allowing the sender to be located, and (4) usually small energy costs. However, visual communication has a limited range, especially in dense vegetation, and is largely ineffective at night and in dark habitats. Visual communication is therefore mostly restricted to mobile animals that can modulate ambient light effectively, that are active by day and live either in surface waters or on and above land. Since visual displays have both indexical and representational capacities, they are well-suited for interindividual communication. In both their aggressive and affiliative displays, monkeys can index one of several individuals by looking directly at that individual during the display. In enlisting aid from others when threatening an opponent, they can identify the opponent by threatening directly at it, all the time glancing briefly back at potential supporters. There are many good examples of visual representation, including all cases of visual mimicry. An interesting example is the mimicry of eggs by dummy eggs on the anal fin of the male cichlid fish *Haplochromis*. Living in freshwater lakes, this species is a mouth-brooder. The female carries the fertilized eggs in her mouth until hatching, and the fry return there for safety in times of danger and at night. Fertilization is assured by a neat trick. During spawning, as the female picks up her eggs with her mouth, the male spreads his anal fin in front of her, displaying the dummy eggs. As the female attempts to pick them up, the male ejects sperm that are drawn into the mouth of the female where fertilization occurs.

Sender Mechanisms

The mechanisms by which visual signals are transmitted may conveniently be divided into extrinsic and intrinsic. Extrinsic mechanisms consist of objects left by the animal in the environment that transmit a message to other individuals, often in the absence of the sender. Male bower birds decorate their bowers, in which they display to females, with flowers and feathers that have unusual color and influence mate choice by females. Intrinsic mechanisms, on the other hand, involve some aspects of the animal itself, which may involve three dimensions: the orientation, the configuration, and the movement of the body, or some part of it. Orientation of the head or entire body away from an aggressor is a common submissive display in many vertebrates. Among many bony fish, the submissive individual swims with its head higher than its tail, and the angle provides a good measure of the intensity of the submissive display. An upright tail position signals a dominant, confident rhesus monkey, while a horizontal or lowered tail indicates lower rank. Configurational changes include alterations in the animal's posture, in the shape and configuration of the body and its appendages, and in color patterns, as in the courtship displays of many teleost fish, including guppies, cichlids, and blennies. Changes in movement patterns can occur during locomotion, such as in the strut of a dominant rhesus monkey, or in the movement of an appendage, such as the tail-wagging of dogs or the claw movements of the courting male fiddler crab. Many visual displays combine two or even three of these dimensions. Greeting and threat in the zebra (Fig. 11-3) combine the orientation of the head toward the conspecific (and of the ears relative to the skull), the configuration of the head brought about by the positions of the ears, and the movement of the jaws as the mouth opens.

Receiver Mechanisms

Light is detected by four basic types of light-receptive organs. Simple light-sensitiva eyespots, which are not capable of forming images but can help the animal determine the direction of a light source, are found in unicellular organisms and many invertebrates; for example, in fly maggots. Compound eyes, composed of numerous, fixed ommatidia, are found in arthropods, including insects, as well as in some polychaete worms. Cephalopods such as octopus and squid possess eyes having remarkable structural similarity with the fourth type, the familiar vertebrate eye. The precise structure and function of the vertebrate eye, as of the other three types, can vary in different species, often in relation to habitat and way of life. Thus, nocturnal animals cannot perceive red light and can therefore be observed during their active phase under red illumination in zoos. Some animals have color vision, while others do not; among those that do, the cones of reptiles and birds contain colored *oil droplets* that absorb certain wavelengths before reaching the pigment layer, whereas cones of mammals such as the human contain different *pigments* that filter out different wavelengths.

Of special relevance for behavior is the effect of selection pressure on the speed with which the visual signal is processed. In Chapter 3, we examined the stimulus filtering mechanism for facial signals in macaques that communicate the identity and emotional state of the sender. These signals require complex neural processing and do not need a rapid, millisecond response: Processing occurs at the level of individual neurons in the upper and lower banks of the superior temporal sulcus. Other visual signals require very rapid responses, notably, those occurring in the context of prey catching and predator avoidance. The following two examples of prey catching in amphibians illustrate how selection pressures result in different filtering mechanisms in two related species. Each ganglion cell in the third layer of the retina receives input from several bipolar cells and a number of photoreceptors, the rods and cones (Fig. 11-4). These constitute the receptive field of the ganglion cell, and there is overlap between the receptive fields of several ganglion cells. In the leopard frog, which feeds on flying insects, stimulus filtering occurs at the level of these ganglion cells. The leopard frog has one type of ganglion cell that is selectively responsive to objects that move irregularly and have convex edges; a second type is selectively responsive to small moving objects that stop moving. The first type is the feature detector for flying insects, which, if stimulated, induces the frog to flick out its tongue to catch the prey. The second type is thought to be a predator detector, since it responds to a stimulus resembling a heron at a distance, moving slowly and then stopping. In contrast to the leopard frog, the European

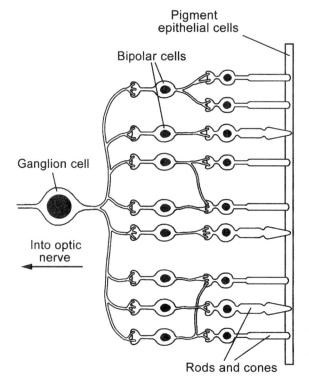

FIGURE 11-4. Schematic cross-section through the retina of the vertebrate eye showing one ganglion cell and its sensitive field. In the leopard frog, stimulus filtering for catching flies takes place at the level of ganglion cells in the retina.

toad catches slow-moving prey such as slugs and worms. The primary stimulus filtering mechanism here is located in the optic tectum of the midbrain, which contains neurons that respond maximally to slow-moving, horizontally elongated objects (Fig. 11-5, middle). The differences between these two species in their prey detection mechanisms are related to the characteristics of their prey and the processing speed needed for an effective response.

In some species, including insects such as fireflies and deep-sea arthropods, communication is by bioluminescence whereby the sender itself generates the light. Compared with communication by reflected light, bioluminescence probably involves much higher energetic costs and is more effective at night than during the day.

Auditory Communication

Signal Properties

The meaningful variables of auditory communication are pulse, or note repetition rate; rhythm, or repetition rate of a series of pulses; frequency, or the pitch of a note; amplitude, or frequency range; and intensity, or loudness of the signal. The importance of these variables depends on the properties of a species' auditory system. Most people, for example, cannot perceive the high frequencies of bat sonar or the postcopulatory vocalization of male rats. High-intensity auditory signals can have a longer range than

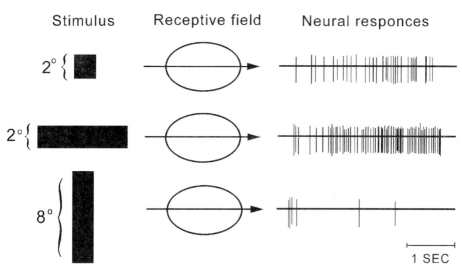

FIGURE 11-5. In European toads, stimulus filtering for catching worms takes place at the level of neurons in the optic tectum (brain) receiving input from their respective receptive fields in the retina. The T5(2) neurons are maximally activated when horizontally elongated objects move slowly across the receptive field (middle); they are thought to be the worm detector. (SOURCE: Modified from Figure 80 in Ewert, 1980. Copyright © 1980 by Springer-Verlag)

visual signals, and their transmission rate and fadeout time are also fast (Table 11-2). Auditory communication has the advantage of not being impeded by darkness or by barriers, which makes it suitable for animals communicating from stationary positions, for example, from within large territories or nesting sites, as well as for nocturnal species and those living underground or in dense cover. With the exception of human language, auditory communication has the disadvantage of generally lacking good indexical and representational capability. While it is moderately easy to detect at least the approximate location of the sender, pinpoint localization of sound requires complex receiver mechanisms such as those in owls and some echo-locating mammals, notably bats and dolphins. It is perhaps more important that the sender cannot readily evaluate whether the intended recipient actually receives the signal. All of us have turned around, along with others, to look for the person behind who just shouted "Hey, you!"; here, vision is needed to determine if the message is received and to locate the sender. This disadvantage makes auditory communication less suited for interindividual distance communication. But there are exceptions. During their territorial phase in the spring mating season, male European robins, which have a variable song pattern, can directly address an individual trespassing onto their territory by imitating its song, leaving the latter in no doubt that the territory owner is aware of its presence. In some pair-bonded species, for example, shrikes and gibbons, the male and female of the pair maintain contact when out of sight of each other by "duetting" or antiphonal singing. In this, one individual begins to vocalize and the other responds in such a way that the two songs are contiguous or overlapping (Fig. 11-6). Some level of representation is contained in the alarm calls of vervet monkeys (*Cercopithecus aethiops*), which have three alarm calls: The first is for

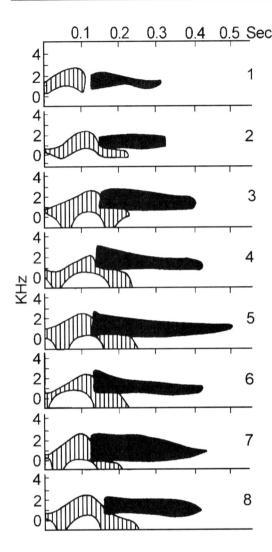

FIGURE 11-6. Sound spectrogram of duetting by a mated pair of shrikes (one bird shaded, the other solid). The close association and overlap of calls allow two individuals to keep in touch by auditory interindividual communication. (SOURCE: Hooker and Hooker, 1969, with permission. Copyright © 1969 by Cambridge University Press)

aerial predators (martial eagles), so listeners will look up; the second is for leopards, so listeners climb trees; and the third is for snakes and prompts listeners to look down. However, the precise location of the predator is communicated visually (indexing) as conspecifics observe where the sender is looking.

Auditory communication is regarded as energetically more costly than visual communication, but that is likely to depend on the type of visual and auditory display. The mating dance of the crane, a visual display, requires considerable energy, whereas the crow of the cockerel is reported to require very little. Auditory signals, like visual ones, lend themselves to communicating changes in motivation because they can readily be graded by increasing the variables of pulse, rhythm, intensity, and amplitude (Fig. 11-7).

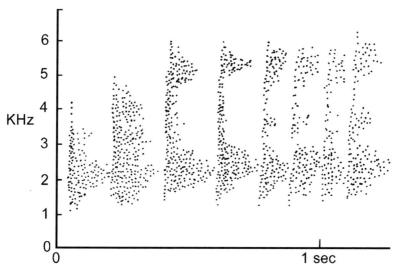

FIGURE 11-7. Intensification of meaning of a signal (urgency of the mobbing call of a European blackbird) is increased gradually by adding higher frequencies. (SOURCE: Andrew, 1961, with permission)

Sender Mechanisms

Like visual signals, auditory signals can be transmitted by extrinsic and intrinsic mechanisms. Extrinsic mechanisms involve the use of an environmental feature as a sounding board in which vibrations are induced by striking it with whole or part of the body. Some will be familiar with being prematurely awakened by a woodpecker drumming on the gutters that act as substitutes for hollow trees, the natural target. Other species produce vibrations on the substrate, on the ground or water surface. Male rabbits advertise their presence to conspecifics by thumping the ground with both hind legs: The receiver may perceive the vibrations by touch rather than by acoustical receptors. Intrinsic mechanisms involve sound production by the animal itself using either non-specialized or specialized organs. Sound production by nonspecialized organs includes the rattling of beaks by owls and storks and the beating on the chest by male gorillas. Organs specialized for sound production are numerous and include the rattle at the tip of the rattlesnake's tail and the various types of the toothed file, together with the structure that is vibrated by its movements to produce the stridulatory sounds of crickets and other insects. The parameters of signals produced by such structures are limited because frequency, amplitude, and intensity are fixed, so that only the pulse and rhythm can be effectively modulated, similar to a nineteenth-century drummer boy signaling commands during the heat of battle.

The syrinx of birds and the vocal chords of mammals, however, can readily modulate all five variables so that a great deal of information may be encoded in the auditory displays of many species. Trying to tease apart which bit of information is encoded by which song variable is extremely difficult, but it has been possible to decode, at least partially, the song of the male indigo bunting (Emlen, 1972). The rise and fall in pitch, note length, internote interval, and possibly the frequency range, are all species-specific and encode species identity. Individual variations in the details of the note structure appear to encode individual identity, and increases in song length and singing rate communicate increasing motivational intensity. Increasing motivational intensity can also be encoded by increasing the frequency itself, such as in human distress calls such as the crying of babies and fear screams of adults, or by adding higher frequencies to the signal, as in the mobbing calls of birds (Fig. 11-7). We do not know what is encoded in the songs of social mammals with large brains capable of complex cognitive processing, namely, dolphins and whales. The song sequences of humpback whales may last 6–7 minutes and are repeated faithfully. Very simple auditory signals containing one overriding message are made by species capable of complex signals in other contexts. The predator alarm call of songbirds, which probably resulted from convergent evolution because its structure is virtually identical in different species (Fig. 11-8), makes the sender of the signal difficult to locate—clearly an advantage that helps to retain the behavior in the repertoire of these songbirds.

The characteristics of auditory displays vary in relation to environmental features in the habitat. Low-frequency sounds carry much farther than high-pitched sounds and are more effective in long-distance communication. This is why modern orchestras

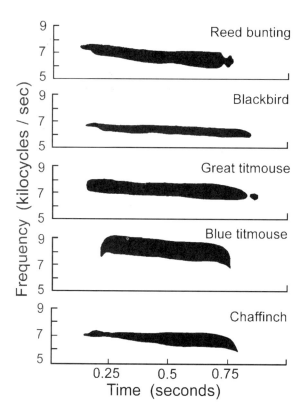

FIGURE 11-8. The similarity of alarm calls in some European passerine birds suggests convergent evolution. (SOURCE: Marler, 1959, with permission from Cambridge University Press)

playing in large concert halls have about four times as many violins as cellos. The booming of the male blackcock's courtship display can be heard by the human over a distance of 10 km, but the range of high-frequency songs by songbirds is about 2 km. The very low-frequency noises of elephants travel several miles, and the ultralow frequencies of some whale songs, which are reflected between the surface and colder, denser, deeper water layers, are thought to travel huge distances, even across oceans. Physical barriers can be bypassed by auditory signals but nevertheless influence the way in which sound is transmitted. Low-pitched sounds are attenuated less than high-pitched ones, and this selection pressure is evident from the frequency range of songs by birds in lowland forests in Panama, where most species have songs in the range of 1.5 to 3.5 kHz, which is least attenuated in this type of habitat (Fig. 11-9). On grassland, there is no such "sound window" and attenuation increases with increasing sound frequency.

Receiver Mechanisms

Sound vibrations have two effects: (1) the particle movement component displaces the molecules or particles back and forth in the surrounding medium, and (2) the pressure component alternately expands and compresses them. Among invertebrates, only the arthropods have sound detectors, some of which respond to particle movement that displaces the hairs of sensory cells, while others detect pressure changes by means of a tympanic membrane whose vibrations also activate sensory cells. An example of the latter is the bat detector of the noctuid moth. The "ears" of the moth are bilateral organs located laterally on the posterior thorax. The tympanic membrane, located within a small pit, faces obliquely sideways and backwards, and covers an air sac. The tympanic membrane responds to pressure changes by activating two auditory neurons (A1 and A2)

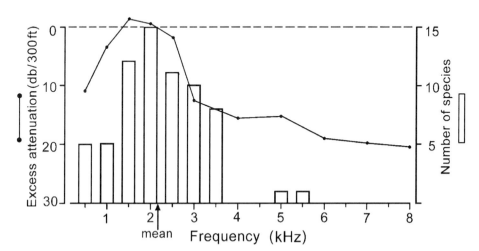

FIGURE 11-9. In a low forest habitat, most bird species have 0.5 to 3.0 kHz songs, a frequency range that is subject to little or no excess attenuation in this habitat. Horizontal interrupted line at 0 decibels (db) gives attenuation rate of 6 db per doubling of distance. (SOURCE: Modified from Morton, 1975. Copyright © 1975 by the University of Chicago Press)

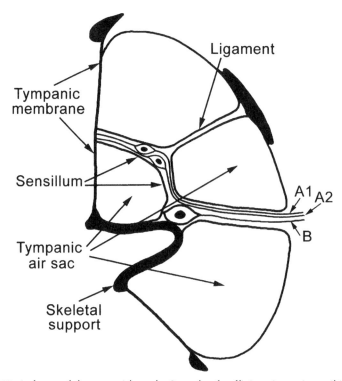

FIGURE 11-10. Left ear of the noctuid moth. Sounds of sufficient intensity striking the tympanic membrane will induce auditory neurons A1 and A2 to fire. (SOURCE: After Roeder, 1963)

within a "sensillum" that is suspended in the air sac and anchored by a ligament and a skeletal support (Fig. 11-10). Compared with A2 cells, A1 cells are very sensitive and can detect bat sounds at 100 feet, long before a bat can detect the moth. They are particularly sensitive to pulses of high-frequency sound used by bats for echolocation. Moths responded to directed playbacks of recorded bat sounds by flying directly away from them, probably by equalizing and decreasing the stimulation of A1 receptors in the left and right ears (Fig. 11-11), and thereby presented the smallest silhouette for the bat sonar to locate. Since a bat rarely flies in straight lines for any length of time, it is likely to locate another prey. But when the bat is close, at about 8 feet, the moth performs very erratic loops and power dives; it is speculated that the less sensitive A2 cells become activated and inhibit some central steering mechanism, causing the wings to beat asynchronously or not at all (Roeder & Treat, 1961). The effect is that the moth tumbles down and may escape the reach of the bat, which is still flying at high speed.

The more familiar vertebrate ears are pressure-sensitive bilateral organs that, as in the moth, permit the approximate location of a sound source by comparing the amount of stimulation reaching each ear. The complexity varies: Most fish have only an inner ear; amphibians and most reptiles have middle and inner ears; and birds and mammals have

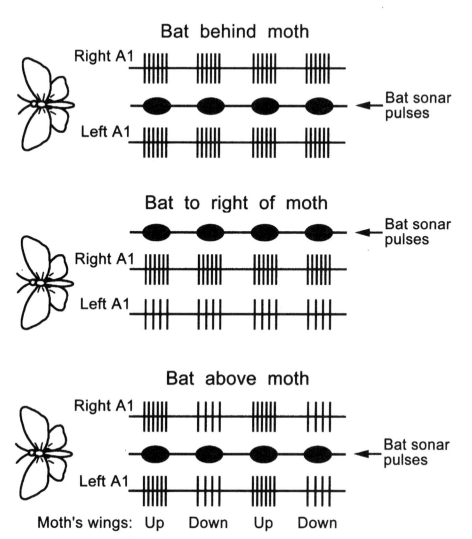

FIGURE 11-11. How moths may determine the location of a bat. Sonar pulses from the rear would result in equal, pulsatile stimulation of both A1 neurons (top), while sonar from the right is thought to produce stronger responses of the right than of the left A1 neuron (middle). Due to partial masking of the ears during the downbeats of the wings, sonar from above (but not from below) would produce alternating stronger and weaker activation of both A1 neurons (bottom). (SOURCE: After Alcock, 1993)

an additional outer ear. Like visual perception, auditory perception varies widely between taxa, primarily in the frequency range that can be detected, a function of the "tuning" of the sensory neurons of a species. In addition, there may be morphological specializations on the head that enhance auditory abilities. A familiar example is the mobile external ear or pinna of most mammals, particularly in prey and predator species.

The pinna is typically funnel-shaped to concentrate sound waves entering it, increasing the probability of hearing fainter sounds (the same principle as the ear trumpet of previous centuries). Each pinna is often independently mobile, so that sounds coming from every direction are monitored. When both pinnae are aligned on the same sound, pinpoint localization of its source is possible; arctic foxes can pounce directly on a lemming beneath deep snow without seeing it. Owls have similar abilities in the dark; the dish-like arrangement of feathers around their forward-facing eyes are also thought to channel sound waves to the ears. The ultimate auditory feats are, of course, the echolocation of bats and the sonar of dolphins and other cetaceans. The individual receiving the signal also produces it by emitting pulses of high-frequency vocalization, and the returning signals are processed after bouncing back off moving or stationary objects. Species so equipped can navigate around obstacles in total darkness while flying or swimming fast in three-dimensional space, and can locate and catch prey, and also avoid predators.

Chemical Communication (Olfaction)

Phylogenetically, chemical communication is the oldest form of communication. Present in slime molds and bacteria, it is very important in communicating between individual cells and organs within complex animals. It is by chemical communication that the nerve impulse is transmitted both from one neuron to another via neurotransmitters and to effector organs such as muscles and glands. Chemical communication is also the means by which the trophic hormones of the anterior pituitary gland modulate the activity of many other glands in vertebrates. In what follows, we can only consider chemical communication between animals, and the main emphasis is on olfactory communication, a subcategory of chemical communication.

Signal Properties

The two meaningful variables of an olfactory or other chemical signal are its chemical identity and its concentration in the air, water, or other medium. Olfactory communication has both advantages and disadvantages (Table 11-2). Depending on its chemical properties, the signal can be effective at a distance, a few miles for insect sex attractants, as well as at close range, in mother–offspring recognition in mammals. It is equally effective in the presence or absence of light, is little impeded by environmental barriers, and has low energetic costs. However, olfactory signals have a low transmission rate and the fadeout time is slow, so that the information content cannot be changed quickly. It is generally difficult to locate the sender of an olfactory signal because of the vagaries of air or water currents, but male moths (*Bombyx mori*) can travel several miles up the concentration gradient in air to locate females. Odor cues, like extrinsic visual signals, have the advantage of being effective in the absence of the sender and can communicate into the future, when the animal producing them has long gone. Olfaction is particularly suitable for providing information about the species, gender, individual identity, reproductive status, and territory boundaries of an animal, but less suitable for communicating motivational changes in aggression, flight, courtship, mating, and parental care.

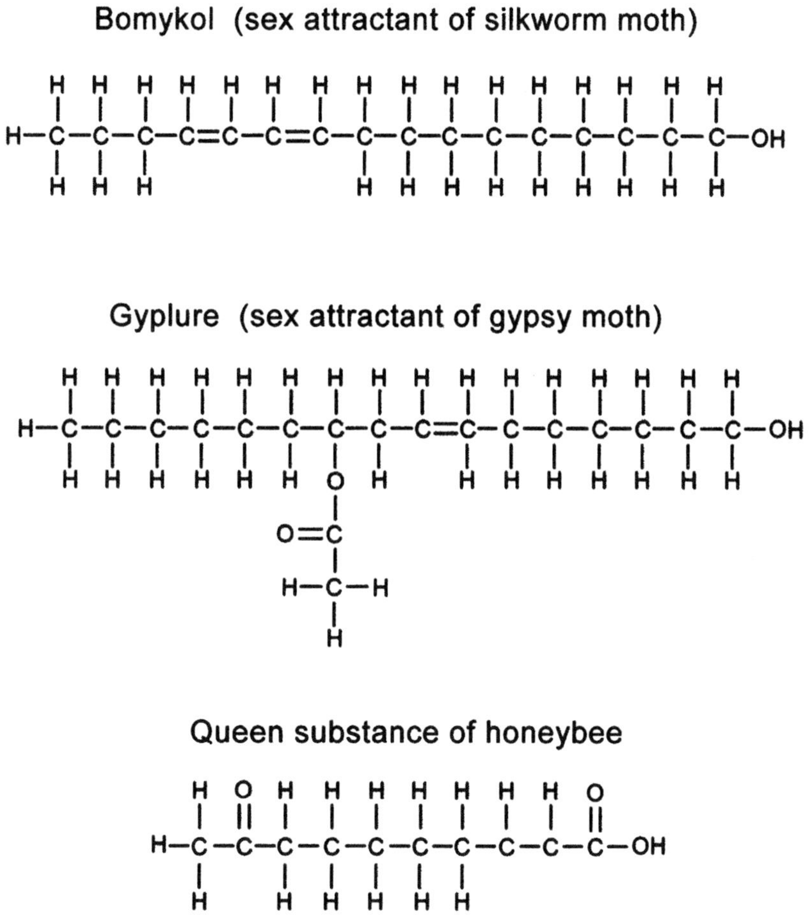

Bomykol (sex attractant of silkworm moth)

$$\text{H}_3\text{C}-\text{CH}_2-\text{CH}_2-\text{CH}=\text{CH}-\text{CH}=\text{CH}-\text{CH}_2-\text{CH}_2-\text{CH}_2-\text{CH}_2-\text{CH}_2-\text{CH}_2-\text{CH}_2-\text{CH}_2-\text{OH}$$

Gyplure (sex attractant of gypsy moth)

Structure: a 16-carbon chain with a $-\text{O}-\text{C}(=\text{O})-\text{CH}_3$ (acetate) substituent on the carbon at position 6 and a $\text{C}=\text{C}$ double bond between carbons 9 and 10, terminating in $-\text{OH}$.

Queen substance of honeybee

$$\text{H}_3\text{C}-\text{C}(=\text{O})-\text{CH}_2-\text{CH}_2-\text{CH}_2-\text{CH}_2-\text{CH}_2-\text{CH}_2-\text{CH}=\text{CH}-\text{C}(=\text{O})-\text{OH}$$

FIGURE 11-12. Species-specific sex pheromones of insects tend to be moderately complex hydrocarbons.

Insect sex attractants are often species-specific, straight-chain hydrocarbons (Fig. 11-12). Some 40 years ago, the chemical structure of the sex attractant of the silkworm moth was identified by Butenandt (1955), and Karlson and Lüscher (1959) coined the term "pheromone" from the Greek *pherein*, meaning "to bear or carry," and *hormon*, present participle of *hormaein*, meaning "to stimulate or excite." Pheromones are described as chemical or olfactory signals that are released to the outside and affect the behavior or physiology of another individual of the same species. A parallel has been drawn between their role in communicating *between* individuals and that of hormones that, released into the bloodstream, communicate *within* an individual. The pheromone concept, developed originally for the reflexive, stereotyped behavior of insects, has now been extended to include the more flexible behavior of mammals. Pheromones are commercially important in the control of insect pests, for example, boll weevils in cotton.

Because pheromones affect behavior as well as physiology, it is difficult to extract and purify these natural glandular secretions, and to assess them individually for behavioral activity after they have been fractionated. Consequently, the precise chemical identities of many mammalian olfactory signals remain largely unknown. Although a handful have been identified satisfactorily, it is easier to classify them by function. There are two main categories of pheromones: primer pheromones and signaling pheromones.

1. *Primer pheromones* result in neuroendocrine changes rather than in an immediate behavioral response. The primer pheromones most investigated are volatile constituents of urine in gonadally intact or castrated, testosterone-treated male social rodents, namely, mice and rats, and effects are also seen in females. It is known that olfactory pathways are involved, since effects are induced by exposing females to male urine and abolished in females made anosmic by olfactory bulb removal. Female mice that are grouped together in the absence of a male have irregular estrous cycles and a high incidence of pseudopregnancy; this grouping phenomenon is the Lee–Boot effect (van der Lee & Boot, 1956). When a male is placed with the females, cycles are both normalized and synchronized, so that all the females become pregnant simultaneously; this is the Whitten effect (Whitten, 1956). Another pheromonal effect results in pregnancy block. If a mated female is exposed to a male of an alien strain within 48 hours of mating, before implantation of the embryos has occurred, the first pregnancy is disrupted and genetic marking shows that the litter is sired by the second male; this is the Bruce effect (Bruce, 1959). These different effects are mediated by the secretion of gonadotropic hormones from the anterior pituitary gland. Another effect is the acceleration of puberty in female mice by their exposure to the odor of adult males during development, which causes greatly increased secretion of follicle-stimulating hormone (FSH), luteinizing hormone (LH), estrogen, and progesterone.

2. *Signaling pheromones* result in an immediate behavioral response by the receiver. Their major functions, as we know them, are similar to those of visual and auditory signals that communicate relatively static information, namely, species, class, deme, kin and individual recognition, social status, territorial boundaries, mate attraction, sexual consummation, and mother–infant recognition. Humans rely less on olfactory communication than do many other mammals, including prosimian and cercopithecoid monkeys, and this is reflected in the reduction of the mass of phylogenetically older brain regions associated with olfaction (Fig. 11-13). Nevertheless, the role of olfaction in our own species has probably been underestimated by scientists, as the magnitude and costs of the perfume industry attest. Mothers can reliably identify their own infants by odor if exposed to their newborns for a few minutes immediately after parturition (Chapter 3, Imprinting). Similarly, within 1–2 days of birth, breast-fed human infants can discriminate between gauze pads from the breasts of their own mothers and those from other breast-feeding mothers. They will turn their heads more often toward their own mothers' gauze pad than toward gauze pads from other mothers; filial imprinting is probably involved (Schleidt & Genzel, 1990). Chemical and olfactory cues are also used as an alarm signal, when they are often referred to by the original German term

COMMUNICATION

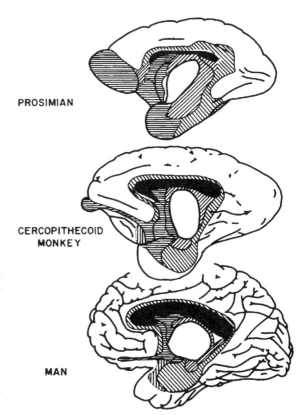

FIGURE 11-13. Reduction in the relative size of the structures concerned with olfaction (hatched areas) from prosimians through cercopithecoid monkeys to the human shown in sagittal sections of brains. Black area is the corpus callosum. (SOURCE: Michael et al., 1976, with permission, after Stephan, 1963)

Schreckstoff. Sea urchins and certain schooling fish move rapidly away from an area containing an injured conspecific that is releasing the signal, and mice avoid an area containing urine from a mouse that has been defeated by another or severely stressed in other ways.

The distinction between primer and signaling pheromones may be less clear-cut than once thought. Although the initial response to primer pheromones is a physiological one, it is invariably followed by a change in behavior. It is also likely that signaling pheromones produce physiological effects, but it is difficult to monitor very short-term changes in physiological parameters such as hormone levels.

Sender Mechanisms

In mammals, the most common pheromonal sources are urine, feces, vaginal secretions, secretions of specialized glands, saliva, and mother's milk. Extrinsic mechanisms, in which the signal is deposited onto an environmental object, are common, for example, the secretions of the ventral glands of gerbils and flank glands of hamsters that are used to mark territories. Intrinsic mechanisms, in which the signal emanates from the individual sending it, may result from odors produced continuously from specialized glands, from the vagina of a mare or bitch in heat, from urine and feces clinging to the animal as it deposits them in the environment, from the boar's saliva, and from glandular secretions that the individual rubs both onto environmental objects and its own body. Many prosimians and New World monkeys have specialized glands distributed about the skin. Ringtail lemurs possess, among others, a gland just above each wrist, which they rub along their tails. When enough secretion is deposited, they wave their tails over their heads at other lemurs in so-called stink fights during intergroup disputes (Chapter 16). A great deal is known about scent glands, and civetone, a cyclic ketone, is a product of the anal glands of civet cats and musk deer; musk is used in making perfumes.

The chemical analysis of scent marks in mammals shows they contain many different compounds, some of which are the products of bacterial action either within the glands themselves or on the body surface after secretions are released. Some are highly volatile and likely candidates for long-distance communication. Others have higher molecular weights, and are nonvolatile and more suited for close inspection. Mammals have separate receiver mechanisms for detecting these two types of compound (see Receiver Mechanisms). We do not know how individual identity, parent–offspring recognition, kin recognition, species identity, and dominance status are encoded in olfactory signals. Because of clear behavioral end points, what little we know comes from sexual signaling, which, together with gender recognition, suggests a hormonal basis.

A known male signaling pheromone is used by pig farmers: Sexually mature boars, which salivate a lot and froth at the mouth when mating, produce a characteristic odor from androstenol in their salivary secretions; men produce it too, in axillary sweat. Androstenol elicits an immobilization reflex in estrous sows that permits the boar to mount and mate. It is used to identify sows that are in heat, and this use has been patented. Another signaling pheromone present in the vaginal secretions of female rhesus monkeys stimulates sexual activity by males and influences their choice of a female partner. Injecting ovariectomized females with estrogen increases, while progesterone decreases, the sexual interest and activity of their male partners. The involvement of an olfactory cue was demonstrated when males were temporarily deprived of olfaction, at which time they failed to respond behaviorally to changes in the hormonal status of the female until their sense of smell was restored. At least part of the signal is encoded in a mixture of volatile, short-chain fatty acids, called copulins, that are produced by bacterial action. Estrogen, by stimulating mucous flow from the vagina, also helps to externalize these acids. Their effectiveness in influencing mate choice by males was demonstrated in small social groups comprising one male with four ovariectomized females. One female received applications of a synthetic mixture of these acids to the perineal region immediately before each behavior test, while another received just the solvent. The changes in mate choice are illustrated in Figure 11-14. Effects were modulated by female dominance rank, and the acids were ineffective when applied to the lowest-ranking female of each group, a phenomenon also observed with hormone treatments. Other anthropoid primates, including the human, produce the same series of volatile vaginal fatty acids in somewhat different proportions, but their communicatory significance in these species is unknown.

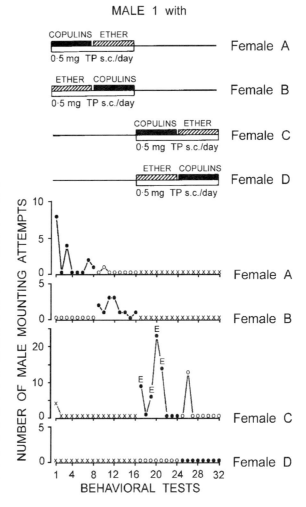

FIGURE 11-14. Experimental design (top) and results (bottom) from one group of rhesus monkeys illustrating the effects of olfactory signals (copulins) on male mate choice. Two of the four ovariectomized females, simultaneously treated with subcutaneous testosterone propionate (TP) to increase their sexual motivation, in turn received applications of either copulins (black bars and closed circles) or vehicle as a control (hatched bars and open circles). Note that copulins were ineffective in the lowest ranking female D, a phenomenon also observed with hormone treatments (Chapter 15). E = ejaculation. (SOURCE: Michael & Zumpe, 1982. Reproduced by permission of the Society for Endocrinology.)

Receiver Mechanisms

As for other sensory channels of communication, the complexity of receiver mechanisms varies with species. The male silkworm moth detects the female's sex-attractant pheromone, bombykol, by means of a single sensory cell in each of its antennae. It changes its direction of flight to equalize the degree of stimulation of both cells and then flies in a direction that increases stimulation of both, that is, up the odor gradient until it finds the female. This is similar to the mechanism by which moths avoid bat sonar. In many vertebrates, there are two parallel receiver subsystems for chemical communication. The familiar one involves the detection of volatile substances by sensory cells in the olfactory epithelium that project to the main olfactory bulb (Fig. 11-15A). The other subsystem detects less volatile substances by a special structure, the

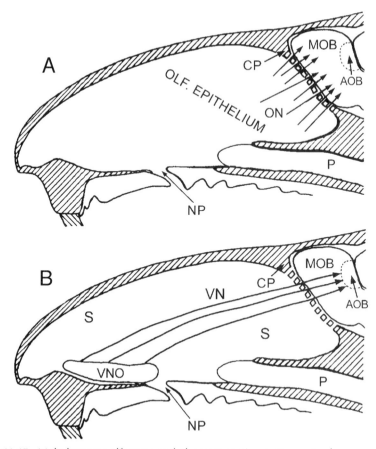

FIGURE 11-15. Male hamster olfactory and chemoreceptive systems. A: chemoreceptors in the olfactory epithelium project via olfactory nerves (ON) to the main olfactory bulb (MOB). B: chemoreceptors in the vomeronasal organ (VNO) project via vomeronasal nerves (VN) through the MOB to the accessory olfactory bulb (AOB). Both sets of nerve fibers penetrate the cribiform plate (CP). Other abbreviations: NP = nasopalatine duct; P = pharynx; S = septum. (SOURCE: Modified from Clancy, 1989, with permission)

vomeronasal organ (organ of Jacobson), whose sensory cells project to the accessory olfactory bulb (Fig. 11-15B). The vomeronasal organ is located bilaterally and medially on each side of the ventral part of the nasal septum near the floor of the nose, a few millimeters back from the external nares. It has a small duct opening into a nasal pit at this point. While sniffing at the genital area of estrous females, males of several species, such as sheep and horses, display a particular facial expression known as "flehmen," in which the upper lip is completely everted over the nostrils. The behavior is thought to involve olfactory and vomeronasal stimulation.

Studies on hamsters, for example, have shown that both the main and accessory olfactory systems are important in sexual attraction and mating. Female hamsters are solitary and very aggressive but must, when in estrus, attract a male. They then mark their territories with vaginal secretions that contain volatile substances, including dimethyl disulphide, which is detected by the main olfactory system and attracts males but does not elicit male copulatory behavior. Vaginal secretions also contain a nonvolatile, higher-molecular-weight fraction that elicits male copulatory activity, but only if the vomeronasal system of the male is intact. The vomeronasal organ is absent in some anthropoid primates but present in the majority of men and women, and there is recent evidence, yet to be confirmed, that it is functional. This controversial topic is not reviewed here, although it excites much interest.

Tactile Communication

Signal Properties

Tactile signals differ from other signals because they are effective only at very short range (Table 11-2). Since tactile communication requires that the sender and receiver overstep each other's individual distances (Chapter 12), it occurs predominantly in the interactions of two individuals that are socially bonded, for example, bees from the same hive, lions in the same pride, parents and offspring, siblings from the same clutch or litter, and prospective or current mates. The other major context for tactile communication is aggression between two individuals, in which the sender's invasion of the receiver's individual distance signals attack and elicits both aggression and flight or submission. Apart from this, tactile communication shares the combined advantages of visual and auditory communication, except for representational capability, and is often associated with them. In most primate species, grooming between social companions or mates initiates and reinforces bonding; this tactile signal is often associated with soft grunting or lip smacking.

Sender and Receiver Mechanisms

Most sender mechanisms are intrinsic, that is, a part of the body is used to touch the body of the receiver, or the receiver uses a part of its body to touch the sender. The former is more familiar to us but the latter also occurs. Honeybees that have located a food source perform either a round dance or a waggle dance on the vertical surface of a comb when they return to the hive. The round dance, during which the bee circles alternately clockwise and anticlockwise, encodes the information that the food source is nearby. During the waggle dance, the bee moves in a figure-of-eight pattern and, when in the "straight run" between the loops, waggles her abdomen and makes a buzzing sound. This dance indicates that the food source is distant. The approximate distance to the food source is encoded in the length of the straight run. The direction of the food source is encoded in the angle by which the straight run deviates from the direction of the sun if the dance is performed outside the hive, or by its deviation from vertical if the dance is performed inside the hive (Fig. 11-16). If the straight run is vertical, from bottom to top, the food source is in the direction of the sun, whereas a vertical run from top to bottom indicates a food source directly away from the sun. These dances are usually performed in dark hives: the other bees cannot see them but they follow the movements of the dancer by touching her with their antennae. The communicatory significance of these

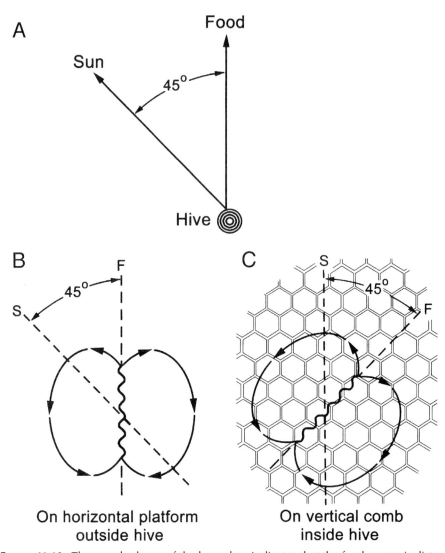

FIGURE 11-16. The waggle dance of the honeybee indicates that the food source is distant. The angle between the sun and the food source (A) is encoded by the angle between the sun and the direction of the straight "waggle" run when performed on a horizontal platform outside the hive (B), and by its angular displacement from vertical when performed on a comb within the hive (C). The distance to the food source is encoded by the length of the straight run. S = actual or symbolic direction of sun; F = actual or symbolic direction of food source.

honeybee dances was worked out by von Frisch (1967), for which he shared the 1973 Nobel Prize for Physiology or Medicine with Lorenz and Tinbergen.

In mammals, there are several types of sensation, many of which have specialized receptor organs. There are two broad categories of sensation: those originating outside the body, known as exteroceptive sensations, and those originating within it, known as proprioceptive sensations. Light touch, superficial and deep pain, pressure touch, temperature, position of limbs, vibration, and the appreciation of form, called stereognosis, are recognized. For light touch, there are both unmyelinated, free nerve endings in the skin and richly around hair follicles, and Meissner's corpuscles and Merkel's discs. For superficial pain, there are also free, unmyelinated nerve endings, and for deep pain and pressure, there are the larger Pacinian corpuscles as well as the Golgi–Mazzoni corpuscles. Krause's end bulbs subserve cold perception and Ruffini's endings subserve heat sensations. The position of limbs and joints is perceived by stretch receptors in the muscles themselves, and by Golgi bodies and Meissner's corpuscles in the tendons and joints. The detection of vibrations and the appreciation of form probably involve a combination of several different receptors. One specialization in receiver mechanisms results from a greater concentration of receptors in certain body areas. In the human, this occurs in the male and female genital areas, around the nipples and breasts, fingertips and palms, and the rhinarium, which comprises the nose, lips, and mouth. Another specialization depends upon the size of the representation in the primary cerebral cortex, where highly sensitive areas such as the mouth and fingertips have more representation than do others. This representation is illustrated for both sensory and motor function in Figure 11-17. In the majority of cases, sensory stimuli reach the spinal cord via the dorsal root ganglia and corresponding structures in the cranial nerves, and then travel rather discretely in their respective spinal columns to the thalamus and cerebral cortex. The sensitivity to tactile signals in communication between sexually bonded individuals is enhanced by hormone mechanisms.

Electrical Communication

The detection of prey by their electrical fields, and intraspecific communication in which the sender modulates its own electrical field, are restricted to a few marine and freshwater species. Two groups of freshwater fishes, the mormyrid elephant-nose fish of Africa and the gymnotid knife fish of South America, are active at night in murky tropical streams and rivers, and use electrical signals to communicate with conspecifics.

Signal Properties

Electrical signals are ideally suited for murky habitats; they are not impeded or even distorted by obstacles, have a fast transmission rate and fadeout time, and the exact location of the sender is detected. The range, about 1–2 meters, is an advantage in reducing "noise" where different species live in close proximity. The information content of electric signals can be varied by changing the shape of the electric field, the waveform of the electric discharge, the discharge frequency, and the duration of the discharge. These variables evidently permit communication of the same messages that are communicated by other sensory channels in other species.

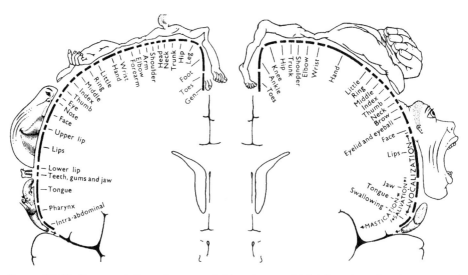

FIGURE 11-17. Representation of sensory (left) and motor (right) function of various parts of the body in the human primary cerebral cortex. The representations of the face and hands are proportionally much greater than those of other regions. (SOURCE: Penfield & Rasmussen, 1950, *The Cerebral Cortex of Man: A Clinical Study of Localization Function.* Copyright © 1950 by the Macmillan Company. Reprinted by permission of the Gale Group.)

Sender and Receiver Mechanisms

All animals produce a weak electrical field when a muscle cell contracts or a neuron fires. This can be exploited by predators such as sharks, which use sound, odor, water disturbance (via lateral line organs) and vision to find swimming prey, and electric fields to detect prey buried in sand. Knife and elephant-nose fish possess an electric organ in the tail that produces a stronger current by virtue of additive effects due to stacking of modified muscle or nerve cells; this is a communicatory, not a shock-producing, organ. Its discharge momentarily renders the tail negative in comparison with the head, producing an electric field whose properties can be varied as described earlier. The signals are perceived by specialized electroreceptor organs embedded in the skin.

In summary, different sensory channels possess different properties that are suited to different habitats and communicatory functions. Sender mechanisms can be either extrinsic or intrinsic and, together with receiver mechanisms, vary widely from the simple to the complex. Important messages are conveyed via at least two sensory channels whose advantages and disadvantages complement each other, and the redundancy maximizes the likelihood that the message reaches the receiver under most environmental conditions or perturbations.

CHAPTER 12
Agonistic Behavior

Most species engage in aggressive behavior in some circumstances and at certain times, and the fact that it is virtually ubiquitous suggests that it must be adaptive despite its hazards. Aggressive behavior has been defined by psychologists as activity that seems intended to convey a noxious stimulus or to destroy another organism, and classifications have included such categories as frustration aggression, fear-induced aggression, and irritable aggression, all based on their supposed underlying motivation. A more function-based classification is now widely accepted in animal behavior, in which aggression is regarded as behavior by which an individual actively attempts to exclude rivals from essential resources such as mates, food, water, and shelter. There are two broad types of aggression: interspecific aggression, largely comprising *predatory* and *antipredatory* aggression, and intraspecific aggression, involving competition between members of the same species. The latter can be subdivided into *territorial*, *dominance*, *sexual*, *parental*, and *parent–offspring* aggression (Wilson, 1975; Moyer, 1976). Predatory aggression, which is motivated by hunger and is part of a different motivational system (Scott, 1972), is not associated with conflicting aggressive and submissive tendencies or the compromises between them. As Lorenz pointed out, a cat killing a mouse may appear as calm as a cow grazing in a pasture; but this in not the case for antipredatory aggression shown by the prey. Intraspecific conflicts, in contrast, typically involve a mixture of aggressive and submissive tendencies, as illustrated by the ambivalent facial expressions and postures of cats (Chapter 2). The terms *agonism* and *agonistic behavior* were therefore coined to describe the entire spectrum of both aggressive and submissive behaviors. Agonistic displays almost always reflect this ambivalence and involve elements of both "fight and flight."

INTERSPECIFIC AGONISM

Aggression as a mechanism for competition for vital resources *between* species is relatively infrequent. In species holding territories that provide necessary food, water, and breeding resources, competition with other species within the territory may result in an increase in territory size, as shown for the song sparrow (*Melospiza melodia*) (Fig. 12-1). The result is a decrease in the number of conspecifics that can maintain a territory within the region of suitable habitat. If a move to another region is not possible, for example on remote islands, divergent evolution may occur, whereby competing species may adapt to take advantage of different ecological niches. A famous example is the diversity of Darwin's finches, so-called because Darwin observed and described them during his visit aboard HMS *Beagle* to the Galapagos Islands in 1835. These volcanic

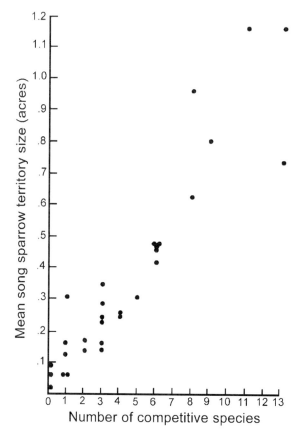

FIGURE 12-1. Increase in the size of song sparrow territories with increasing numbers of competing species in different localities of the Pacific Northwest and Wyoming. (SOURCE: Yeaton & Cody, 1974, with permission)

islands, lying in the Pacific about 600 km west of Ecuador, are populated by the descendants of relatively few species of birds, reptiles, and insects that drifted or were blown there from the continental mainland thousands of years ago. Here, a founder species, probably a seed-feeding ground finch, evolved by adaptive radiation into a number of different species, each of which was morphologically (Fig. 12-2) and behaviorally specialized to fill its own feeding niche, including fruit, bud, seed, and insect eaters, and even a species using twigs to extract insects from tree bark.

There are, however, situations in which different species come into conflict over specific food sources, especially if food is scarce, when they are not evenly distributed in the environment but are clumped in a few locations. This situation will be familiar to those fond of watching wildlife television programs from Africa: lions, hyenas, jackals, and vultures all compete for access to a recent kill. Less familiar are observations that some of these species come into deadly conflict in contexts not directly associated with feeding competition. Lions and hyenas hold large territories that overlap each other, and both species hunt the same ungulate prey, typically antelope, wildebeest, and zebra. It has been documented that male lions will search out and kill adult hyenas, and that

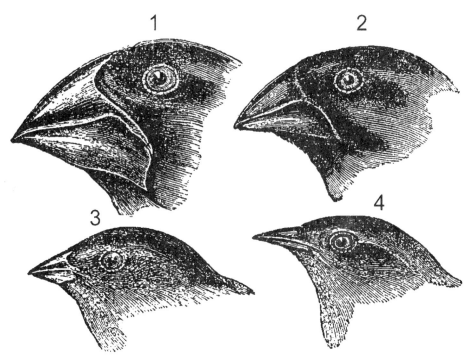

1. Geospiza magnirostris
2. Geospiza fortis
3. Geospiza parvula
4. Certhidea olivacea

FIGURE 12-2. Four species of Darwin's finches with graded beaks adapted for different foods. (SOURCE: Darwin, 1839, with permission)

hyenas will kill lionesses and cubs given the opportunity. Lions also kill cheetahs and cubs entering their territories. Although this behavior is lethal, it has some of the features of intraspecific aggression. While not occurring in direct competition for access to a kill, such deadly interspecific aggression is probably related ultimately to the need to compete for seasonally limited prey.

Predatory Aggression

Probably the most frequent form of interspecific aggression is the behavioral sequence used by predators to kill their prey. This aggression, rarely associated with signs of emotion, is triggered by hunger or the presence of prey (food) and often involves weapons not used in intraspecific aggression. Thus, rattlesnakes inject venom when they bite prey but never use their fangs in intraspecific disputes (Fig. 12-3). Predatory aggression, at least in some species, has a neural substrate that differs from that of intraspecific aggression. In cats, for example, the stalk, crouch, pounce, and killing bite in predatory aggression can be elicited by electrical stimulation of the *lateral* hypothalamus, while fighting with conspecifics can be evoked from the *ventral* and *medial* hypothalamus (Flynn, 1967). Intraspecific conflicts can be just as intense and injurious in herbivores as in carnivores (discussed later). For all these reasons, ethologists tend to place predatory aggression in a separate category from aggression of other types.

FIGURE 12-3. Intraspecific conflicts between male rattlesnakes (*Crotalus ruber*) consist of wrestling matches and never involve the venomous bite used in predatory aggression. The fight ends when the winner pushes the loser to the ground (bottom). (SOURCE: Modified from Shaw, 1948, with permission)

Antipredatory Agonism

Unless caught by surprise, prey usually flee from a predator or, under certain circumstances, may attack it. As in predatory aggression, antipredatory "defensive aggression" may involve the use of weapons not employed in intraspecific conflicts. Giraffes, for example, defend themselves from predators by kicking and slashing with their sharp hooves but use their short, blunt horns in intraspecific (usually male–male) aggression. Hediger (1934) was the first to show that animals have species-specific flight distances that can be modified only somewhat by experience. Smaller species generally have shorter flight distances than do large species, and well-protected species such as armadillos, porcupines, and skunks have shorter flight distances than do more vulnerable species. As noted for vervet monkeys (Chapter 11), a single species may have different vocalizations and patterns of flight for different types of predator (snakes, leopards, and eagles in the case of vervet monkeys). Chickens flee into cover from flying raptors but fly upward themselves to escape from ground predators such as foxes. When escape is prevented, some animals will attack the predator: the cornered rat effect. Antipredator attack occurs when the predator oversteps the prey's species-specific critical distance (Hediger, 1950), which is much shorter than its flight distance. Attack also occurs if the animal first becomes aware of a predator or intruder when already within its critical distance; this is how most people are inadvertently bitten by snakes. Animal trainers know about this and never violate the critical distance of dangerous animals, which it is the objective of training to reduce. Another type of antipredatory aggression is mobbing, in which several individuals of a prey species gang together to attack a single predator. This can be seen, for example, when a group of songbirds mobs an owl or hawk. In addition to species-specific flight and critical distances, most animals also have an even shorter individual distance, which is important in intraspecific interactions.

INTRASPECIFIC AGONISM

Individual Distance

If the individual distance is violated, the response is either to threaten or to avoid the intruder. Typically, individual distances are quite short, often only slightly longer than the individual's reach, and it appears that their function is to provide just enough space to maneuver unhindered should the need for sudden action arise. Individual distance in birds is clearly apparent in the fall, when flocks of migrants congregate on telegraph lines before departure. The birds distance themselves from each other so symmetrically that they look like beads on a necklace. If an individual attempts to land between two others that are already close together, both will repel the intruder, which must find a larger gap. We are sometimes uncomfortably aware of encroachments on our individual distance when approached at a cocktail party by someone with a much shorter individual distance than our own. The person comes too close, and we find ourselves backing up until we collide with the furniture or the wall in an attempt to keep our distance comfortable. Because intrusions upon individual distances arouse agonistic responses, it is not surprising that the closeness required by mating activity stimulates agonism that must be reduced by courtship if mating is to proceed satisfactorily.

Intraspecific Aggression

The majority of intraspecific agonistic episodes occur in the context of direct or indirect competition between individuals or groups for vital resources such as food, water, shelter, mates, and a place to breed. Since intraspecific aggression occurs in many different contexts, there has been debate as to whether it makes sense to talk of a unitary aggression drive in the way we do of a unitary sexual drive. Some authorities take the view that there is no biological, genetically based predisposition to behave aggressively, and that aggressiveness is primarily a consequence of experience and learning. J. P. Scott (1960), for example, found that he could produce aggressive and nonaggressive male mice depending on the experiences to which he subjected them; mice became more aggressive if they were allowed to win fights with smaller opponents, whereas they became nonaggressive if they lost fights with larger opponents. Another hypothesis, formulated by Dollard and others (1939), states that aggressive behavior is a consequence of frustration that arises when striving for a goal—a view reflected in some human educational programs without entirely satisfactory results.

In contrast, Freud, and then Lorenz, saw aggression as having evolved because of its adaptive advantages in a variety of circumstances. While experience undoubtedly affects the amount and intensity of aggression shown by the individual, there is evidence suggesting that the potential to behave aggressively is genetically programmed in many species: Some species are inherently more aggressive than others.

Evidence for the Genetic Control of Aggressive Behavior

Evidence for a major genetic component of aggression is described as follows:

1. Morphological features act as releasers for aggression. Certain fish that inhabit coral reefs (e.g., butterfly fish, *Chaetodontidae*) are highly territorial when young; the territory provides both food and shelter. Butterfly fish are brightly colored with bold patches and stripes of white, yellow, red, and black. The species-specific color pattern of each fish acts as a releaser for territorial aggression by conspecifics, but not by other species, indicating coevolution of the morphological characteristic and the behavioral response.

2. Artificial selection and inbreeding results in changes in aggressiveness within a few generations. We are all familiar with the many species that have been domesticated to reduce their aggressiveness, but some, including pit bull terriers, fighting cocks and fighting bulls, have been bred to increase it. It has been shown that inbred strains of aggressive and nonaggressive mice retain their behavioral characteristics even when pups are cross-fostered from birth, providing strong evidence for a genetic component.

3. There are neural substrates for aggressive behavior. Studies in which different regions of the brain were lesioned, electrically stimulated, or examined by recording neural activity have shown that discrete regions of the limbic system, notably, the hypothalamus and amygdala, are involved in agonistic behavior.

4. Animals perform appetitive behavior, such as running a maze, crossing an electric grid, or pressing a lever, to engage in aggression that appears to be a consummatory response. Like other juvenile coral reef fish, damsel fish (*Pomacentridae*) establish individual feeding territories that are maintained until sexual maturation. The territorial boundaries are defended aggressively against intruders. Studies with *Microspathodon chrysurus* demonstrated that these fish would negotiate an L-maze consisting of an opaque tube to enter a small bottle, an aversive environment but one that provided an opportunity to threaten at and fight with a conspecific in an adjacent aquarium in a simulated territorial boundary fight (Rasa, 1971).

5. Aggressive behavior can be used as a reward or "positive reinforcement" in a learning task in the same way that food is used as reward in an operant conditioning paradigm. The damsel fish study also demonstrated that the opportunity to engage in intraspecific aggression acted as reinforcement for the learned task of entering an opaque tube with the small bottle at its end. Performance of this task could be extinguished by withholding the reward, namely, the opportunity to engage in aggressive behavior.

6. Aggressive behavior appears spontaneously during ontogeny. In albino mice, for example, aggressive behavior first appears at 28 days of age, and tractable, easily handled infant rhesus monkeys spontaneously start threatening and biting at a few months of age, even when they are hand-fed daily.

7. Species-typical agonistic behavior patterns develop in isolation-reared individuals deprived of the opportunity of learning them from conspecifics, and this may be observed in mice, rats, and rhesus monkeys.

Reducing the Costs of Intraspecific Aggression

It appears from the foregoing discussion that engaging in intraspecific aggression is rewarding if it does not result in repeated loss, pain, or injury. It is difficult to see how it could be adaptive if it took the form of injurious and potentially deadly fights. Two main mechanisms reduce the costs of intraspecific aggression. The first is the form of the behavioral interactions themselves, and the second is that certain types of aggression, notably, territorial aggression and dominance aggression, actually reduce the frequency of aggressive behavior in the long run.

Intraspecific, unlike interspecific, aggression does not generally cause serious injury; it resembles jousting, that is, a ritualized test of strength and endurance that permits the loser to withdraw, perhaps to fight a more evenly-matched opponent another day. Lethal weapons such as teeth and claws are either not used or targeted toward a well-protected part of the body, for example, the thick manes of lions and baboons or the thick layers of protective blubber of sealions and elephant seals. Many ungulates have horns or antlers that are either permanent, as in bighorn sheep, cattle, antelopes, giraffes, and reindeer, or seasonal, being shed after the mating season, as in deer and moose. They are used by opponents to push against each other during contests, but the avoidance of serious damage is well illustrated in species whose females do not have horns, such as the Nilgau antelope. Males have short, sharp, unbranched horns used to butt against each other head-on (Fig. 12-4), but they never butt each other in the flanks, which would cause serious injury. In contrast, females fight each other with the same head-butting move-

FIGURE 12-4. Male Nilgau antelope (top) trying to push each other's heads to the ground. Male oryx (bottom) butting each other's horns. Neither species ever butts with horns against the opponent's flanks. (SOURCE: Eibl-Eibesfeldt, 1975, with permission)

ments, but these are often directed at the other female's flank region without, of course, producing serious injury (Fig. 12-5).

Intraspecific contests usually terminate with one opponent yielding, either by leaving the scene more or less hurriedly or by making a submissive gesture that inhibits further aggression (see below). When flight is prevented, for example, by the presence of dependent offspring or in captivity, the "cornered rat" effect results in a fight, and the loser may not be able to terminate the contest before being seriously hurt or killed. Painful stimuli such as tail pinch or electric shock can elicit immediate aggression in many species; at this point, the loser may escalate the fight with injurious defensive aggression containing elements of fear, which is virtually identical to the response of prey to attack by a predator.

Categories of Intraspecific Aggression

As noted, aggression occurs in a wide variety of behavioral contexts that presumably reflect its different functions, and it is typically categorized by function. Two categories of intraspecific aggression, territorial and dominance aggression, are argu-

FIGURE 12-5. Female Nilgau antelope, lacking horns, often butt at the vulnerable flanks of another female. (SOURCE: Eibl-Eibesfeldt, 1975, with permission)

ably more prevalent and more important than the others. Territoriality and dominance can occur independently but they often occur in combination, depending in part on the interaction between the sociality of the species and the population density. Thus, increased sociality promotes the formation of dominance hierarchies, while increased population density promotes territoriality; where both sociality and population density are high, dominance orders exist both between group members and between groups.

Territorial Aggression

Individuals and groups can share the resources of the habitat with others, as do nomadic tribes in Asia and Africa, or they can utilize overlapping home ranges, like some primates and grazing herds in Africa. They can also defend smaller pieces of the habitat, namely, their own territories, against all other conspecifics. Both male and female hamsters have individual territories that they defend against all others; monogamous pairs of gibbons defend their territories against even their own adult offspring; and social groups of wolves, lions, and hyenas defend their territories against other groups. Territories generally provide access to food, water, shelter, and breeding sites. Territoriality is likely to develop if adequate resources are available in an area that can be defended effectively. Territories can be permanent year-round. In seasonally breeding species, for example, many songbirds, territories are abandoned at the end of the mating season when the birds flock, often before migration. Territoriality can be restricted to one developmental phase; some fish, for example, damsel and butterfly fish, are territorial as juveniles but not as adults. When a feeding territory cannot be defended effectively because the food is sparse and widely dispersed, territories may be small and contain only one important resource, for example, a suitable nesting site in herring gulls and cliff-breeding seabirds, including guillemots and puffins. This is similar to nomadic tribes laying claim to and defending a reliable water source in the Sahara. Between species with large territories that provide almost everything and species with small territories that provide a single resource, there are species that defend two different territories for two different purposes. Herons and storks, for example, have small nesting territories that are clumped closely together but when hunting for food, they distribute themselves widely on large, individual fishing territories. Finally, territories may have a temporal component analogous to time sharing an apartment at a vacation resort. Several male cats may share much of the same territory but use it at different times of day, so that they usually avoid each other.

One effect of territoriality is that individuals or groups are spaced out over the available habitat, which limits the number of competitive encounters with conspecifics. Neighbors in established territory become familiar with each other and with their common boundaries, which results in decreasing aggressive intensities and eventually in periodic patrolling and occasional mild threat: this has been called the "dear enemy effect." Tinbergen showed that an animal is always the victor near the center of its territory and is increasingly likely to be the loser as it encroaches onto the territory of a neighbor (Fig. 12-6). Territorial boundaries are often demarcated in marine and freshwater fish by environmental features such as rocks or corals, patches of sandy bottom, and vegetation. In terrestrial species, similar environmental features are used but often are supplemented in mammals with feces, urine, and scent marks for which specialized scent glands have evolved (Chapter 11). Some species also physically mark the environment; for example, bears tear at the bark of certain trees with their claws and deer thrash the undergrowth with their antlers. Environmental objects may even be rearranged; the European red squirrel lines up pine cones near its territorial border. Nevertheless, territorial boundaries are not so much a "line drawn in the sand" as a strip of no-man's-land in which both neighbors are fearful of each other and reluctant to attack. This tranquil state of affairs promptly ends when a third individual attempts to establish a new territory from pieces of existing territories. Consequently, although the establishment of a territory depends on agonistic behavior and is potentially hazardous, it is worthwhile because, in the long term, the territory provides not only assured access to resources but also greatly reduces aggressive competition for them.

Dominance Hierarchy

Dominance hierarchies occur in species that live in individualized groups, in which animals recognize each other individually (Chapter 10). Aggressive competition between group members and between different groups results in the development of intragroup and intergroup dominance hierarchies. Schjelderup-Ebbe (1922) was the first to draw attention to the "pecking order" of domestic chickens, which is linear and stable when fully established. Figure 12-7 illustrates the development of a dominance hier-

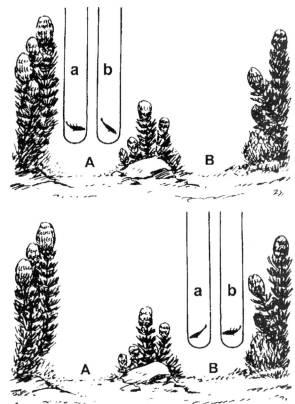

FIGURE 12-6. Two male sticklebacks (a and b) living in 2 territories (A and B) were moved between territories in glass tubes. When in A, male a tries to attack b, which tries to escape (top), and vice versa (bottom). (SOURCE: Eibl-Eibesfeldt, 1975, with permission)

archy in 6 roosters that were strangers before being placed together. Establishing a dominance hierarchy, like establishing a territory, is associated with much fighting. This can be exhausting and injurious but, once established, rank order greatly reduces the frequency and intensity of aggressive exchanges; it is then maintained by mild threats and increasingly perfunctory submissive gestures.

In species with higher neocortical development, dominance hierarchies tend to be more complex and less linear than those of chickens. Alliances between two or more lower-ranking individuals may give them a temporary advantage over a higher-ranking individual. Rank order can differ with the resource being contested, so that one individual may be dominant in competition for food and another may be dominant in competition for a mate. Differences depend on the level of motivation for access to a particular resource, and this interacts with the basic dominance relationship.

Dominance hierarchies typically involve less agonism under natural conditions than in captivity for two reasons. First, animals in nature spend much time hunting for food and feeding, so those of low rank have the chance to avoid the vicinity of higher-ranking individuals. Second, in species such as macaques and baboons, the females live in the social group all their lives and inherit their dominance status from the mother. This social inheritance obviates much of the intense fighting associated with establishing the

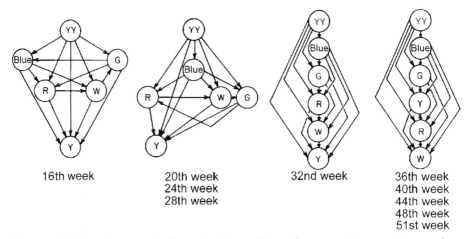

FIGURE 12-7. Development of a linear dominance hierarchy among 6 roosters (arrows show direction of aggressive behavior). (SOURCE: C. Murchison, 1935, The experimental measurement of a social hierarchy in *Gallus domesticus*: IV. Loss of body weight under conditions of mild starvation as a function of social dominance. *J. Gen. Psychol.* **12**, 296–311. Reprinted with permission of the Helen Dwight Reid Education Foundation. Published by Heldref Publications, 1319 Eighteenth St., NW, Washington, DC 20036-1802. Copyright © 1935)

hierarchy. Daughters attain a rank just below that of their mothers, with the youngest daughter outranking the next older sister that, in turn, outranks *her* older sister. This arrangement protects the infant and results in a rank order both within and between female members of different matrilines.

There are sometimes costs for low-ranking individuals, as illustrated by studies on feeding behavior and reproductive success in Japanese macaques in the wild (Soumah & Yokota, 1991, 1992). Throughout the year, a group received supplementary food in the form of wheat placed in a small area every afternoon. Higher-ranking females largely monopolized this preferred food and obtained more calories than did lower-ranking females (Fig. 12-8). Consequently, the latter spent a larger proportion of their time foraging and feeding, and less time resting and grooming (Table 12-1). This difference correlated with decreased reproductive success in low-ranking females, which were older when they had their first infant, had longer interbirth intervals thereafter (Table 12-2), and a higher cumulative infant mortality rate (17%) by the time infants were age 8–12 months compared with high-ranking females (1%) (Fig. 12-9). This resulted in fewer offspring per lifetime and diminished reproductive success. The marked negative effect of low rank would be unlikely to occur under completely natural conditions, but the study made its point. Similar negative effects of low rank on female reproductive success were found among chimpanzees in the Gombe National Park, Tanzania (Pusey *et al.*, 1997). Data spanning 35 years showed that, compared with subordinate females, dominant females had significantly higher infant survival, faster maturing daughters, and more rapid production of young.

There may also be costs associated with high rank. Data were collected between 1967 and 1992 from the females of five social groups of olive baboons in Tanzania. Olive

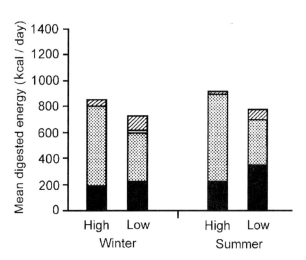

FIGURE 12-8. High-ranking female monkeys obtain more calories from provisioning (stippled areas) than do low-ranking females. Black areas = energy intake from natural food before provisioning; hatched areas = energy intake from natural food after provisioning. (SOURCE: Soumah & Yokota, 1991, with permission from Karger, Basel)

baboons have the same multimale–multifemale social organization as Japanese macaques, and females usually retain their rank for life. Lower rank in olive baboon females was associated with the same phenomena described in Japanese macaques. Their lifetime reproductive success was no lower than that of dominant females, because the latter had more miscarriages, and some showed decreased fertility or were completely infertile (Packer et al., 1995). The stress associated with maintaining high rank was thought to play a role in this, and it was argued that the reproductive costs of high rank may act as a selective constraint on traits promoting agonistic competition in females.

Sexual Aggression

Sexual aggression can be defined as threats and attacks used to secure and keep mates. For reasons discussed in Chapter 13, it is primarily a male phenomenon, namely, intermale aggression. Sometimes this consists only of threatening facial expressions,

Table 12-1. Activity Budget (%) of Female Japanese Macaques by Season and Rank

	Winter		Summer	
	High rank	Low rank	High rank	Low rank
Feed	35	47	21	31
Forage (search for food)	9	11	11	12
Move	8	9	15	17
Groom others	10	5	7	4
Groom self	2	2	1	1
Rest	36	26	45	36

SOURCE: Soumah & Yokota, 1991. With permission from Karger, Basel.

Table 12-2. Rank and Reproductive Success in Japanese Macaques

Rank	Age at first birth	Interbirth interval
High	6.09 yrs	1.65 yrs
Medium	6.43 yrs	1.81 yrs
Low	6.90 yrs	1.98 yrs

SOURCE: Data from Soumah & Yokota, 1992, with permission.

body postures, and vocalizations. It also includes actual fighting between males for access to females, as in elephant seals and in deer and many other ungulates; territorial aggression between males where the territory provides breeding resources, as in sticklebacks, many songbirds, and gibbons; and much intermale dominance aggression in social groups, including many primates such as langurs, talapoin and patas monkeys, macaques and baboons. Sexual aggression can also involve threats and attacks directed at the mate. In rhesus monkeys, for example, at the start of the fall mating season there is a large increase in intermale aggression as well as in aggression directed by males at females. The male hamadryas baboon will attack and punish a female from his harem if she wanders too far or attempts to associate with another male. Castration decreases male sexual motivation and behavior, and also decreases male aggression. For these

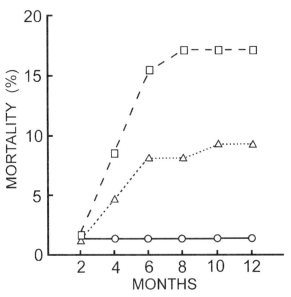

FIGURE 12-9. Cumulative infant mortality (1987–1988) in high-ranking (circles), middle-ranking (triangles) and low-ranking (squares) female Japanese macaques. (SOURCE: Soumah & Yokota, 1992, with permission)

reasons, it has been used worldwide for centuries in domestic animals to make them more tractable: the world over, bullocks not bulls are used for pulling carts and plows. The greater aggressiveness of males compared with females may be due primarily to sexual aggression. Although less commonly reported, sexual aggression is also displayed by females. Female hamsters and gibbons, for example, are as aggressive toward other females encroaching on their territories as males are toward other males, and female rhesus monkeys will threaten other females more often during their fertile midcycle period than at other times. Other forms of aggression that occur in a sexual context include aggressive interactions between a male and female at the start of courtship and those occurring when there is a conflict between the interests of the male and female, for example, when the male attempts to mate but the female is unreceptive and threatens or attacks him.

Parental Aggression

Parental aggression is directed toward an intruder by a parent in the presence of its offspring. In defense of young, a parent—in mammals, usually the mother—will attack another animal to which it would not otherwise respond aggressively. Female lions, tigers, and bears with cubs, for example, will threaten or attack an approaching male, even their recent mate, in their cubs' defense (males are a serious threat to cubs). Parental aggression can also be interspecific. In a phenomenon known as parental antipredatory aggression, parents defend their young against predators and nonpredators if they approach too closely. State and National Parks invariably post warnings to keep a distance from animals with young.

Parent–Offspring Aggression

Parent–offspring aggression usually takes the form of aggressive, disciplinary actions by a parent toward its offspring. Perhaps more common in mammals than in other taxa, it is usually associated with weaning (Chapter 15). Conflict develops between the mother and her maturing offspring near the time of weaning, when the mother responds with increasing aggression in response to the offspring's attempts to suckle. Similar kinds of behavior can be observed in birds. Families of starlings, for example, often hunt for earthworms on lawns in late spring. As parents pull worms out of the ground, their offspring close in, begging loudly to be fed; initial avoidance of the fledglings by parents is gradually replaced by pecks and chases. There is prolonged postweaning dependence in some species, notably carnivores hunting dangerous prey and species with complex social organizations, for example, primates, whose young take time to learn the necessary skills. Cheetah cubs, for example, remain with the mother for a couple of years after weaning. When they have made their first substantial kill, the mother aggressively rejects them. Female bears respond in similar fashion to partly grown cubs when they are big enough to fend for themselves. Sometimes this behavioral change is initiated by the onset of heat or by a new pregnancy. In multimale–multifemale groups of primates there is the usual maternal weaning aggression, but thereafter the situation is different because the mother–daughter relationship persists throughout life. Sons are not clearly rejected by their mothers, but all adolescent males are punished by adults when they engage in loud and disruptive fighting; they become peripheralized and eventually learn to remain silent during agonistic interactions.

Sibling Aggression

Aggression between siblings, not usually dignified as a separate category, is nevertheless a common occurrence among species in which parental care includes feeding and protection of altricial young after hatching or birth. Competitive aggression between littermates can be restricted to mild pecking, biting, or pushing to secure first access to food or to a nipple. The aggressive intensity seems to be related to both a sibling's ability to monopolize food and the number of young in a litter; it can be lethal and the norm for a species. Among birds, siblicide occurs regularly in great egrets, which feed their young on fish small enough to be taken by one chick. Because egrets start to incubate after the first egg is laid, there is asynchronous hatching and the oldest chick has a size advantage, so it can kill younger siblings either directly or by pushing them out of the nest. In contrast, siblicide is rare in great herons, which bring fish that are too large for any one chick to monopolize. Among mammals, a striking example of sibling aggression occurs in the spotted hyena. Groups are female-dominated; females have very high testosterone levels and only the highest-ranking female breeds. Twins are born with fully erupted teeth. Captivity studies have shown that, within minutes of birth, they engage in a vicious biting fight that usually results in the death of one twin. There is, of course, a great deal of competition between human siblings, but mostly not quite of this order.

Other Forms of Intraspecific Aggression

Xenophobia. Literally, xenophobia means "fear of strangers" but it translates behaviorally as defensive aggression toward strangers, that is, both fear of, and aggression toward, them. Phylogenetically, xenophobia is very old, occurring in the majority of vertebrates, and it is evoked both by conspecifics and by individuals of other species. For conspecifics, the adaptive function of xenophobia is readily apparent in territorial species, which respond aggressively to an intruder in order to maintain their territory. Xenophobia is also adaptive in species living in closed, anonymous, or individualized social groups: here, an intruder is likely to upset the social order. Xenophobia can also occur toward group members whose appearance or behavior falls outside the norm. For example, a light-plumaged hen in a dark-plumaged flock of chickens generally becomes the lowest-ranking individual because it elicits aggression from all other flock members in what amounts to "scapegoating"; a hen threatened by a higher-ranking hen redirects its own aggression toward the light-plumaged hen. Several chimpanzees in the wild were infected by poliomyelitis some years ago, and survivors developed atypical gaits or peculiar modes of locomotion. This attracted both fear and repeated threat response from healthy group members. Xenophobia is, of course, virtually ubiquitous among humans and must have been adaptive in the past. It is now a divisive and destructive human trait responsible for much scapegoating and persecution of minorities that differ from the "norm"; it is, of course, the biological basis of all racism.

Infanticide. Field studies on a number of different mammalian species have documented the occurrence of infanticide sufficiently often not to dismiss it as an aberration. Infanticide occurs when one male, or several allied males that may be kin, overthrows the resident male or males in a social group containing several females and young. Shortly after the takeover, the new males systematically kill all dependent offspring, that is, infants that have not yet been weaned. Such incidents have been observed repeatedly in langur monkeys, lions, and other species (Hrdy, 1977). The proximate mechanism remains unknown but its adaptive advantage may be twofold. First, lactation inhibits ovulation, and when infants die, lactation ceases and the females become sexually receptive much sooner than would otherwise be the case. This provides earlier and more reproductive opportunities for the new males. Second, the new males will not have to expend energy protecting infants that were sired by other males and therefore could not ultimately contribute to the new males' inclusive fitness. In humans, infanticide occasionally occurs in similar situations, namely, by a new stepfather (Chapter 17), and is a serious crime. In certain parts of the world, it is more prevalent in some rural populations because of economic necessity, and daughters appear to be more at risk than sons.

War and Genocide. We are all aware of this special case of intergroup aggression that is devastatingly injurious and may result in the demise of an entire society. Regarded as a human phenomenon until recently, war and genocide may also occur in other species, but it is rare and difficult to document, requiring years of field study. Jane Goodall and colleagues described guerrilla warfare and genocide among the chimpanzees in the Gombe National Park. Chimpanzees live in groups centered around the adult males, many of them related, and the groups are rather stable over time. The organization is the opposite of that in macaques and baboons because males remain in the group, whereas females typically emigrate from their natal community around puberty. In one incident, the males of one community systematically ambushed and killed the males of another community (guerrilla warfare), eliminating it (genocide). In another incident, a female and her daughter eventually killed several females and their offspring within the group (genocide). It has been documented in different geographical locations that small parties of male chimpanzees will periodically set out on "raids" into the home range of a neighboring community and ambush, attack, and kill a male of that community if they happen to encounter him alone. This is reminiscent of similar behavior among some human tribes.

Intraspecific Submission and Flight

Neither submission nor flight appears to be rewarding; the observation that repeated defeat acts as a negative reinforcement for aggressive behavior supports this. It has been fully documented that the loser of a fight, or the lowest-ranking individual in a social group, may die as a result of stress within a few days if not removed from the stressful situation. There may be no external injuries but death results from renal ischemia and uremia (von Holst, 1972; van den Höövel, 1973).

Whether the loser of a contest uses submissive displays or flight depends in part on the social organization and ecology of the species. If species live in individualized groups, either to provide defense against predators (baboons and macaques) or to facilitate hunting dangerous prey (wolves and hyenas), flight from an aggressor within the group is not an option since the individual cannot survive on its own. Submissive or "appeasement" displays that inhibit further aggression are therefore performed by the losers of conflicts and also by lower-ranking individuals. As first pointed out and illustrated in dogs by Darwin (1872), submissive behaviors typically involve postures and actions that make the animal look smaller and less formidable, that is, the antithesis of those used by dominant and aggressive individuals. Dogs about to attack are straight-legged, bare their teeth, have a raised tail, and the hair on the back of the neck (hackles) is raised. Submissive dogs crouch, tuck their tails under, and flatten their hair (Fig. 12-10). Dogs and wolves losing fights may terminate the aggression by exposing the side of the neck or by rolling over onto their backs, exposing vulnerable body parts, and this inhibits further attacks by the winner. Lorenz speculated that the biblical injunction to "turn the other cheek" might have been misunderstood, and that its true meaning is that such a submissive act would inhibit the aggressor from striking again. In highly mobile species that live either in anonymous aggregations as a defense against predators (schools of fishes, large herds of ungulates) or dispersed throughout their habitat, withdrawal or flight is the rule, and these species may have no submissive behavior patterns in their repertoire.

COMPARISONS BETWEEN INTERSPECIFIC AND INTRASPECIFIC AGONISM

Interspecific and intraspecific agonism appear to differ considerably in both their behavioral expressions and their functions. Much of this difference stems from the fact that predatory aggression involves no elements of fear, which is undoubtedly related to the fact that it would not be adaptive to prey on species that cannot usually be overpowered. The likelihood of winning the competition between predator and prey, the major form of interspecific competition, will be overwhelmingly in the predator's favor or it will not prey upon that species. The responses of the loser are very similar in interspecific competition and intraspecific conflicts when the loser is injured and in mortal danger, namely, outright flight or, if escape is not possible, damaging defensive aggression. On the basis of such observations, it has been proposed that all interspecific and intraspecific aggression may best be divided into two basic categories that can be distinguished by both behavior and their causes and functions: property-protective aggression that is mild, ritualistic, and usually nondamaging, and self-defensive aggression that is all-out and injurious (Rasa, 1976); this distinction is a useful one.

HUMAN AGGRESSION

Humans show, although not in identical form, the same general types of aggression that we have described for nonhuman species. Human interspecific aggression is usually directed at species providing food, when vast numbers of crustaceans and fish are

FIGURE 12-10. Contrast between threat posture (top) and submissive posture (bottom), as illustrated in the dog. (SOURCE: Darwin, 1872)

harvested, and this also applies to domesticated food sources such as poultry, pigs, sheep, and cattle. But there is also a good deal of killing for sport, from foxhunting to shooting "game" animals. Sometimes this pushes a species toward and even over the edge of extinction, because populations cannot regenerate at the speed with which they are destroyed (e.g., whales). This represents a form of genocide, although more properly the term should be reserved for exterminating our own species. Genocide has been widespread throughout human history, and underlying economic, historical, and religious factors are usually involved. The *others* are scapegoated, everything "bad" is attributed to *them*, while everything "good" is attributed to *us*. These mechanisms apply to quite small as well as very large groups. Nations indulge in nationalism, generally regarded as destructive because the interests of one's own nation transcend the rights of all others. Patriotism, which simply describes the love of one's own country is, on the other hand, usually regarded as constructive. Xenophobia plays a major role in human affairs, and to help combat this, educationists encourage young people in particular to travel and work overseas in different cultures before they become too narrow and inward looking in their views.

War between peoples, as well as gang warfare, is intraspecific aggression for power and control of resources. But we are all familiar with the sexual aggression associated with mate acquisition, although nowadays this is more subtle than the ritualized, medieval joust for a maiden's hand. Territorial aggression is also frequent; not only does this take the form of disputes between neighbors that result in violence and litigation, but it is also seen in the academic disputes over who will teach this or that course in a college. Numerous other examples will leap to mind. Aggression in humans is complex, and we should mention so-called passive aggression. This is a form of attack in which both victim and perpetrator may be only vaguely aware that aggression is involved at all. It may take the form of letting others down by using infuriating tricks such as always being late, being on time but on the wrong day, or being chronically overdue with assignments. The attack is hidden but it is there nevertheless; help and assistance is simply withheld when needed, or given when no longer useful. Certain forms of overt

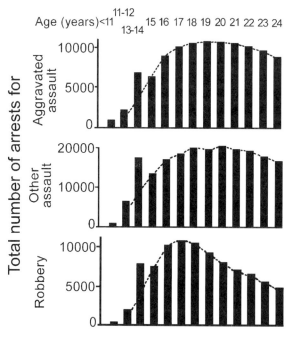

FIGURE 12-11. Changes during and after puberty in numbers of arrests for violent crimes (the majority by males) in the United States during 1977. Data from the FBI Uniform Crime Reports. (SOURCE: Michael & Zumpe, 1990, in *Control of the Onset of Puberty* (ed. M. Grumbach). Copyright © 1990 by Lippincott-Williams & Wilkins)

aggression are peculiar to humans, and some are associated with drug and alcohol abuse. More recently, serious intraspecific aggression has resulted from chronic steroid abuse by would-be athletes, and puberty in the male is associated with a marked increase in violent crimes (Fig. 12-11). Some forms of aggression are characterized by poor impulse control and occur in various psychiatric syndromes and in metabolic and neurological disorders; these are outside our present scope, but brain damage can be diffuse, as in arteriosclerosis, or localized, as in space-occupying lesions and tumors. Epilepsy can be associated with violence of which the sufferer may be relatively unaware (fugue states), and temporal lobe epilepsy, which can be effectively treated, is a prominent example.

CHAPTER 13
Sexual Selection

ASEXUAL AND SEXUAL REPRODUCTION

Reproduction is of two kinds, asexual and sexual. Asexual reproduction, namely, reproduction in the absence of fertilization resulting from the fusion of two gametes, is largely confined to simple life forms such as viruses and bacteria, and consists of the production of identical copies of the parental type. This is achieved by binary fission in unicellular protozoans and by budding in simple metazoans such as hydra. In more complex organisms, including a few insect and even vertebrate species, asexual reproduction results from the development of offspring from unfertilized eggs, a process known as parthenogenesis. Propagation of one's genes is accepted as the main driving force behind evolutionary change insofar as traits that enhance one's ability to survive and reproduce will be favored over those which do not. Sexual reproduction offers the possibility of genetic variation because of the recombination of genes inherited from both parents, and this provides the basis upon which natural selection can mold the process of evolution (Chapter 4). Yet by sexual reproduction, only half of an individual's genes pass to its offspring, whereas asexual reproduction, including parthenogenesis, ensures that all parental genes are passed on. Another disadvantage of sexual reproduction is that the recombination of genetic material from the egg and sperm may break up successful parental genotypes and produce offspring that are homozygous for a deleterious trait encoded in recessive alleles, resulting in diseases such as cystic fibrosis, sickle-cell anemia, phenylketonuria, and Tay–Sachs disease in humans. Parthenogenesis requires only females, and sexual reproduction can be regarded as "wasteful" in that it produces unneeded males. There are also the energetic costs required for the specialized structures and behaviors associated with finding and courting the opposite sex, competing with the same sex, as well as dealing with conflicts when males and females attempt to pursue different sex-specific strategies for maximizing their own reproductive success (see below). Mating can also be hazardous because of injury and death from intraspecific aggression during mate competition, and because close proximity facilitates transmission of parasites and many diseases, including the familiar human examples of gonorrhea, syphilis, and AIDS.

In view of these disadvantages, why is sexual reproduction so prevalent among complex plants and animals? The answer is elusive; but it is not just that it is more fun. Most authorities agree that the overriding advantage in a biological sense is that sexual reproduction results in phenotypic variation, and this maximizes the likelihood that at least some offspring are better adapted and more successful than their competitors; more importantly, some offspring could survive and thrive in radically different environmental conditions. For example, if one wolf cub in a litter has a pelage thicker than that of its siblings, it is more likely to survive a dramatically harsh winter and reproduce. Williams (1975), the major proponent of this view, likened reproduction to a lottery in which the chances of winning are increased by buying a series of tickets with different numbers (sexual reproduction) instead of a series of tickets with the same number (asexual reproduction). He compared the life cycles of species such as aphids and liver flukes that alternate between asexual and sexual phases. Parthenogenesis typically occurs under constant, predictable conditions, whereas sexual reproduction precedes dispersal into unpredictable environments. Thus, throughout late spring and early summer, female aphids on a suitable food source, for example, a prized rosebush, reproduce asexually, producing several generations of wingless and occasionally winged daughters. But later in summer, females produce both winged sons and winged daughters that disperse to overwintering sites. Sexual reproduction in the fall results in fertilized females whose eggs survive the winter and hatch in early spring as winged founder females.

The prediction that asexual reproduction is linked to highly predictable ecological conditions is also confirmed by unusual cases in which a vertebrate has evolved from a sexually to an asexually reproducing form. Certain species of whiptail lizards consist exclusively of genetic females. They share most of their reproductive physiology and behavior with females, as well as the mounting behavior of males, of closely related species that reproduce sexually (Crews, 1987). These parthenogenetic species live in the harsh but highly predictable environment of the desert grasslands of southwestern North America, where there is little competition from their sexually reproducing cousins. It might be that these two factors combined to tip the balance in favor of the simpler mode of asexual reproduction.

SEX DETERMINATION

In birds and mammals, sex determination is genetic. In mammals, females carry two X chromosomes and males carry an X and a Y chromosome. Consequently, eggs always carry an X chromosome, while sperm carry either an X or a Y chromosome, and the sex of the offspring is determined by whether the egg is fertilized by a sperm carrying an X chromosome, producing a female, or by one carrying a Y chromosome, producing a male. Genetic material carried by the Y chromosome induces a testis in the primitive, undifferentiated fetal gonad. In the human, this occurs very early in embryonic development, some 4–6 weeks after fertilization, when the woman may be scarcely aware that her cycle has been disrupted by pregnancy. The production of testosterone by the fetal testes is responsible for the cascade of differentiation that eventually produces a phenotypic male (Chapter 5). In the absence of the Y chromosome, the fetus develops an ovary and differentiates into a female; so sexual differentiation is not symmetrical and, in the absence of any hormonal influences, the basic somatotype in mammals is female; certain genetic errors in humans confirm this. In birds, the situation is reversed: The female carries two different sex chromosomes, termed Z and W, while males carry two Z chromosomes. In each taxon, the basic somatotype is the homogametic sex that develops in the absence of any hormonal influence (Jost, 1971).

Sex determination may, however, be independent of genetic programming. In some reptiles, including the gecko and alligator, the sex of offspring depends on the temperature at which eggs are incubated; higher temperatures result in males and lower temperatures in females. In snapping turtles, both high and low temperature extremes produce females and intermediate temperatures produce males. Sex reversal, either from male to female, known as protandry, or from female to male, called protogyny, occurs in some invertebrates and in several families of teleost fishes. If a male blue-headed wrasse dies or is removed experimentally from its group of females, the dominant female transforms into a male: There is a marked color change and sperm are produced instead of eggs. This type of sex change is sometimes termed sequential hermaphroditism and is irreversible.

Some invertebrates, including earthworms and snails, are true hermaphrodites, with one individual producing eggs and sperm concurrently. Copulation may involve the simultaneous exchange of sperm between two individuals. In at least one genus of freshwater snail capable of storing sperm, however, the male role is initially adopted by all individuals. It is the relative sizes of the two individuals that determine which inseminates the other; the smaller inseminates the larger because the latter does not contest the insemination attempt (DeWitt, 1996). This is thought to be adaptive because fecundity is positively correlated with size, so that the fitness of both individuals is greater if the larger of the two produces fertilized eggs.

SEX RATIO (SR)

The ratio of males to females at birth in the white U.S. population is 106:100. The deviations from 100:100 in zygotes may be due either to differential production and efficiency of X- and Y-carrying sperm or differential mortality of male and female zygotes. The SR at fertilization is called the primary SR, that at birth is the secondary SR, and that of adults is the tertiary or operational SR; the latter varies with age because of differential mortality. Family planning has no effect on SR. In the majority of species, the ratio of males to females is approximately 1:1; any bias favors the sex that primarily competes for mates, which is usually males, but it can be females in species with sex role reversal. Fisher's Theorem (Fisher, 1930) proposed that any decrease, for example, as a result of a climatic change, in the numbers of either sex in a randomly mating population will automatically select for the production of the less numerous sex.

THEORETICAL CONSIDERATIONS

Darwin was the first to point out in *The Descent of Man and Selection in Relation to Sex* (1871) that sexual selection exists because males and females exert selective pressures on each other as they acquire mates. But the question of why this should be so was not satisfactorily explored until Bateman and Trivers proposed two key ideas that together suggested a plausible explanation that has generated testable hypotheses. These ideas also gave rise to a third idea, which proposes that mammalian mothers might manipulate the sex of their offspring so as to maximize the sex most likely to have greater reproductive success.

Bateman's Principle

Working with *Drosophila* having a variety of morphological genetic markers, Bateman (1948) demonstrated quantitatively that there is greater variability in the reproductive success of males than of females. He explained his findings in terms of a sex difference in the energy invested in gametes. His reasoning can be summarized as follows: The production of eggs requires more energy than that of sperm because eggs contain an energy store, whose size varies in different species, needed to nourish the developing embryo. Given that males and females devote the same amount of energy to producing gametes, males can produce far more sperm, and potentially more offspring in a lifetime, than females can produce eggs. Moreover, in certain mammals, including the human, the stock of oocytes is limited and fixed before birth. This makes eggs a more limited resource for which males compete, winners having greater reproductive success than losers. Variance in reproductive success, assessed here by numbers of mates, is smaller in females than in males (Fig. 13-1) because there is no shortage of males to fertilize their eggs. Since females may have several males competing for them, it can be further argued that it would be advantageous for females to choose the highest quality male as a mate because an error in this respect would have a more damaging effect statistically on the female's than on the male's lifetime reproductive success. Bateman's Principle has led to the predictions of male mate competition and female mate choice, namely, that males will generally compete with each other for access to a fertile female, while females will be selective in choosing the "best" male.

Trivers's Theory of Parental Investment

Trivers (1972) extended Bateman's Principle by proposing that a sex difference in total parental investment, which includes investment in gametes, results in a higher variance in male than in female reproductive success (Fig. 13-2). He reasoned that females generally invest more time and energy than do males in raising their offspring. In female mammals, for example, there are the physiologically determined costs to the female of pregnancy and lactation, and in many birds, of incubating eggs and rearing

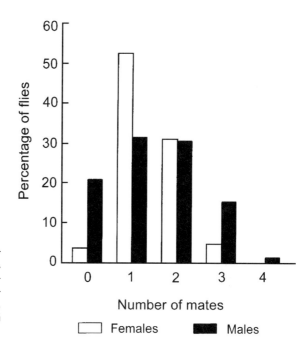

FIGURE 13-1. Bateman's Principle illustrated in fruit flies. There was more variance between males than between females in numbers of mates. (SOURCE: Bateman, 1948, with permission)

chicks. Some species, including the stickleback among fish and the Northern jacana among birds, have polyandrous mating systems consisting of one female with several males (Chapter 15). In these species, there is a sex-role reversal, with males incubating eggs and rearing offspring. In agreement with Trivers's Theory of Parental Investment, the females of polyandrous species exhibit mate competition, while males exhibit mate choice. Although environmental variables affect the adaptive value of parental care for both males and females, the advantages will usually be greater for the female than for the male for the following reasons:

1. The female has fewer lifetime reproductive opportunities than does the male, so parental care is more important for her.
2. While caring for his young, the male loses more opportunities to produce additional offspring than the female.
3. The offspring in which the female invests are certain to be her own except when there is brood parasitism. The male has no such "paternity certainty" and risks reducing his own reproductive potential by investing time and energy in offspring that may not carry his genes.

The male maximizes his reproductive success by mating with as many females as possible, whereas the female does so by choosing the "best" available male and by restricting mating to the most fertile period. This can produce a conflict of interests between the sexes, which is clearly apparent in mammals, where females provide the majority of parental investment in the form pregnancy and lactation, and male mate competition is often fierce. A conflict between the sexes is virtually undetectable in

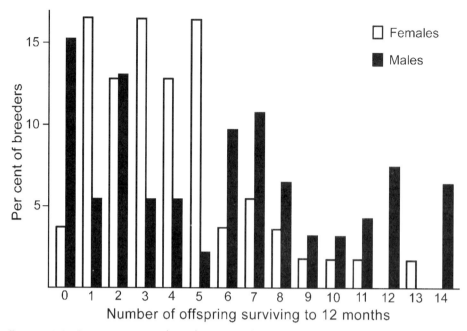

FIGURE 13-2. Bateman's Principle and Trivers's Theory of Parental Investment illustrated in lions. There was more reproductive variance, measured by the number of offspring reaching 1 year of age, between males than between females. (SOURCE: Packer et al., 1988. Copyright © 1988 by the University of Chicago Press)

species showing no parental care, for example, in marine fish such as herrings, that simply discharge their gametes into the environment, leaving both fertilization and the survival of offspring to chance.

It is worth emphasizing that while males generally compete more than females for mates, and females are more selective than males in their choice of a mate, this refers only to statistical averages. Assuming an average sex ratio of 1:1, logic alone suggests that if most female birds and mammals choose males with the "best" possible genes, they will sometimes find themselves in conflict with each other in obtaining access to the "best" male or the only one available at the time. Similarly, there will be situations where males would have no need to compete unless they were choosing certain females in preference to others. For example, when several female rhesus monkeys are observed with only one male in an enclosure, there is female mate competition when the highest-ranking female maintains almost exclusive sexual and social access to the male by intimidating the other females. But there may be male mate preference when the male interacts more with one of the subordinate females when they are alone together (Chapter 16).

Trivers–Willard Hypothesis

Based on Bateman's Principle and Trivers's Theory of Parental Investment, Trivers and Willard (1973) together proposed that mammalian mothers might manipulate the SR of offspring as a function of prevailing conditions to produce the sex that gives the greatest increase in inclusive fitness for a set amount of investment. The Trivers–Willard Hypothesis does not indicate how this manipulation occurs but, if true, it might be mediated by physiological mechanisms, for example, by stress hormones, or by behavioral mechanisms, for example, parental neglect. In general, high-ranking mothers would theoretically maximize their reproductive success by having sons, as these would be highly competitive compared with other males and could therefore produce more offspring than would daughters, with their more limited reproductive potential. Conversely, low-ranking mothers could attain greater reproductive success by having daughters, since even poor daughters are likely to reproduce, whereas poor sons are less likely to do so. The same reasoning would apply to mothers in environments with either plentiful or scarce resources. The theory has been supported by data from the nonplacental opossum (Austad & Sunquist, 1986), whose young are born at a very early stage of development. Free-ranging females receiving sardines as food supplements produced 1.4 sons for each daughter, whereas unsupplemented females produced equal numbers of sons and daughters. In contrast, old females in poor physical condition produced 1.8 daughters for every son. The hypothesis has received little support from data in a variety of other species, but in some human cultures differential parental support of sons and daughters is in agreement with the hypothesis (Chapter 17).

Based on these theoretical considerations, it is thought that in most species, intrasexual selection is primarily a male phenomenon resulting from selection pressures brought to bear on males by mate competition with other males, while intersexual or epigamic selection is primarily a female phenomenon resulting from selection pressures brought to bear on males by females choosing the "best" males. We consider this below. In practice, it is difficult to distinguish between the roles of intrasexual and intersexual selection in the development of specialized morphological or behavioral trait in males. For example, greater size, strength, and aggressiveness could evolve in males because it gives them both a competitive edge in mate competition and makes them more attractive to fertile females.

INTRASEXUAL SELECTION

Competition among Males

Before and during courtship, and even after mating has occurred, males may compete to enhance their reproductive success by a variety of behavioral and physiological mechanisms that involve competition for copulations, sperm competition, and lowering the reproductive success of competitors (Table 13-1).

Competition for Copulations

Low Mating Threshold. Females typically do not mate except during their rather restricted period of fertility, around the time of ovulation. Males are ready to mate with a fertile female throughout the year or, in seasonally breeding species, throughout the mating season. This behavioral predisposition has a physiological basis insofar as testosterone levels, which facilitate both male sexual activity and spermatogenesis, are

Table 13-1. Mechanisms for Intrasexual Selection (Male Mate Competition)

Competition for copulations via
  Low threshold for mating
  Monopolization of females by excluding other males from
    The vicinity of receptive females
    The harem of females
    High status in multimale groups
    Areas with resources attractive to females
    Display sites attractive to females
  Coping with higher-ranking males by
    Submission and temporary postponement of reproduction
    Courtship in the absence of dominant males
    Sneak copulations
    Rape

Competition for sperm usage via
  Guarding behavior by
    Prolonged guarding of a female or harem
    Temporary guarding of an inseminated female
  Reduction of repeat copulations by an unguarded female by
    Insertion of a mating plug after copulation
    Behavioral or physiological signals that reduce female attractiveness or receptivity
  Interference with sperm from a previous mating

Lowering reproductive success of competitors via
  Sexual interference by
    Interruption of another male's courtship
    Female mimicry (inducing other male to waste time, energy, or sperm)
  Attempts to injure competitors by
    Assault
    Assault on other males' mates or offspring (i.e., infanticide)

SOURCE: Modified from Alcock, 1979, with permission.

rather constant and high all the time, whereas estrogen levels, which facilitate female sexual activity, are high episodically, around the time of ovulation (Chapter 5). Male mating motivation can be so high that males may attempt to mate an inappropriate individual, or even an inanimate object, with minimal stimulation or even as a vacuum activity. In some domestic animals, such as cattle bred for certain characteristics that include robust reproductive potential, males can be induced to mount an appropriately sized support and ejaculate into an artificial vagina for sperm collection for subsequent use in artificial insemination.

Monopolizing Females. Males can compete for copulations by preventing other males from gaining access to females in different ways. The male may simply chase other males away when they approach a receptive female, or drive her away from them; this can readily be observed in ducks during their spring breeding season. Or males may fight each other for exclusive access to a harem of females as the mating season approaches, a phenomenon that will be familiar from television films of various ungulates, including bighorn sheep, deer, and moose, as well as sea lions and elephant seals (Fig. 13-3, top). This form of direct, aggressive competition obviously exerts a strong

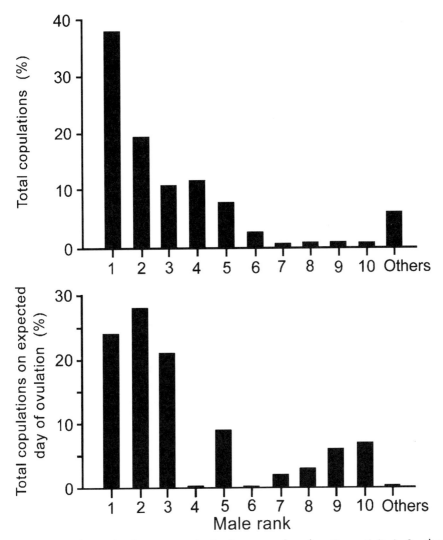

FIGURE 13-3. Relationship between male dominance rank and mating activity in Southern elephant seals (*Mirounga leonina*) (top) (SOURCE: Modified from McCann, 1981, with permission from the Cambridge University Press) and baboons (*Papio cynocephalus*) (bottom) (SOURCE: Modified from Hausfater, 1975, with permission from Karger, Basel)

selection pressure, promoting male size, strength, and endurance, as well as formidable weaponry such as antlers and tusks. A more unusual harem variant occurs in the non-seasonally breeding hamadryas baboon of Ethiopia, whose males often recruit immature, prepubescent females into their harems. Probably because these females are either immature or fertile for only a few weeks every year, and because harems are permanent and do not break up and re-form annually, there is little fighting between males for the possession of females.

Where social groups contain several males and females, for example, in some macaques and baboons (Fig. 13-3, bottom), high dominance rank confers an advantage in obtaining access to all resources, including fertile females; this may not be expressed in much overt fighting, because low-ranking males do not challenge and will avoid mating while a higher-ranking male is in the vicinity, even if the latter is constrained experimentally and physically prevented from interfering (Chapter 16).

The types of behaviors described so far involve guarding one or more females from contact with other males. A similar effect results when males compete for exclusive access to areas that are attractive to females, either because they contain high-quality resources such as food, water, shelter, and suitable nesting sites, or because they are favored display sites. In the former case, a male typically establishes a territory, which is defended against other males permanently or only during the mating season. This is a common phenomenon in many songbirds, including monogamous and polygynous species; in some cases, for example, the red-winged blackbird, the size and quality of the "real estate" held by the male appear to determine whether he has a single mate or several. In other species, including the monogamous gibbon, both the male and the female defend the territory against like-sex, but not opposite-sex, conspecifics; gibbons even eject their own offspring when they mature, although parents will help offspring establish their territories. Finally, males may attempt to exclude other males from courtship display sites that are attractive to females. An appreciable number of species, including grouse and peacocks among birds and hammer-headed bats among mammals, have lek mating systems (Chapter 15). The males congregate year after year on traditional display sites termed *leks*, where they perform their spectacular courtship displays while females look on and then approach and copulate with particular males. Within each lek, males establish small, exclusive territories for themselves, and there is evidence that females may choose their mates by the position of their territories on the lek. For example, peacocks holding a territory near the center of the lek appear to copulate with more females than do males near the periphery.

Coping with Higher-Ranking Males. If males less powerful than their competitors cannot monopolize females, they can nevertheless attain high numbers of copulations over a lifetime by one or more of four behavioral devices. They may simply submit to higher-ranking males and postpone their sexual activity until they have achieved higher rank. They may court and mate with a female while the dominant male is temporarily absent, or they may engage in so-called sneak copulations while the dominant male is out of view or not paying attention. These latter behaviors require cooperative females and, since the advent of DNA "fingerprinting," there have been many studies demonstrating multiple paternity in the same brood among birds with monogamous mating systems, showing that cooperative females are not lacking. Finally, lower-ranking males may rape unwilling females in those species in which rape is physically possible. In many species, however, the male cannot fertilize the female unless she is cooperative and adopts a receptive posture (Chapter 5).

Competition for Sperm Usage

Guarding Behavior. Prolonged monopolization of females to prevent contact by other males will effectively guard females after as well as before insemination. Many anthropoid primates, including macaques, most baboons, langurs, and vervet monkeys, have multimale–multifemale social organizations. During a female's most fertile period near midcycle, temporary pair bonds are formed with a male; these may last a few hours, a few days, or as long as 2 weeks. During these longer consortships, which extend into the postovulatory period, the pair repeatedly copulates and the male follows the female closely when she travels and feeds. Such consortships are nutritionally costly for the male because he must forage less and has shorter feeding bouts than at other times; these costs may be outweighed because the consort male is more likely to be the sire of the female's offspring.

Reduction of Repeated Copulations by Unguarded Females. Copulations by other males may be minimized by physiological adaptations. One of these is the deposition of a copulatory plug, consisting of a thick, viscous material in the vagina of the female. In guinea pigs, this plug blocks subsequent insemination, and in prosimians, such as the potto, the large plug hardens rapidly into the consistency of wax: the equivalent of a medieval chastity belt. Repeat copulations by an inseminated female can also be prevented by behavioral or physiological signals that render her either unattractive or unreceptive to subsequent males. In some insects, including *Drosophila* and a species of neotropical butterfly (*Heliconius erato*), the first male to inseminate transfers antiaphrodisiac substances to the female during mating. About 50% of the sperm in butterflies and moths consists of "dud" sperm, that is, smaller sperm that carry no nuclear material ("apyrene" sperm, as opposed to normal "eupyrene" sperm). It has been suggested that, among other possibilities, these serve as packing material to fill out the female's spermatheca, a storage organ that can hold sperm from several males, to the point that she becomes unreceptive to further mating.

Interference with Sperm from a Previous Mating. These mechanisms give an advantage to the first male to mate with a female, and it will be no surprise that mechanisms giving an advantage to the last male have evolved in some species. In many insects, the last male typically has the advantage in terms of paternity, perhaps in part because his sperm is released first from the spermatheca of the female as her eggs pass down the oviduct during egg laying. But there can also be active chemical and physical interference with the sperm deposited by the predecessor. Male damselflies have special backward-directed hairs on their penis by means of which they scoop the sperm of other males from the female's spermatheca before depositing their own. Male dunnocks (*Prunella modularis*), a common hedge sparrow in Europe, repeatedly peck at the cloaca of a female before mating with her until she everts it (Fig. 13-4), which sometimes results in the ejection of sperm from a previous mating.

Lowering the Reproductive Success of Competitors

Sexual Interference. Harassment by a high-ranking male of a lower-ranking male's courtship and mating activity occurs in many species, including macaques, most baboons, and several other anthropoid primates. Interference may be active, as when a dominant male threatens or attacks a subordinate male; in male rhesus monkeys, testicular regression and low plasma testosterone levels may follow a defeat. Inter-

FIGURE 13-4. Sperm competition in the dunnock. The male pecks at the female's cloaca (a), which may result in the ejection of sperm from a previous mating with another male, before mating with her (b). (SOURCE: Modified from Davies, 1983, Polyandry, cloaca-pecking and sperm competition in dunnocks. *Nature* **302**, 334–336. Reprinted by permission. Copyright © 1983 by Macmillan Magazines Ltd.)

ference may simply depend on the presence of the dominant male; subordinate male talapoin monkeys show behavioral suppression together with testicular regression and low testosterone levels in the absence of overt aggression. In some salamanders, males interfere with courtship and mating of other males in several ways. One of these consists of mimicking the behavior of females so that the competitor not only wastes the spermatophore he is induced to deposit, a form of sperm competition, but also the time and energy he has invested in courtship.

Injury of Competitors. Males may increase their own reproductive success by harming their competitors or the mates or offspring of their competitors. Fights between males for access to females and territories can result in injury and death. In chimpanzees, communities of sometimes closely related males will attack and even kill females that enter their home range, unless the females have the marked swelling of the anogenital region that characterizes sexual receptivity in this species, in which case males take turns to mate. Infanticide by males that have successfully invaded a social group and defeated the resident male has now been documented in many species, from tree swallows to lions and monkeys (Chapter 10). Also recall from Chapter 11 that exposure of a newly mated female mouse to pheromones in the urine of a strange male before implantation blocks pregnancy; this Bruce Effect is an intrauterine case of males preventing the survival of the progeny of another male.

Competition among Females

Several of the phenomena in male mate competition occur in mate competition among females. Dominant female rhesus monkeys excluded lower-ranking females from mating in a laboratory setting where there was no opportunity for female mate choice, since groups comprised one male and four females. This is a clear case of female mate competition similar to that in males (see Chapter 16). In addition, female baboons harass lower-ranking females during pregnancy and the early postpartum period, and the ovaries of adolescent female marmosets remain suppressed as long as the females remain in their natal group. An alternative interpretation of these two phenomena, in females as well as in males, might be resource competition and inbreeding suppression, respectively, namely, aspects of natural rather than sexual selection. This illustrates the difficulties in distinguishing between sexual and nonsexual selection pressures, and there are similar difficulties in distinguishing between intrasexual and intersexual selection.

INTERSEXUAL (EPIGAMIC) SELECTION

In intersexual selection, it is generally female choice between mates that determines the evolution of "preferred" traits in males. This raises some questions. What male qualities are important in female mate choice? Which trait does the female use as the marker for the preferred quality? How do females evaluate that marker in a male? How does the marker, and the female's preference for it, evolve?

Desirable Male Qualities

The male qualities generally thought to shape female mate choice are (1) the ability to provide sufficient sperm, for example, resulting in the avoidance of a male whose sperm is depleted from a previous mating; (2) the ability to provide essential resources such as food, water, shelter, and protection from predators; (3) superior parental care for the defense of, and provision for, offspring; and (4) general good health.

Markers of Male Qualities

Which trait a female uses as a marker for the quality of choice is difficult to establish with any certainty. In the red-winged blackbird, the intensity of the male's courtship predicts his subsequent effort in feeding the young, and the size of the red wing patch is positively correlated with the amount of effort devoted to defense of the nest. Females may therefore use courtship intensity and size of the wing patch as markers for parental care. In mammals, odor cues may be important markers for the physical quality of potential mates. Both male and female meadow voles (*Microtus pennsylvanicus*) spent more time investigating odors from the anogenital area, urine, and feces of opposite-sex individuals fed high-protein diets than from those fed low-protein diets (Ferkin *et al.*, 1997).

Selecting males with "good" genes comprises several phenomena: The mate should be a male, a conspecific, and in some cases, a conspecific from the same ecological niche. Because of the hazards of both inbreeding (leading to increased homozygosity for deleterious recessive alleles) and excessive outbreeding (resulting in the breakup of a successful parental genotype) the mate should be neither too closely nor too distantly related, resulting in so-called assortative mating. The traits for assessing such genetic characteristics are typically encoded in the precise form of courtship displays and the physical appearance of the male. In many species, males and females disperse before sexual maturation, but in those that do not, including the human (Chapter 17), long-term familiarity can be a marker for close kinship. It produces a general lack of sexual interest between brother and sister that inhibits mate choice and results in inbreeding avoidance. In humans, prohibition of sibling and parent–offspring marriages is widespread and is extended to first-cousin marriages in some cultures, while ethnic or religious mixed marriages between individuals with clearly apparent physical or behavioral differences are often strongly resisted by family members.

On a different level, females may choose males by traits reflecting good general health, the absence of injury and disease, and the ability to hold large territories or achieve high rank. Less fluctuating asymmetry between the left and right sides in color patterning, feather size, tail shape or, indeed, facial features, reflects better protection against random accidents during development and hence is thought to characterize a better genome. There is evidence indicating that females tend to choose mates with less asymmetry over those with more. Female zebra finches that chose males wearing symmetrical arrangements of colored leg bands invested more parental effort and produced more offspring surviving past the period of parental care than did females choosing males with asymmetrical bands. This effect may be mediated by higher androgen levels in eggs laid after mating with a more attractive male. Likewise, humans tend to find symmetrical faces more attractive than asymmetrical ones.

Evaluation of Male Qualities

Given that females select a marker for a given trait, how do they evaluate males with respect to that trait? Do they have an internal template, genetically fixed or acquired during development, for example, by imprinting, against which they match the prospective mate, or do they compare the available males and choose on the basis of differences between them? This is a problem that is even more difficult to address experimentally than choice itself, and it is likely that different mechanisms have evolved in different species and circumstances.

Evolution of Male Traits and Female Preferences for Them

Three principal models for the evolution of intersexually selected traits have been proposed. Fisher's runaway selection model (Fisher, 1930) states that if a male trait associated with a reproductive advantage is preferred by a female, she will have greater reproductive success; her sons are more likely to carry the trait and her daughters the

SEXUAL SELECTION

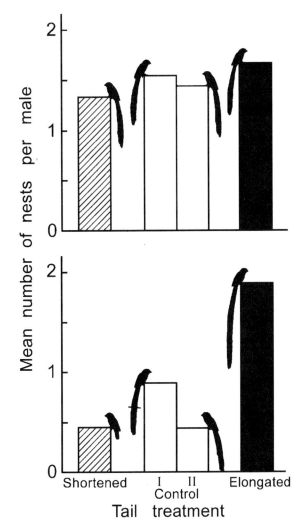

FIGURE 13-5. Intersexual selection for tail length in male long-tailed widow birds (*Euplectes progne*). Before tail manipulation (top), there was no difference in male reproductive success as measured by numbers of active nests in the territory of each male. After tail manipulation (bottom), reproductive success was significantly greater in males with elongated tails (solid bars) than in all other males. (SOURCE: Modified from Andersson, 1982, Female choice selects for extreme tail length in a widowbird. *Nature* **299**, 818–820. Reprinted by permission. Copyright © 1982 by Macmillan Magazines Ltd.)

preference for it. Thus, both the trait and the preference continue to be selected in a runaway fashion, resulting in highly exaggerated traits, such as the tail of the long-tailed widow bird (Fig. 13-5) and the peacock, to the point at which they become an impediment by physically handicapping the male or attracting predators.

Zahavi's handicap model (Zahavi, 1975) suggests, instead, that only males with superior general health, agility, and strength will be able to express many secondary sexual characteristics to the utmost because traits such as long tails, heavy antlers, bright coloration, and the ability to establish high rank and maintain large territories all incur energetic costs and expose their bearers to increased predation and other risks. There is evidence for this. The relationship between tail size, male mating success, and survival was examined in free-ranging blue peacocks (*Pavo cristatus*) at Whipsnade Zoo in

England (Petrie, 1992). In the winter of 1990, two foxes got inside the perimeter fence and killed 5 of 33 peacocks, which displayed at four different lek sites and showed very large variation in reproductive activity, from 0 to 60 copulations in the previous mating season. Whereas 4 of the 11 males that had not mated were killed (36.4%), only one of the 22 males that had mated was killed (4.5%), yet the surviving males had significantly longer tails with more eyespots on them than did the males killed by predators.

The third evolutionary model takes into account that the health, and therefore quality as a mate, of an animal can vary over time and need not be genetically fixed. Hamilton and Zuk (1982) proposed that sexually selected traits have evolved as a barometer of the animal's current state of health. They pointed out that, in birds, those species whose males have the brightest, showiest plumage are also those that are particularly likely to suffer from infestation with blood parasites, and similar observations have been made recently in other vertebrate taxa. Males in poor health are less likely than healthy ones to have long horns, bright colors, or ornate feathers, so females choosing males with well-developed sexually dimorphic traits are also choosing healthier males. The results of the study on blue peacocks do not distinguish between Zahavi's handicap model and the Hamilton and Zuk model, that is, whether females avoided mating with the males killed by predators because of smaller tail size as an indicator of poorer genetic constitution or as an indicator of disease. Whatever the proximate cause, it seems likely that poorer displays may signal a poorer-quality individual, and that females discriminate between prospective mates accordingly.

Mate Choice by Males

Little attention has been given to intersexual selection in males, although it is readily observed in macaques and other species both in captivity and in the wild. Mate choice by males is also likely to be a factor in polyandry, in which sex roles are reversed (Chapter 15), in monogamy, and in assortative mating.

CHAPTER **14**

Courtship and Mating

Reproduction encompasses behaviors that range from the identification and courtship of a suitable mate to the successful rearing of offspring. Behavioral mechanisms determine the social, mating, and parental care systems of a particular species, and depend on complex interactions between anatomy and physiology on the one hand, and environmental factors, notably, habitat and predation, on the other. For example, high predator pressure may promote greater defense of young and, in turn, the extent to which the male can effectively contribute to this defense helps to determine the mating system (Chapter 15).

These social and environmental selection pressures have importance for the evolution of species-typical social, parental, and mating behavior, but evidence from many different species suggests that these pressures may also have a direct effect on individuals by modifying their behavior according to current circumstances. We are all familiar with "wild," elusive animals such as deer that cease to show flight responses to dogs or people when maintained in captivity, where they are protected from predators. Constraints imposed by other features of the habitat, including the terrain, amount of cover, and the type and distribution of food and shelter may result in different mating systems for different individuals of the same species, and the social organization can vary in different populations. We describe some primate examples of this variability in Chapter 16, but emphasize here the dynamic interactions between the variables that shape the patterns of reproductive behavior.

FACTORS IMPORTANT FOR THE ONSET OF COURTSHIP AND MATING

Seasonal Factors

Most terrestrial and freshwater species, and also some marine species, live in environments where there are seasonal fluctuations in day length, temperature, rainfall, and food. These exteroceptive factors have an impact on the animal's environment and on the resources it provides. Exteroceptive stimuli also directly affect the physiology of the individual via well-known pathways involving distance receptors such as the retina and olfactory epithelium, the brain, particularly the hypothalamus, and the pineal and pituitary glands. Consequently, many species breed seasonally and their mating is timed for births in the season optimal for raising young. At a given latitude, the change in the photoperiod (Chapter 7) is unvarying, and is a potent factor in triggering the onset of gonadal activity and courtship behavior in seasonal breeders (Chapter 5). Spring mating is typical of many species including carnivores, rodents, and nonruminating ungulates such as the pig, horse, zebra, and rhinoceros. On the other hand, mating in fall is typical in ruminating ungulates such as the cow, sheep, goat, camel, bison, and yak. Several primate species that are seasonal breeders do so in the fall. In each case, young are born during the most prolific season, generally early summer. Some species, for example, female bats, store sperm after mating, and other species use different reproductive devices that obscure the relationship between mating and birth: The wolverine, for example, mates in winter but delays implantation until spring. Many tropical and subtropical animals do not have a breeding season. The cynomolgus monkey (*Macaca fascicularis*, which inhabits areas of southeast Asia near the equator, breeds throughout the year, but the closely related rhesus monkey (*M. mulatta*), which mainly inhabits northern India, has a well-marked mating season in the fall. In highly inbred laboratory rats and mice, the ancestral mating seasonality is lost because of selection against seasonality and shielding from photoperiod and temperature changes. Domestication in dogs, for example, results in some loss of seasonality but other canids (e.g., fox, wolf, and jackal) all mate in the spring.

Hormonal Stimulation

Other extremely important proximal factors for courtship and mating are, of course, the well-known hormonal changes occurring in both males and females. As described in more detail in Chapter 5, in mammals, the great increase in the secretion of gonadal steroids is responsible for the great increase in sexual activity around puberty in both sexes. Thereafter, under the influence of exteroceptive factors at the onset of each mating season, the testes and ovaries emerge from a period of quiescence, gametes are produced, and the gonadal steroids stimulate courtship and mating. In males, sexual activity is sustained in a rather constant fashion by the steady secretion of androgens by the testes until senescence (this ignores short-term diurnal variations). In female mammals, on the other hand, the production of estrogens and progesterone by the ovaries is episodic and associated with cycles of sexual activity. These cycles also continue throughout life except in those few species that live long enough to have a menopause. Now that anthropoid monkeys and apes survive to advanced ages in captivity, a menopause is becoming recognized in these species also.

Social Stimulation

In addition to seasonal environmental changes and hormonal factors, social stimulation by conspecifics plays an important role in determining the onset of courtship and mating by synchronizing the reproductive physiology and behavior of the male and female. The famous eighteenth-century Scottish surgeon, John Hunter (1728–1793; Hunter's canal, Hunter's chancre), had a parrot of unknown sex. Feeling it was bored, he provided the bird with a mirror. A few days later, the parrot laid an egg! Apocryphal, perhaps, but it illustrates forcefully the role of social stimulation that is also evident in the centerfold cult among some men. Where courtship is prolonged, there is opportunity for positive feedback effects between the two sexes, whereby increased excitement in one partner increases excitement in the other. The crucial role of social stimulation is apparent in lekking species (Chapter 15), for example, the black grouse. Male grouse congregate in large numbers on traditional lek sites, where they engage in spectacular displays and emit loud booming vocalizations. Although just three or four males are sufficient to attract some females to the lek, this number is apparently too small for reproduction; populations below a certain critical number of individuals become extinct. Social stimulation is also a factor in rhesus monkeys. When estrogen-treated receptive females were released into a free-ranging social group during the summer nonmating season, when males and females were sexually quiescent, the mating activity of resident males was stimulated immediately, and the resident, untreated females were also affected because their infants were born some 6 weeks earlier than usual (Vandenbergh & Drickamer, 1974). The precise mechanisms remain speculative, but those involved in the courtship of other species, for example, ring doves (*Streptopelia risoria*), are well understood (see below).

Social stimulation can also be important in parthenogenetic species that have evolved from sexually reproducing species, for example, the whiptail lizard (*Cnemidophorus uniparens*). If females are mounted prior to ovulation in male-typical fashion by another female, they produce more eggs than when not mounted, and the hormonal changes of a given female determine when she is mounted and when she mounts other females (Fig. 14-1).

FUNCTIONS OF COURTSHIP

Various courtship displays have been described in previous chapters, and their functions can be grouped into six categories.

Species (and Strain) Identification

Species-specific courtship displays are important in preventing abortive mating attempts with an inappropriate individual from a different species. This is especially the case where there are few morphological differences between individuals of different species, or if individuals have no opportunity to "learn" how conspecifics should look, sound, and smell, for example, by an imprinting process similar to that described for ducks. Consequently, species identification is most likely to be encoded in the courtship displays of animals with little or no parental care, and in those that are widely dispersed and need to attract opposite-sex individuals from afar. Among invertebrates, for exam-

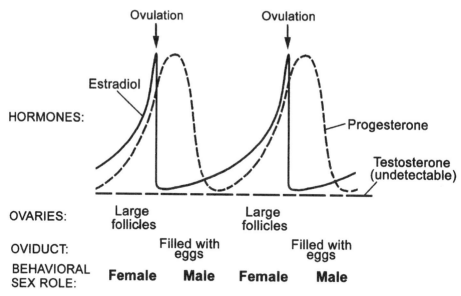

FIGURE 14-1. Relation between circulating hormone levels, ovarian state, condition of the oviduct, and male and female mating behavior in parthenogenetic whiptail lizards. Females show female-typical sexual behavior when preovulatory and with large yolking follicles, and show male-typical mounting shortly after ovulation or when gravid with eggs in the oviduct. (SOURCE: After Crews, 1987)

ple, the temporal patterns of the chirps of male crickets (Chapter 4), the bioluminescent light flashes of male fireflies, and the claw waving of male fiddler crabs, all differ between closely related species and are responded to appropriately only by conspecific females. Among vertebrates, we have already discussed species-specific variations in homologous courtship displays of ducks (inciting by female sheld ducks and mallards, Fig. 2-11) and the role of imprinting on the song by which male zebra finches attract females (Chapter 3).

Gender Identification

Gender-specific courtship displays are necessary for reasons similar to those for species identification. In some salamanders, for example, if a male is mounted by another male, the former adopts a head-down display that is not shown by mounted females, and this elicits a dismount from the other male. On occasion, however, a mounted male fails to signal his gender in this manner, inducing the mounting male to deposit a wasted spermatophore.

Aggression Reduction between the Male and Female

Although it might seem strange, the initial interactions between conspecifics of opposite sex are generally agonistic. In the male, the agonistic displays may lead on by

COURTSHIP AND MATING

redirection to fighting with a competing male rather than with the female, but the agonistic behavior of the male must be converted into the first phase of courtship; unless this happens successfully, mating cannot ensue. Early in courtship, females therefore tend to be subject to attacks by males. Furthermore, the close proximity required by internal fertilization especially, but also by external fertilization in which eggs and sperm are deposited in a small area of substrate, violates individual distances and generates agonistic tendencies in both sexes. Courtship displays initially contain strong agonistic elements that are clearly derived from attack and flight behaviors such as male threat, female submission, male appeasement, and female approach, but as agonistic tension wanes, they are replaced by components of sexual behavior. Female mammals are often mated by dominant males and even solicit such mating, but without the aggression-reducing function of courtship, violent resistance may develop. The absence of courtship may also be a factor in the intense violation experienced by women raped by strangers, and in the phenomenon of date-rape, when the agendas of the male and female differ sharply.

Individual Recognition

Even the most stereotyped courtship displays contain small variations that are unique to the displaying individual, and courtship gives potential partners an opportunity specifically to identify each other as individuals and not just as opposite-sex conspecifics. This is important where pair bonds ensure that mates remain together to rear young after hatching or birth. Geese form lifelong pair bonds and can live some 70 years; even after many decades, the partners typically perform the "triumph ceremony" on being reunited after an absence (Fig. 14-2). Individual recognition can also be adaptive in nonbonding species, for example, when it allows individuals to identify potential mates with which they have not already mated. Typically, this is adaptive for males, but it could also be adaptive for females.

Behavioral and Physiological Synchronization between the Male and Female

Partners are rarely at precisely the same motivational state at a given moment, and intensive courtship behavior by the more motivated partner can rapidly bring the less motivated partner to the same motivational level. This important function is readily identified, because courtship often consists of a chain of activity whereby the behavior of one partner is the releaser for the other partner's response which, in turn, releases another behavior pattern by the first partner, and so on to consummation. If the chain of interlocking FAPs remains unbroken, the behavior of the two sexes is synchronized and mating ensues; because there are also physiological changes in the partners, the mating results in fertilization. A good physiological example would be reflex ovulation in those species in which ovulation is induced by coitus, such as rabbit, cat, lion, and ferret. The relationship between courtship displays and hormonal changes has been studied in detail in, for example, canaries and ring doves. In birds, many authorities have contributed to knowledge, but the work of Lehrman (1959) on ring doves was a seminal contribution to our understanding of the interactions between external factors, such as the presence of a

FIGURE 14-2. The triumph ceremony of graylag geese (*Anser anser*) initiates the pair bond and subsequently occurs as a greeting. The male threatens a third individual E (1, 2), chases it off (3), and rushes "triumphantly" back toward the female that comes to meet him (4). Both partners make rolling and cackling vocalizations and show neck postures, typical of threat, that are directed past each other (5, 6). (SOURCE: Fischer, 1965, with permission)

male, nesting material, and eggs, and internal factors involving the pituitary and the endocrine glands. Increasing day length activates pituitary gonadotropin secretion, and thence the growth of ovarian follicles, the oviduct, and shell glands. High estrogen levels in the female result in nest building and, in some species, almost synchronous sexual behavior (copulation). The courtship activity of the male, together with the presence of nest-building material, further enhances progesterone secretion, which encourages egg laying and incubation. In turn, the stimulus provided by the presence of eggs causes prolactin secretion that, together with the estrogens, results in edema and defeathering of a part of the chest called the incubation or brood patch that warms the eggs during incubation. The areas of the brood patch correspond to the ventral apteria, areas of skin with only down feathers (or no feathers at all) that lie between tracts of contour feathers. Prolactin inhibits further gonadotropin secretion and stimulates the growth of the crop in the ring dove, and the production of crop "milk" for feeding the young in birds with previous breeding experience.

Signaling Competitive and Parental Abilities

As described earlier, courtship also signals to the prospective mate that the displayer possesses good qualities of general health, strength, competitive ability, or the capacity to provide the resources and defense required to raise offspring. With regard to parental care, it is interesting to note that fledgling-like begging by the adult female and feeding by the male are conspicuous components of the later stages of courtship in many monogamous birds in which both the male and the female must feed the young to ensure their survival. The ability to provide adequate resources for rearing young certainly seems important for female mate choice in the human also (Chapter 17).

MATING CATEGORIES

External and Internal Fertilization

The primary function of mating is, of course, to bring the ovum and sperm together so that fertilization occurs. Infertile matings, namely with nonovulatory females, also play a role in establishing and maintaining social bonds in some species, certainly in bonobos (*Pan paniscus*) and humans. Mating patterns are determined in part by whether fertilization is external or internal. In marine and freshwater habitats, fertilization is often external, although internally fertilizing species are common. With external fertilization, mating may simply involve the synchronized shedding of massive numbers of eggs and sperm into the water, where a very few sperm will contact and fuse with a few eggs, and there is no parental care whatsoever; examples include herrings and wrasses. Other species have complex courtships, after which the female deposits a discrete number of eggs on a small patch of substrate; thereafter, the male deposits sperm immediately above the eggs. This type of mating is generally associated with extensive parental care by either or both parents, as in many cichlid fishes.

In terrestrial species, unless the adults temporarily return to water to breed, fertilization *must* be internal because the gametes could not survive desiccation; all birds and mammals have internal fertilization. Sperm are transferred from male to female either by close apposition of their genital duct openings, as in worms, or of their cloacas, the chamber in birds and reptiles into which digestive, urinary, and genital ducts all open. In mammals, sperm is transferred by the insertion of an erect penis into a vagina. For the act of insertion, the term *copulation* is used, a word derived from the Latin *copulatus*, past participle of *copulare*, meaning "to unite or couple." Both males and females have a phallus, the penis and the clitoris, respectively; in females, the latter is an homologous structure that can be erectile but does not function as a conduit. Only the virgin human female has a hymen. In some Middle Eastern and North African cultures, the hymen, clitoris, and labia of middle- and upper-class prepubertal girls are surgically ablated and the vagina is sewn up; this custom seems to be sustained at the instigation, strangely enough, of mothers who were similarly mutilated. The custom persists because it is believed to ensure the daughter's marriageability.

Copulatory Patterns in Mammals

The precise form of the mating pattern in mammals has been categorized on the basis of four variables of male copulatory behavior (Dewsbury, 1972): (1) whether there is a copulatory lock, in which the penis becomes engorged within the vagina, resulting in a mechanical tie that prevents the male and female from separating for minutes or hours; (2) whether there is single or multiple thrusting of the penis within the vagina during an

Table 14-1. Different Copulatory Patterns in Some Mammals

Common name	Lock	Multiple thrusting*	Multiple intromissions	Multiple ejaculations
Dog	+	+	−	+
Wolf	+	+	−	+
Cat	−	−	−	+
Golden mouse	+	−	−	+
House mouse	−	+	+	+
Montane vole	−	+	+	+
Meadow vole	−	+	−	+
Norway rat	−	−	+	+
Mongolian gerbil	−	−	+	+
Bison	−	−	−	+
Black-tailed deer	−	−	−	−
Elephant	−	+	−	+
Bonnet monkey	−	+	−	+
Rhesus monkey	−	+	+	+

*After intromission.
SOURCE: After Dewsbury (1972).

intromission; (3) whether there is single or multiple intromission before ejaculation occurs; and (4) whether there is single or multiple ejaculation during a single mating episode. As shown in Table 14-1, closely related species have different copulatory patterns and the functions of these four variables for the most part remain unknown, but in reflex ovulators, a single intromission may not be sufficient to induce ovulation. Reflex ovulation occurs in rabbit, cat, ferret, mink and raccoon, while spontaneous or rhythmic ovulation occurs in cow, sheep, mare, bitch, rat, and the human.

BISEXUAL BEHAVIOR

Nonhuman Animals

We have mentioned the sexual transformations occurring in certain fishes and that sex determination is temperature-regulated in several reptiles. Almost complete morphological transformations have been produced in amphibia and birds either by adding heterologous hormones to the water in which they are maintained or by direct injections into incubating eggs. In mammals, experimental sex reversal of the gonads themselves is more difficult but can be produced in marsupials that are in a very undeveloped stage at birth and can be given hormone injections while in the pouch. It is clear from this, and from a mass of other data, that a bisexual potential exists in the mammal that can involve the gonads themselves; however, the bisexual potential of the mammalian gonad is lost very early in embryogenesis. The best naturally occurring transformations occur in the well-known freemartin condition in cattle (Chapter 5, Organizational Effects during Development). Although the gonads can be extensively transformed in this "experiment of nature," it has been difficult to achieve this experimentally in placental mammals.

Pregnant female rats, guinea pigs, rabbits, and monkeys have been treated with androgens during gestation, and sexual differentiation in the female embryo has been altered; there is little effect in male embryos. Although treatment does not transform the gonads themselves, marked changes are produced with relative ease in all the accessory structures and in the external genitalia, and genetically female fetuses are masculinized.

In a wide range of mammalian species, bisexual patterns of behavior are a naturally occurring, spontaneous phenomenon. Bisexual behavior is the manifestation or expression of sexual behavior that is usually characteristic of the opposite sex, in addition to the behavior of the same sex. Normal estrous female rats and guinea pigs will quite frequently show mounting activity, a characteristic of the male, directed toward other females and males when in maximum heat. If two female cats confined in a pen are at the height of estrus, one may mount the other in a manner characteristic of the male and make a series of male-like pelvic thrusts, yet both females will be fully receptive if an active tom is introduced. There are similar examples of bisexual behavior by farm animals, particularly pigs and cattle. In primates, bisexual behavior has been described in tree shrews, squirrel monkeys, vervet monkeys, and talapoin monkeys; in rhesus, cynomolgus, pigtail, and Japanese macaques; in several species of baboon; and, less frequently, in the great apes. In macaques and baboons living under natural conditions, males and females also show bisexual behavior patterns in nonsexual contexts, namely, in greeting and to signal dominance by male-typical mounting and submission by female-typical presentation. In some captive pairs of male and female rhesus monkeys, females mount males and do so more frequently when most receptive, near the expected time of ovulation. The mounting pattern can be bizarre and is accompanied by pelvic thrusting (Fig. 14-3). Some males do not tolerate this behavior and push the female away. The following two generalizations are probably true for mammals: (1) Mounting by females is greatest when sexual motivation is highest and males are not readily available; and (2) mounting patterns by females occur more frequently than do lordosis patterns by males.

Humans

Attention is given to this topic because of both its intrinsic biological importance and its current social impact. Our sexual proclivities can cause trouble, and strong ethical and religious views are held by different societies at different times as to what is and is not acceptable. Bisexuality or homosexuality among humans has been reported since antiquity in many cultures. Ford and Beach (1951), in their well-known monograph, reported its overt occurrence, neither rarely nor secretly, in 48 of 76 human societies; but this takes us into the domain of cultural anthropology. Homosexuality is a vast subject about which much has been written and it would serve no useful purpose to treat it superficially here. The original Kinsey data (Kinsey *et al.*, 1948), which showed that up to 30% of adult American and British males had some homosexual experience, now seems overdone. Overt homosexuality has been confirmed at 4–5% in an adult, college-educated, white, male population (Gebhard, 1972). As definitions have become more precise, the reported incidence has dropped. A family pedigree of 50 predominantly heterosexual and 50 predominantly homosexual men showed that heterosexual men have the same number of homosexual brothers as predicted from the natural

FIGURE 14-3. Different mounting patterns of female rhesus monkeys on males and other females. (SOURCE: Michael et al., 1974)

prevalence data, while homosexual men have four times as many homosexual or bisexual brothers as predicted (Pillard & Weinrich, 1986). There were no significant differences of any kind for females. Studies with identical and fraternal twins, as well as DNA-linkage studies, provide further support for the view that genetic factors may contribute to sexual orientation in a subset of homosexual men (Chapter 4). These data do not exclude a role for potent environmental influences operating within a family but suggest that male and female homosexuality are different traits.

Homosexual bonding and behavior occurs in may nonhuman primates, but as far as we are aware, the total avoidance of sexual activity with the opposite sex, as can occur in the human, is not observed in any other species. This avoidance occurs in some forms of human homosexuality as well as in transsexualism, a condition unique to human beings and one that has now been successfully delineated both from homosexuality and from transvestitism (cross-dressing). The transsexual phenomenon involves reversals in both gender role and gender identity, and persons with this condition not only wish to dress and pass as the opposite sex but are so intolerant of their own anatomic sex that they relentlessly pursue the objective of physically altering their bodies. This is achieved both by the use of hormones and by reconstructive surgery that, in some cases, is sufficiently successful to permit a happy resolution and, when society permits, a relatively normal life. Success rates are better for genetic males than for genetic females.

Unlikely as it may seem, the study of sex in the recent past has been difficult for scientists in academic institutions because of stigmatization, which made it easier to discredit or ignore results. The Committee for Research in Problems of Sex was established in 1921 under the aegis of the Division of Medical Sciences, National Research Council, to facilitate sex research and make it more acceptable. Despite scientific advances and changes in public opinion, the work of Masters and Johnson (1966), who pioneered studies on the physiology of the human sexual response, initially encountered great opposition that to some extent continues. Two factors have helped to legitimize the field: (1) the genetic and twin studies already mentioned, and (2) the discovery of structural differences between the brains of males and females. Sexually dimorphic nuclei in the medial preoptic area of the rat are three to eight times larger in males than in females; other species, including gerbils, guinea pigs, and ferrets, but by no means all, show a similar difference. In the human, the splenium of the corpus callosum and the anterior commissure, which are involved in interhemispheric transfers, are larger in women than in men when measured in the midsagittal plane (Chapter 5).

Other traits are also sexually dimorphic in the human. It is well documented that the verbal ability of girls is superior to that of boys, and this difference is established by 5 years of age. Lateralization of function between the hemispheres, and particularly of the language regions, occurs earlier in girls than in boys. However, cross-cultural studies from widely separated parts of the world show that spatial abilities are better in boys than in girls. The dominant hemisphere controls both handedness and language. In right-handed people, the dominant hemisphere is the left one, and this is the site controlling speech, whereas spatial ability is located mainly in the subdominant right hemisphere. Gender differences in the lateralization of function are thought to underlie differences in the severity of signs and symptoms following strokes, as well as being a factor in the speed and extent of recovery. Men have larger brains than women, but the difference diminishes when corrected for body size. Even so, males have about 8% greater brain weight when corrected for differences in height from the second year onwards. In the human preoptic area, there is also a sexually dimorphic nucleus that is twice the size and contains twice the number of neurons in men than in women (Allen & Gorski, 1990). Although approximately of equal size in boys and girls up to about age 4 years, thereafter the sex difference develops due to a decrease in cell numbers in females (Hofman & Swaab, 1991). Some of these size differences in the anterior hypothalamus have been related to sexual orientation in comparisons between the brains of heterosexual and homosexual men (LeVay, 1991), but postmortem data are controversial and the findings are difficult to interpret.

CHAPTER **15**

Parental Behavior and Mating Systems

Trivers (1972) defined parental investment as "any investment by the parent in an individual offspring that increases that offspring's chances of surviving (and hence reproductive success) at the cost of the parent's ability to invest in other offspring." This includes investment in the survival of gametes and embryos as well as in the parental care of offspring after hatching or birth, until they become independent. In many species, parental care is virtually absent: In some fish, huge numbers of eggs and sperm are released into the water during spawning; fertilization and the survival of eggs and hatchlings are left to chance. In other species, the female, the male, or both provide parental care by (1) ensuring a food supply; (2) regulating the immediate environment, for example, fish fanning eggs to provide oxygenated water and birds incubating eggs; and (3) protecting the offspring from predators; and, in animals with complex brains, by (4) serving as models for acquiring social and hunting skills. Provision of food can involve quite bizarre mechanisms, including the cannibalization of eggs and younger embryos within the mother's oviduct by other sand shark embryos, so that usually only one embryo remains. Female digger wasps lay an egg into a paralyzed, but not dead, honeybee upon which the larva feeds. In an African species of social spider, the mother makes the supreme sacrifice by being totally devoured by her young. Protection from predators can also take an unusual form. In an African mouth-breeding cichlid fish, *Haplochromis*, the female picks up her eggs with her mouth immediately after laying them and simultaneously picks up male sperm during attempts to pick up the "dummy" eggs on the anal fin of the male. Fertilization, incubation, and hatching all occur within her mouth, to which the fry continue to return at night and when threatened.

MODELS OF THE PARENT–OFFSPRING RELATIONSHIP

From their own childhood experiences, and from watching their pets, people tend to think of parental behavior as a one-way process from parents to offspring. The first of three models for the parent–offspring relationship is based on this view.

The Parental Provision Model

In rats, the mother builds a nest and broods the young to keep them warm. She suckles them, providing both nutrition and antibodies to fight infection; as in human babies, rat pups cannot produce their own antibodies during the first weeks after birth. Since pups cannot defecate and urinate spontaneously, the mother stimulates these processes by licking their anogenital regions. When necessary, the mother also uses her mouth to transport each pup to a new nesting location. In many social carnivores, including wolves, hyenas, and lions, and in anthropoid primates such as macaques and baboons, mothers enforce social roles and skills, and they protect their young from aggression by other group members either directly by intervention, or indirectly by maternal rank. They also protect them from predators and serve as models for the acquisition of hunting and foraging skills. In many birds, for example, zebra finches, both parents help feed the young and, during a critical period, male fledglings learn the father's song and become sexually imprinted on their own species (Chapter 3). In all of these cases, it seems that there is a one-way flow of resources from parent to offspring until the latter become fully independent.

The Mutual Benefit (Symbiosis) Model

The second model differs from the first by proposing that not only the offspring but also the parent benefits immediately from the relationship. For example, when a mother rat licks the anogenital regions of her pups to stimulate excretion, she benefits by drinking the urine that provides her with essential salts and with about two-thirds of the water she needs for lactation. When either mothers (Babický et al., 1970) or pups (Gubernick & Alberts, 1983) were injected with radioactively labeled water or saline, and radioactivity was measured 24 hours later in either the pups or mothers, it was shown that mutual water transfer rates increased until day 15 postpartum and decreased rapidly after weaning around 3 weeks postpartum (Fig. 15-1).

The Conflict Model

The third model is based on Trivers's Theory of Parental Investment (Chapter 13) and predicts a fundamental conflict between parent and offspring, at least by the time of weaning (Fig. 15-2). The offspring carry only 50% of the parent's genes. Around the time when offspring can fend for themselves, they have nothing to lose and everything to

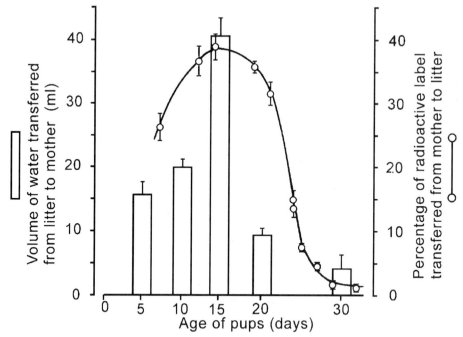

FIGURE 15-1. The amount of radioactively labeled water taken up by rat mothers from the urine of their pups (histograms) is mirrored by the percentage of radioactive label transferred in milk from mothers to their pups (open circles). In both cases, water transfer between mothers and pups peaks at about 15 days postpartum. (SOURCES: Histograms: Reprinted from Gubernick & Alberts, 1983, Maternal licking of young: Resource exchange and proximate controls. *Physiol. Behav.* **31**, 593–601. Copyright © 1983, with permission from Elsevier Science. Open circles: Babický *et al.*, 1970, with permission)

gain by attempting to obtain more care. In contrast, the parent could double its reproductive output by ceasing parental care in order to reproduce again (Fig. 15-2, ratio of benefit to offspring:cost to mother < 1.0). This results in conflict between mother and young until the costs to the mother, in terms of delayed subsequent reproduction, double in comparison with the benefits to the offspring (Fig. 15-2, ratio = 0.5). Benefits decrease as they mature, and cease entirely when their inclusive fitness gains little from continuing parental care but a great deal more from new kin in the form of siblings. Studies with several species, including cats (Fig. 15-3) and macaques, have provided support for this model. The time frame is rather constant for a given species, but environmental influences can change it. In fact, the model implies that a parent would terminate care prematurely, namely, abandon its young, if it found itself in unusually unfavorable conditions that imposed excessive costs. It would then increase its reproductive success by attempting to survive so as to reproduce under better conditions. Bonnet monkey mothers subjected to a variable foraging demand in the experiment described in Chapter 9 responded as predicted by the conflict model: They rejected attempts by their infants to

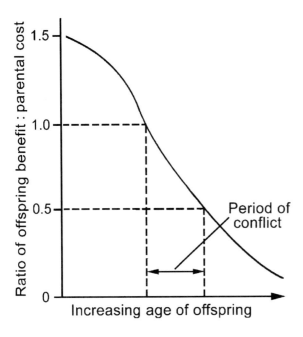

FIGURE 15-2. Trivers's model of parent–offspring conflict. Benefits and costs are measured in units of reproductive success of the offspring and comparable units of reproductive success of the future offspring of the parent, respectively. (SOURCE: After Trivers, 1974)

make contact with them, effectively abandoning them, which precipitated the termination of the study.

These three models have been useful for generating testable research questions, and the results so far indicate that the models are not mutually exclusive and that their usefulness in describing parent–offspring interactions varies between species.

EVOLUTION OF PARENTAL CARE

The theoretical basis for understanding the evolution of parental care was developed between the late 1950s and early 1970s by a number of authors including Williams (1966, 1975), Hamilton (1966), and Trivers (1974).

Sex Differences in Parental Care

The evolution of parental care in both sexes is thought to depend partly on the relative costs and benefits under different environmental conditions (Fig. 15-4). Because females typically have fewer gametes and invest more in each than do males (Chapter 13), they produce fewer offspring per lifetime. Consequently, a single reproductive error resulting in the death of offspring is more costly for females than for males, so females have more to gain by providing parental care under all conditions. Where ecological resources such as food and shelter for the young are plentiful and predator pressures on offspring are low, immature offspring are more likely to survive without parental care

PARENTAL BEHAVIOR AND MATING SYSTEMS

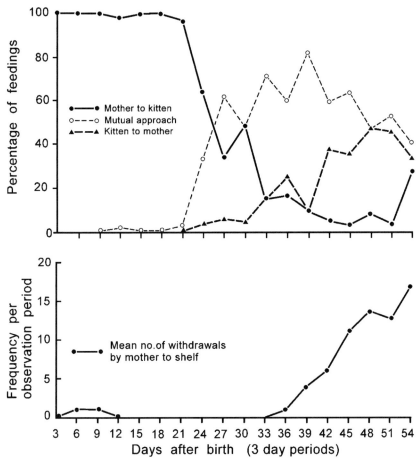

FIGURE 15-3. Weaning conflict in cats is demonstrated by a gradual change from the mother's to the kittens' feeding initiation (upper panel) and a concomitant increase in the mother's avoidance of her offspring by withdrawing to a shelf (lower panel). (SOURCE: Modified from Schneirla et al., 1963)

than where food and shelter are scarce and predation is high. Consequently, under good environmental conditions, the fitness costs of parental care incurred by the delay in subsequent reproduction exceed the fitness benefits conferred by ensuring higher offspring survival rates in both sexes (Fig. 15-4, N). Under intermediate conditions, it benefits the female but not the male to provide parental care (Fig. 15-4F), while in difficult conditions the benefits exceed the costs for both parents to provide care (Fig. 15-4B). There are, however, many exceptions.

While female care predominates in species with internal fertilization, male care is often associated with external fertilization. Three different hypotheses have been proposed for this.

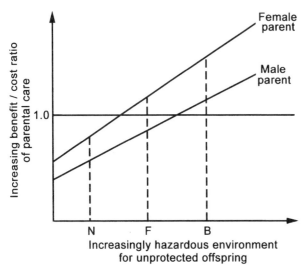

FIGURE 15-4. Ecological variables affecting the chances of survival of offspring influence the benefit–cost ratio of parental care, which is higher for females than for males in most species. N = neither parent cares for offspring; F = only the female cares for offspring; B = both parents care for offspring.

Certainty of Paternity Hypothesis

This hypothesis is based on Trivers's Theory of Parental Investment (Chapter 13), which proposes that males are more likely to invest parental care in young when they are genetically related than when they are not. In species with internal fertilization, males, unlike females, might not be related to the mate's offspring because another male could deposit sperm and fertilize the eggs either before or after the putative father. In species with external fertilization, the certainty of paternity is greater because the male typically deposits sperm on the eggs immediately after the female lays them, so there is less risk of males caring for unrelated offspring. Theoretical models have not supported this hypothesis, and it seems that natural selection for male parental behavior is not influenced by paternity itself.

Gamete Order Hypothesis

It has been suggested that it would be adaptive for a parent to desert its offspring and leave the other parent to care for them. The parent releasing its gametes before its mate does so has the first opportunity to depart. This is usually the male among internally fertilizing species such as birds and mammals, which results in female parental care, although males also contribute in many species because of other selection pressures. In externally fertilizing species such as fish and amphibians, the female releases her gametes first, which generally results in male parental care; for example, female three-spined sticklebacks depart, leaving the male to take care of the eggs.

Association Hypothesis

Formulated by Williams (1975) and supported by data from many families of teleost fishes and amphibians, the association hypothesis proposes that care is given primarily by the parent most closely associated with the offspring when they are hatched or born. One might think that this would always be the female, and this is usually the case, in internally fertilizing species because the male may have long since left the area. External fertilization often occurs in male-defended territories and the male remains near the eggs, whereas the female may not. In fish, this provides the male with some added benefits of parental care because the territory itself and the eggs within it attract other females for spawning. Many fish continue to grow throughout their lives and the number of eggs, and hence fecundity, of females often increases exponentially with increasing body size. Periods of parental care, during which a female cannot feed at maximal rates, would effectively reduce her body size and thus her lifetime reproductive success. Consequently, parental care is more costly for female fish than for females of other taxa. Although the association hypothesis is supported in fish and amphibians, for many other species we do not know why parental care evolved primarily in one sex and not in the other.

In some animals, both parents give parental care more or less equally. This is believed to depend either on unusually high predator pressures, where survival of at least some offspring depends on the vigilance of both parents, for example, in African and South American freshwater cichlid fishes (*Cichlidae*), or on scarce food supplies or dangerous prey, where one adult alone is incapable of feeding the brood, as in many birds. The highly arboreal, small, and vulnerable marmosets and tamarins give birth to twins; this is unusual among primates, which typically give birth to a single infant. The female is incapable of rearing twins on her own, and infants are carried on the backs not only of parents, more often the father than the mother, but also of other close relatives, until they almost reach adult size. It is possible that both twinning and a very high level of male parental care, unusual among mammals, coevolved in response to high predator pressure.

Selection Pressures for Parental Care

The main ecological factors and intervening biological variables thought to be important for the evolution of parental care are illustrated in Fig. 15-5 (Wilson, 1975).

Stable, Structured Habitats

Stable, predictable, and structured habitats promote competition between conspecifics for limited resources, resulting in the defense of nest sites and breeding and feeding territories, namely, philopatry (see Chapter 1, K selection). Parental care is greatly facilitated when resources are within familiar or defended areas. Such habitats are also associated with greater prenatal investment in terms of larger eggs, longer gestation times, and, consequently, larger offspring at birth. Probable coadaptations of such early investment are a prolonged period of immaturity, larger adult size, since bigger is better when competing for resources, and a longer life. These characteristics promote parental care; the fitness of the parent is more likely to increase if, by caring for the offspring, the parent ensures that an already competitive individual becomes even more so. By analogy, our parents, having discovered we are quite bright and likely to be competitive as adults, thought there might be an additional edge if we became even

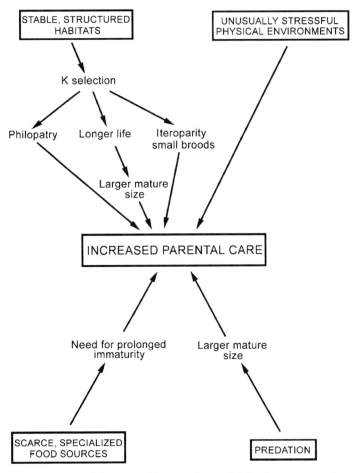

FIGURE 15-5. Environmental factors and intervening variables that promote the evolution of parental care. (SOURCE: E. O. Wilson, 1975, *Sociobiology: The New Synthesis*. Harvard University Press: Cambridge, MA. Reprinted with permission of the publisher. Copyright © 1975 by the President and Fellows of Harvard College)

brighter and decided to invest in us for an additional 4 years by paying university tuition fees. The expected positive correlation between size and parental care generally holds among insects, birds, and mammals.

Related to the phenomena just described is the fact that species in stable, structured habitats tend to produce small broods repeatedly, termed *iteroparity*; in extreme cases, there is a single offspring every year or two, as in many large mammals. In its opposite, termed *semelparity*, there is just one massive reproductive effort during a lifetime; mayflies, for example, produce a *single* brood of thousands during their short lives. David Lack first developed the idea that the probability of the parent caring for its young increases with decreasing brood size in iteroparous species, and that the more the parent cares for the brood, the more precisely controlled the brood size will be. This sprang from his observation that the number of songbird fledglings was reduced if the number of eggs in a clutch deviated by just one or two eggs from the species norm. He argued that too many eggs would exceed the capacity of the parents to rear their offspring and result in undernourishment and high mortality in the brood as a whole, whereas too few eggs would fall below the optimal capacity of the parents to rear them. Data from barn owls also support this notion. Their main food is small rodents, whose populations can fluctuate dramatically from year to year. In a given pair of owls, the number of eggs varies with food availability, with none at all being laid in a poor year.

Unusually Stressful Physical Environments

While stable, predictable habitats promote parental care, so do harsh environments, for example, those with exceptionally stressful temperature or humidity extremes. Unlike other members of the beetle family *Staphylinidae*, one species (*Bledius spectabilis*) inhabits the intertidal mud on the northern coast of Europe, where it often encounters oxygen shortages and high salinity. This is associated with an unusual degree of maternal care. The mother provides a burrow for her larvae, which she protects from intruders, and periodically brings them fresh algae to feed on. Plethodontid salamanders have penetrated the land environment more fully than other salamanders, laying their eggs in humid sites on land, in the ground, or in pieces of wood, where the mother often protects them until they hatch into adult-like form; they bypass the typical aquatic larval stage. In at least one species, when broods were deprived of maternal care, embryos were smaller and less than half the offspring survived compared with broods whose mothers were present.

Predation

Parental care is also promoted in habitats with high predator pressures, as shown by mouth-brooding African cichlids described earlier. Moreover, high levels of predation can result in larger mature size—bigger is better in competing with predators—which, as already noted, makes increased parental care adaptive.

Scarce, Specialized Food Sources

Extended parental care is highly adaptive in species that specialize in foods that are difficult or dangerous to find, capture, or retrieve, notably, birds of prey and some mammalian carnivores. Among birds, albatrosses, condors, and eagles are the slowest to breed. Royal albatrosses and condors may not begin breeding until they are about 9 years old. A single offspring is raised over a 1- to 1.5-year period before the parents breed again. This extensive parental support is necessary to allow the offspring to mature to adult size and strength, and to acquire the necessary hunting skills to fend for itself. Some eagles, for example, range over enormous areas encompassing thousands of square kilometers in search of prey, which requires good homing skills. During the breeding season, movement is considerably restricted; the male typically hunts for himself as well as for the female and chick, until the latter is almost fully-grown. Bringing large prey, as large as a small antelope in the case of the crowned eagle, back to the nest requires stamina and strength. Long periods of immaturity and dependence also characterize large canid and felid carnivores, including wolves, hyenas, and lions. These species have complex social organizations whose members cooperate in hunting large and dangerous mammals such as wildebeest, large antelopes, and zebras. The long period of parental care facilitates physical maturation and provides time for the young to learn cooperative hunting skills and both their social roles within the group and the social skills needed for group living.

MATING SYSTEMS

The evolution of mating systems cannot be understood without taking parental investment and care into account, which is why we have dealt with the latter first. Bateman's Principle and Trivers's Theory of Parental Investment (Chapter 13) predict that because females invest more heavily than males in offspring and therefore produce fewer offspring than males, there is a fundamental conflict of interest between males and females when it comes to mating strategies. Typically, a male can fulfill his greater reproductive potential by taking several female mates or sneaking copulations with females other than his own mate. The conflict is reduced when a male can contribute substantially to parental care, for example, in birds where males as well as females may incubate eggs and feed fledglings, and is increased where males contribute relatively little parental care. In mammals, only the female nurtures and protects the developing fetus, and only the female suckles the developing offspring, often for a prolonged period. As described below, the relative contributions of maternal and paternal care, together with selection pressures exerted by the environment, influence the mating strategies of males and females and hence the evolution of different mating systems.

Definition of Mating Systems

Mating systems are not easy to define. They are the result of compromises between conflicting male and female mating strategies that maximize the reproductive success and, ultimately, the inclusive fitness of both sexes. This is quite distinct from the copulatory pattern itself and should not be confused with it. Different authorities use different operational definitions, typically applying one or more of three broad criteria: (1) the number of partners with which copulation occurs; (2) the presence or absence of pair-bond formation involving cooperative parental care; and (3) pair-bond duration. Defining the pair bond has its own set of problems, for there is no universally accepted definition. The first and apparently simplest of these three criteria, number of copulation partners, is very difficult to assess in the natural habitat, and the increasing use of paternity exclusion analyses, particularly in birds, has demonstrated that mixed paternity in a brood is the norm for several species previously regarded as monogamous. This indicates that both partners copulate with individuals other than their mates, although they cooperate only with each other in raising a brood. More successful males in species with polygynous mating systems have several mates, but less successful ones may have only one mate, and the same is true of females in species with polyandrous mating systems. For these reasons, the term *mating system* should perhaps be replaced with the term *breeding system*, which is more accurate when comparing different species. It should be kept in mind that, within species, different individuals may have different

Classification of Mating Systems

There are several classifications of mating systems. That discussed in more detail below uses three terms from other classifications: monogamy ("single marriage"), namely, one male with one female; polygyny ("many females"), namely, one male with several females; and polyandry ("many males"), namely, one female with several males. The term polygamy ("many marriages") is the opposite of monogamy and includes both polygyny and polyandry. Some classifications include the category promiscuity, with its pejorative overtones, and this is defined as multiple matings by at least one sex in the absence of any prolonged associations between the sexes (Wittenberger, 1979).

Table 15-1 summarizes the behavioral and morphological features that characterize each of the major categories of mating systems. The order in which features are listed from left to right approximates the likely evolutionary chain of causation. Male parental ability indicates whether or not phylogenetic and environmental factors enable males to provide resources, care, or both to the female and her offspring. Male parental ability determines which male behavior will maximize male reproductive success; and male behavior, in turn, affects the degree and type of morphological sexual dimorphism. Female mate choice is influenced by the male's behavior or morphological characteristics (although, as noted earlier, it can also influence them), and itself helps to determine the mating system.

Polygyny

As already noted, male mammals generally contribute a smaller proportion of parental care than males of other taxa because female mammals invest so heavily during pregnancy and lactation, and this correlates well with estimates that about 90% of mammalian species are polygynous, including humans (Chapter 17). Polygyny can be divided into two categories on the basis of whether or not males provide resources for females and their offspring.

Pure Dominance Polygyny. Pure dominance polygyny occurs when males do not contribute any parental care, be it by protection, by nurturance, or by giving females access to environments with abundant food and shelter. Instead, males demonstrate their superior quality and competitiveness by sequestering or attracting females away from other males.

Among mammals, the females of many ungulates, including deer, sheep, cattle, wildebeest, horse, and zebra, band together for protection from predators and the males fight each other during the mating season for sole possession of the largest possible herd. In other species, including elephant seals and sea lions, females congregate in large numbers on beaches during the breeding season, and males have opportunity to fight each other for exclusive access to groups of females. This is sometimes termed *female defense polygyny*.

Table 15-1. A Classification of Mating Systems

Male parental ability	Male behavior	Morphological sexual dimorphism	Female mate choice	Mating system
Males do not provide useful assistance for mate and offspring.	Males compete with each other to demonstrate their quality or monopolize females.	Males are very much larger, stronger, and more spectacular than females.	Competitively superior male	Pure dominance polygyny
Males monopolize resources on behalf of mate and offspring and/or provide parental care.	Some males assist female more than others do.	Males are somewhat larger, stronger, and more spectacular than females.	Better helper	Parental investment polygyny
	Most males assist females.	Males and females appear similar.	Unmated male	Monogamy
	Male's parental care exceeds that of female.	Females are somewhat larger, stronger, and more spectacular than males.	Females primarily invest in egg production; sex-role reversal.	Polyandry

Males of other species do not sequester females but instead compete with each other by means of spectacular displays on small areas known as *leks*. This is most common among ground-nesting gallinaceous birds such as peacocks and grouse. A few mammals, including the hammer-headed bat, also have leks. Leks are typically located in areas frequented by females and are often used for many generations. Each male competes with others to carve out a small territory for himself, upon which he performs spectacular visual and auditory displays when females are nearby. Females then visit the individual territories and mate with the male or males of their choice, after which they leave the lek until the mating season the following year. Typically, a few males perform most of the copulations and many, over 50% in some cases, do not copulate at all in a given year. Among the black grouse, sexually successful males are usually 2–3 years old, dominant over unsuccessful males, occupy territories in the center of the lek, have higher testosterone levels during the mating period and a lower parasite load, and are more likely to survive the next winter. This is sometimes called *lek polygyny*.

Since males compete intensively with each other for females, either aggressively or with highly ritualized displays that are often associated with morphological characteristics that exaggerate them, there is considerable sexual dimorphism in species with pure dominance polygyny. Males are very much larger than females and have more spectacular and ornamental plumage in birds, specialized appendages such as the large claw of the fiddler crab, and large horns in many ungulates. Both intrasexual and epigamic selection may account for many of these male characteristics. It is difficult to distinguish between the effects of these two processes, but there is evidence for some species that females prefer to mate with males having an experimentally exaggerated sexually dimorphic feature (Chapter 14). In humans, exaggerations include military or other uniforms with epaulettes that widen the shoulders ("for the military chest seems to suit the ladies best, there is something about a soldier that is fine, fine, fine!"). There are indications that females of many species respond to supranormal releasers (Chapter 3), and that these features, together with the females' preferences for them, probably coevolved.

There are exceptions, notably the bower birds of New Guinea, which have a lek polygyny similar to that of grouse. Depending on the bower bird species, males build impressively large and complex structures or bowers in their lek territory that are then decorated with brightly colored objects, either natural (flowers, leaves, etc.) or man-made (Coca-Cola bottle tops, etc.). The preferred color is species- or locale-specific and a function of female preference. Sometimes males must search far to retrieve objects of the appropriate color if they cannot steal them from a neighbor's bower. The males themselves are remarkably plain and there is little sexual dimorphism. In comparisons between species, there is a negative correlation between the appearance of males and the appearance of their bowers; the plainest males build the most striking bowers.

Parental Investment Polygyny. Parental investment polygyny occurs when some males provide more parental care than others or sequester more resources for females and their offspring. Under such conditions, a female might maximize her reproductive success by choosing a male that is better able to support her and the young, even when the male already has several mates, provided that his resources are adequate for all. This has been examined in several species of birds in which males defend territories that can differ greatly in size and quality, and where males with larger or richer territories are polygynous. In red-winged blackbirds, there is no systematic difference in the reproductive success of females as a function of the number of other females sharing the territory of a male; it appears that females do not mate with a male unless his resources are adequate for them. Similar observations have been made for other bird species, but in mammals, for example, yellow-bellied marmots (Fig. 15-6) and humans (Chapter 17), female reproductive success decreases with increasing harem size, whereas male reproductive success typically increases when more than one female is available. There is currently no explanation for the difference between birds and mammals. This type of polygyny is sometimes known as *resource defense polygyny*.

Although some males have only a single mate and others none at all, the variance in male mating success is somewhat less with parental investment polygyny than with pure dominance polygyny. How this affects lifetime reproductive success, especially in longer-lived species, is not known because it is so difficult to determine in the field; it is possible that each male is successful during the prime of life, as appears to be the case for black grouse (discussed earlier). Sexual dimorphism in size and ornamentation is also generally less extreme than with pure dominance polygyny.

Monogamy

If most males provide parental care, a female and her offspring would probably fare better with a male that does not already have a mate, because he would devote all his effort to them alone. Over 90% of bird species exhibit monogamy but, as noted earlier, this does not mean that copulations themselves are necessarily restricted to a single mate. The pair bond may last for a single reproductive cycle or for the entire breeding season, but in some cases, for example, greylag geese and swans, the pair bond may last the entire life span of some 70 years. The high incidence of monogamy among birds is probably related to the fact that male birds, unlike male mammals, can contribute substantially to parental care by sharing nest building, egg incubation, and feeding. Monogamy is rare in mammals and often associated with specializations in hunting or retrieving dangerous or scarce prey. Monogamy may comprise a single male–female

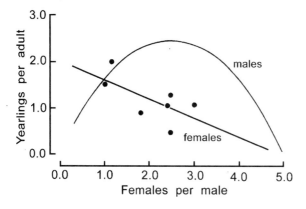

FIGURE 15-6. The reproductive success of female yellow-bellied marmots (*Marmota flaviventris*) decreases with polygyny (points around regression line), whereas that of males is maximal when they have exclusive access to two or three females. (SOURCE: From Downhower & Armitage, 1971. Copyright © 1971 by the University of Chicago Press.)

pair, as in foxes, or occur in the context of a highly organized social group, as in wolves, where mating is generally restricted to the dominant male and dominant female of the pack.

In monogamous mating systems, females choose primarily unmated males rather than the largest, strongest, or most ornamented ones. Consequently, there is less sexual dimorphism; males and females tend to be more similar in size and appearance and difficult to distinguish in the field.

Polyandry

Polyandry is less common than the other types of mating systems and usually occurs when males can assume all parental duties, enabling females to produce several broods in quick succession. Because the sperm of one male is generally adequate to fertilize all the female's eggs, and the male loses mating opportunities when assuming parental care, polyandrous mating seems to have disadvantages for both sexes; two hypotheses have been proposed for its evolution. The first (Graul *et al.*, 1977) suggests there may be advantages for both sexes where food supplies fluctuate dramatically. Under poor conditions, with a female's energies depleted by egg production, parental care by the male would be adaptive, and when conditions improve, the female could produce another clutch, either tending it or leaving its care to another male. The second hypothesis (Maxson & Oring, 1980) suggests that predator pressure is the important environmental variable. If predation is high, both partners could increase their chances of raising young if the male takes care of the first brood so that the female can produce another. This latter hypothesis has been supported by data from spotted sandpipers, birds that are subject to intense nest predation. Females will mate with new males if they appear, but otherwise care for their young. In areas with high predation, the birds crowd into the safest refuge areas, increasing the ability of females to acquire multiple mates, and the number of offspring reared increases with the number of males monopolized by the female.

In addition to parental behavior, there is generally sex-role reversal in other behaviors as well. Females rather than males compete with each other for mates and resources, while males tend to be more selective in choosing prospective mates; females tend to be larger, stronger, and more ornate than males. In another bird species, the Northern jacana, the female weighs 145 gm, whereas the male weighs only 89 gm. The female defends her territory whether or not young are present, takes the dominant role in courtship, and does not aid her male partners in brood care. In addition to this "true" polyandry, which is always associated with the sexual dimorphism described earlier, there is another form of polyandry, *resource defense polyandry*, which is not. This occurs, for example, in hummingbirds, when females copulate with males in exchange for valuable resources that are guarded by males. There is no sex-role reversal, and since males mate with several females and are not monopolized by them, this mating system is better regarded as a combination of polygyny and polyandry.

CHAPTER **16**

Nonhuman Primates

In this and the next chapter, we consider the behavior of nonhuman primates and humans, respectively. To facilitate an appreciation of the phylogenetic relationships between the various species from which the behavioral examples are drawn, Table 16-1 gives a simple classification of living primates. Latin names of suborders, families, and genera are italicized and, where applicable, common English terms are given in parentheses. Taxonomic classifications vary with different authorities, but the principle is the same: Different species (not given here) of the same genus are more closely related than are different genera of the same family, and different families of the same suborder are more closely related than are different suborders. Many prosimian species, notably the *Lemuridae*, *Indriidae*, and *Daubentoniidae*, are confined to Madagascar, where there are no anthropoid primates or "true" monkeys (*Platyrrhinae* and *Catarrhinae*). It is thought that anthropoid primates diverged from prosimians after Madagascar was separated from the rest of Africa, and generally outcompeted the prosimians elsewhere.

ETHOLOGY

Compared with nonprimate species, except perhaps for some marine mammals such as whales and dolphins, most primates have greatly increased neocortical development. The neo- or new cortex has six layers (Brodman's classification) and, in phylogenetic terms, contrasts with the archi- or old cortex. The latter has one or two layers and is concerned with smell, among other things. In anthropoid primates, the neocortex is convoluted and folded into gyri separated by sulci, and this convolution increases

Table 16-1. Simplified Classification of Living Primates (Order: Primates)

Suborder	Family	Genus
Prosimii (prosimians)	*Tupaiidae* (tree shrews)	*Tupaia, Dendrogale, Urogale, Ptilocercus*
	Lorisidae (lorises)	*Loris* (slender loris), *Nyctocebus* (slow loris), *Arctocebus* (golden potto), *Perodicticus* (potto)
	Galagidae (bushbabies)	*Galago*
	Lemuridae (lemurs)	*Microcebus* (mouse lemurs), *Cheirogaleus* (dwarf lemurs), *Phaner, Hapalemur, Lemur, Lepilemur* (sportive lemur)
	Indriidae	*Propithecus* (sifakas), *Avahi* (woolly lemur), *Indri* (indris)
	Daubentoniidae	*Daubentonia* (aye-aye)
	Tarsiidae (tarsiers)	*Tarsius* (tarsiers)
Platyrrhinae (New World monkeys)	*Callithricidae* (marmosets)	*Callithrix* (marmosets), *Cebuella* (pygmy marmoset), *Leontideus* (tamarins), *Saguinus* (tamarins)
	Callimiconidae	*Callimico* (Goeldi's marmoset)
	Cebidae	*Aotes* (owl monkey), *Callicebus* (titis), *Pithecia* (sakis), *Chiropotes* (sakis), *Cacajao* (uakaris), *Alouatta* (howlers), *Saimiri* (squirrel monkeys), *Cebus* (capuchins), *Ateles* (spider monkeys) *Brachyteles* (woolly spider monkey), *Lagothrix* (woolly monkey)
Catarrhinae (Old World monkeys and apes; humans)	*Cercopithecidae*	*Macaca* (macaques), *Papio* (baboons, mandrill, drill), *Theropithecus* (gelada); *Cercocebus* (mangabeys), *Cercopithecus* (vervets, guenons, talapoin), *Erythrocebus* (patas)
	Colobidae	*Presbytis* (langurs, leaf monkeys), *Pygathrix* (Douc langur), *Rhinopithecus* (snub-nosed monkey), *Simias* (pig-tailed langur), *Nasalis* (proboscis monkey), *Colobus* (colobus monkeys)
	Hylobatidae (lesser apes)	*Hylobates* (gibbons), *Symphalangus* (siamang)
	Pongidae (great apes)	*Pongo* (orangutan), *Pan* (chimpanzees), *Gorilla* (gorilla)
	Hominidae	*Homo* (human)

cortical surface area and is contrasted, for example, with the smooth or lissencephalic cortex of the rat. As one moves from the more primitive prosimians to the human, there is a relative reduction in the brain structures concerned with olfaction and increased bulk of the neocortical structures (Chapter 11, Fig. 11-13). This is associated with greater cognitive abilities, behavioral flexibility, and often complex social relationships, particularly in the Old World monkeys and apes. These characteristics enable individuals to modify their behavior according to the environment and social influences such as dominance rank, for which the memory of previous social interactions is needed. Some nonhuman primate species are approaching extinction because of habitat destruction and human predation. Many are hard to observe in their natural habitats because they are nocturnal, small, or living in the canopy of dense forest. They are difficult to maintain in captivity because their social organizations and nutritional requirements are not sufficiently known. Moreover, larger species are long-lived (in excess of 20 years), and decades of continuous observations on known individuals were needed to obtain an understanding of the social organizations and behavior of just a few species, notably, some macaques, baboons, chimpanzees, and gorillas.

For these reasons, and because of ethical considerations, many of the manipulations used to identify FAPs, releasers, and other ethological phenomena in nonprimates are less feasible in monkeys and apes. Nevertheless, comparisons of closely related primate species, and of populations of the same species living in different geographical locations, suggest that many of the phenomena described in nonprimates can also be found in our closest living relatives.

Fixed Action Patterns (FAPs)

Neonatal primates show nipple-searching behavior, whereby the hungry infant presses its face against the mother's chest and moves it from side to side until encountering the nipple. This behavior is shared by all primates including the human and is a FAP (Chapter 2). When traveling, older infants may ride on the mother's back, for instance, in baboons, but at other times the mother facilitates feeding by holding the infant ventroventrally against her abdomen or chest, and the infant clings to her hair with both hands and feet. This clinging response is also an FAP and may be related to the Moro reflex of the human neonate. Facial expressions and vocalizations are almost certainly FAPs because they take a similar form in different species and, at least in rhesus monkeys (*Macaca mulatta*), develop in infants reared in isolation from conspecifics. In macaques, for example, an intense open-mouth threat involves a stare at the opponent with erect ears and lowered eyebrows while the mouth is opened widely but in such a way that the teeth remain covered by the lips. This may be accompanied by forward jerks of the head and low-pitched grunts. In contrast, during the fear grimace of intense submission, the jaw is closed or barely open but the lips are retracted and the corners of the mouth are drawn so far back that all the teeth are clearly visible. The ears are laid back and the eyebrows are maximally raised, and there may be high-pitched screaming vocalizations. These same behaviors occur in baboons, which have a quite different-shaped jaw, resembling a dog-like muzzle. Some macaques (*M. fascicularis*) and baboons (*Papio anubis*) have rather dark facial skin except above the eyes. When they raise their eyebrows, this pale skin stands out like a pair of beacons, and is seen at a distance of many yards. The adaptation makes greetings between social companions clearly visible at a distance; when one individual peacefully approaches another, it looks directly at the other, rapidly and audibly smacking its lips while flattening its ears and maximally raising its eyebrows. This is reminiscent of the "eyebrow flash" shown by humans when greeting each other at a distance (Chapter 17).

FAPs are very probably involved in sexual behavior. All monkeys mate dorsoventrally and the male approaches and mounts the female from the rear. This pattern is also typical for most apes in the wild, but ventroventral copulation has been observed in captivity. In the pygmy chimpanzee or bonobo (*Pan paniscus*), however, copulation typically occurs face-to-face both in the wild and in captivity. There is also species specificity in copulatory patterns, even within a single genus. In rhesus monkeys, ejaculation occurs at the end of a series of mounts, each with an intromission and multiple thrusts. In cynomolgus monkeys, some ejaculations occur at the end of a single mount, with intromission and multiple thrusts, but sometimes there are several such mounts without ejaculation before there is one with ejaculation. In bonnet monkeys (*Macaca radiata*), copulation typically consists of a single mount, with intromission and multiple thrusts terminating in ejaculation. The significance of these differences between closely related macaques remains obscure.

Conflict Behaviors

Various types of conflict behavior occur frequently in many primate species, especially in those living in large social groups where the behavior of an individual is often constrained by its age, sex, and rank. Intention movements occur in the conflict between approach and avoidance, threat and submission, and in situations involving tactile contact. A female rhesus monkey paired with an unfamiliar male, for example, may tentatively brush the male's hair in an intention movement to groom, whereas, with a familiar mate, she would groom with confidence. Repetitive scratching, yawning involving the prominent display of the canines, and exaggerated chewing movements or "chomping" are the most frequent forms of displacement activity in primates and can readily be observed in zoos in many species ranging from prosimians to apes. This behavior occurs in various contexts, for example, in a standoff between two males that have just threatened or fought each other. Redirection of aggression is very common; an individual threatened by a higher-ranking social companion may threaten a nearby lower-ranking individual, the scapegoat phenomenon, which may result in the high-ranking individual joining in with threats toward the low-ranking animal. Affiliative behavior can also be redirected. When one individual grooms another and the latter departs, the groomer may begin grooming a nearby companion (redirected) or itself (self-directed) for a while before moving away. Ambivalence and alternation are often apparent in the highly mobile faces of Old World primates, particularly in agonistic contexts. In macaques, there are many ambivalent gradations between threatening and submissive facial expressions described earlier, from an intense open-mouth threat, where the corners of the mouth and the lips are slightly retracted so that the tops of the teeth are just visible, to a fear grimace, in which the corners of the mouth are not fully retracted and the jaw is partially open. Alternation may also occur when the animal changes rapidly and repeatedly between the more aggressive and more submissive facial expressions and postures, as in dogs and cats.

Ritualization

It cannot be stated with certainty that any behavior in higher primates is phylogenetically rather than culturally ritualized. However, the two examples given here involve motivational changes, which suggest phylogenetic ritualization. The first concerns the threatening-away behavior shown by captive male–female pairs of rhesus monkeys. Threats are directed not at the other monkey but away from it, toward some irrelevant feature in the environment. Threatening-away occurs in a sexual context, just before and during, but not immediately after, a mounting series terminating in ejaculation. As described in Chapter 3, hormone-induced changes in the sexual motivation of the male and female showed that the frequency of threatening-away behavior in both sexes was positively correlated with the level of sexual interest in the partner and reflected sexual rather than aggressive motivation. This behavior occurs in the wild and in related species such as cynomolgus monkeys. In the main form of female sexual invitation shared by most nonhuman primates, the female stands quadrupedally or crouches in front of the male with her anogenital region directed at him; this is the familiar sexual presentation posture. There is evidence that threatening-away has undergone further ritualization into three additional sexual invitations by female rhesus monkeys that can initiate male mounting by the use of three other movements: a slap of the hand on the ground (hand-reach), a rapid downward jerk of the head (head-duck) and an upward jerk of the head (head-bob). These movements, which are more discrete and exaggerated versions of movements that occur in threatening-away, are made while the female is sitting close to the male with her back or side to him. Like threatening-away, they are made only in a sexual context, occurring most frequently when the estrogen levels of the female are high; if they do not elicit mounts by the male, they are ignored. Females differ in the proportions with which they make the three gestures, irrespective of the responses of the male partner (Fig. 16-1). Most females show all three movements under appropriate conditions.

Another example of what is probably phylogenetic ritualization comes from reconciliation in bonobos. In some socially living mammals, including nonhuman primates, two individuals will come together shortly after a dispute and direct affiliative behavior toward each other. This occurs particularly frequently when the two individuals are important to each other, such as kin or partners in dominance alliances, and the function of the reconciliation, as implied by the term, appears to be to repair the relationship that was damaged by the prior agonistic interaction. In most nonhuman primates, reconciliation usually involves submissive and greeting behavior followed by touching and perhaps mutual grooming. In bonobos, however, reconciliation takes an unusual form: the two individuals hug ventroventrally and rub their anogenital regions against each other. Unlike true mating activity, this behavior occurs in all age and sex classes, including between males, between females, and between close kin.

Releasers

In our previous discussion of releasers and innate releasing mechanisms (Chapter 3), the releaser was recognized because it consistently elicited the same, appropriate behavioral response. But even in the case of the classical aggression releaser, namely, the red belly of the male stickleback, the behavioral response is situation-dependent: It is milder toward a territory neighbor than toward an unfamiliar intruder. In primates, behavior is far more flexible and situation-dependent than in sticklebacks or geese, making it difficult to identify releasers by the apparent inevitability of the behavioral response. Only in humans can one ask about changes in attitudes and feelings and circumvent the problem (Chapter 17). There is reason to believe, however, that releasers exist in monkeys and apes. One certain releaser in rhesus monkeys is a soft-textured surface, normally the mother's belly hair, that releases the clinging behavior of the neonate. Famous studies in which infant rhesus monkeys were raised in social isolation on various artificial mother surrogates (Harlow & Zimmermann, 1958) showed that the latter had to be covered in terrycloth to elicit clinging (Fig. 16-2).

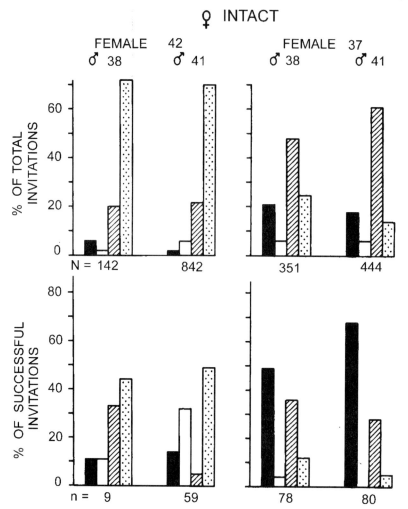

FIGURE 16-1. There were individual differences in the proportions of presentations (black), head-bobs (white), hand-reaches (hatched), and head-ducks (stippled) made by two female rhesus monkeys, but each female made the same proportions of invitations to different males, irrespective of the males' responses to them. (SOURCE: Michael & Zumpe, 1970b, with permission)

It seems probable that facial expressions reflecting emotion are releasers, if not for overt behavior, then for a change in mood, as suggested by the finding that infant rhesus monkeys at about 3–4 months of age stop pressing a lever to see photographs of a threatening conspecific that elicits fear (Chapter 3). Compared with true monkeys and apes, prosimians such as the ring-tailed lemur (*Lemur catta*) have rather immobile faces, limiting the extent to which threat, submission, affiliation, and so forth, can be communicated by facial expressions. Instead, these prosimians have a variety of scent glands

FIGURE 16-2. Neonatal rhesus monkey clinging to mother surrogate covered in terrycloth. (SOURCE: Harlow & Zimmerman, 1958, with permission)

and use their secretions widely in olfactory communication. Ring-tailed lemurs live in social groups in which females are dominant over males. In aggressive exchanges between two adult males, each faces the other, tucks his long tail between his legs, bringing it up in front of the chest (Fig. 16-3), and wipes it along scent glands on the inner arms. The male then stands up and arches the heavily anointed tail over his back and head, with the tip pointing toward the opponent. The tail is then rapidly quivered vertically, wafting the odor forwards. The opponent does precisely the same thing, and the animals dance about waving the tails in each other's faces. These disputes have been called "stink-fights" (Jolly, 1966). The tails are white with broad, black rings along their lengths; when waved about, they provide an additional visual signal. It seems likely that the combination of olfactory and visual signals serve as releasers in agonistic interactions.

Releasers are also involved in eliciting sexual behavior by males in some species in which the circumgenital skins of both sexes are distinctively colored. In females, the color of this "sex skin" fades and intensifies with hormonal status and may be associated with marked swelling, so that the color is brightest and the swelling is largest at midcycle, when fertility is maximal. In some macaques such as bonnet monkeys, baboons, and chimpanzees, the swelling can be very large, resulting in a protuberance

FIGURE 16-3. Ring-tailed lemur with tail between its legs.

that is clearly visible at a distance. This is a releaser for male mounting because, in these species, males virtually never mate with a female unless there is at least a partial sex skin tumescence. Maternal behavior may also depend in part on releasers. Very young infants often have a much lighter or darker pelage than do their older conspecifics. In many species, young infants attract much attention from older siblings and adults, especially females, which often try to take the infant from the mother in order to cradle it themselves. These observations suggest that certain properties of the infant, including its size, proportions, color, odor, and vocalizations, act as releasers for caregiving behavior by older conspecifics, as has been shown for humans (Chapter 17).

Sensitive Periods, Imprinting

Learning during development plays a large role in primates, especially in social anthropoid primates, because of the relatively long time elapsing between birth and puberty. Immediately after birth, the infant is a mostly passive participant in the interactions of the mother with her kin and with all other group members, giving it the opportunity to learn much about social relationships and social structure within the group. Starting a few weeks after birth, when the mother begins to allow it off her body to explore, initially for brief periods, the infant begins to interact actively with other infants in play. One function of play is thought to be the opportunity for so-called cognitive rehearsals for behaviors important in adult life, including agonistic interactions (chasing, play-fighting) and mating behavior (play-mounting and thrusting, mostly by males, and accepting mounts and thrusting, mostly by females). But more important than the opportunity for early learning is the development of a social bond with a living, responsive mother, mother surrogate, or peers soon after birth (Harlow & Harlow, 1962). Infant rhesus monkeys separated from the mother immediately after birth and placed with artificial mother surrogates in social isolation failed to develop normal play, social, aggressive, and sexual behaviors when introduced in adolescence to normally raised peers. They responded to friendly approaches by withdrawal or aggression and, when they became adults, were incapable of normal social and sexual interactions. The few females that were, nevertheless, impregnated by an experienced male either ignored or maltreated their own infants. These studies indicate that there is an important, early sensitive period for social bonding in rhesus monkeys that is essential for normal adult behavior. No studies of this kind have been conducted with other primate species. We have, however, described the profound effects on mother–infant bonds in bonnet monkeys when the mothers were subjected to a variable foraging demand, and how this permanently affected the physiology of the stress and coping responses of offspring when they became adults (Chapter 9). Another clear example of a sensitive period early in development is the acquisition of symbolic communication, that is, language, in great apes and humans, described in the last section of this chapter.

SOCIOBIOLOGY

Social Systems

Social organization is both a product of relationships between individuals and a factor influencing individual relationships. Genes can code for an individual being either "large" or "small" but not for an individual being "larger than" or "smaller than" another, namely, for a relationship between two individuals. Since genes do not code for direct relationships, social organization is further removed from direct genetic control than most other morphological and behavioral traits. It is perhaps no accident that this has been pointed out emphatically by such primatologists as Kummer and Bernstein, because there are no predictable relationships between social systems and phylogeny in primates, even when selection pressures such as food specialization and predation are taken into account. At present, we do not understand why certain primates have certain social systems, while related species have quite different ones. Eisenberg and colleagues (1972) distinguished between five types of social system in primates based on the degree of tolerance between and within the sexes (Table 16-2). These are each described below using representative species.

Solitary

The adult orangutan is essentially solitary; males appear to have home ranges overlapping those of females that travel either alone or with their immature offspring. Neither males nor females appear to seek out and tolerate frequent interactions with

Table 16-2. Social Systems in Some Nonhuman Primate Taxa

	Solitary	Male–female pair	Unimale group	Age-graded male group	Multimale group
Tupaiidae	Tree shrews	—	—	—	—
Lorisidae	Loris, potto	—	—	—	—
Lemuridae	Mouse, sportive, and dwarf lemurs	—	—	—	Ring-tailed lemur
Indriidae	—	Indri	—	—	Sifaka
Daubentoniidae	Aye-aye	—	—	—	—
Callithricidae	—	Marmosets, tamarins	—	—	—
Cebidae	—	Owl and titi monkeys	Capuchin	Howler, squirrel, and spider monkeys	—
Cercopithecidae	—	—	Patas, sacred and gelada baboons, drill, mandrill, grey-cheeked mangabey	Toque macaque, talapoin, vervet, white-collared mangabey	Macaques, savannah baboons, vervet
Colobidae	—	—	Langurs, colobus	Langurs	—
Hylobatidae	—	Gibbons, siamang	—	—	—
Pongidae	Orangutan	—	—	Gorilla	Chimpanzees

SOURCE: Excerpted with permission from Eisenberg et al., 1972, The relation between ecology and social structure in primates. *Science* **176**, 863–874. Copyright © 1972 by the American Association for the Advancement of Science.

adults of either sex. Mating seems to occur at chance meetings and, according to captive studies, is either of a rape-like nature, when the female is not fully receptive, or is gentle and repetitive, when the female is fully receptive near midcycle.

Male–Female Pair

Among New World monkeys, marmosets and tamarins live in groups consisting of a single male–female pair, together with their offspring. These small animals are unusual among primates because they typically give birth to twins rather than to a single infant, and the twins can only survive when the father is present and provides extensive parental care. He carries both twins, and later one twin, more often than does the mother: compared with adults, marmoset infants are proportionally much larger than infants in species with greater size. Adolescent offspring are not actively expelled from the group but they do not mate with their parents and eventually leave to form a new group. In fact, ovarian activity is suppressed in mature daughters living within the family group but starts rapidly when the daughter is removed from the family or exposed to a novel adult male.

Among Old World primates, gibbons and the siamang have a similar social system. Each sex is highly intolerant of same-sex adults. The pair defend a territory from which the adult male expels his sons as they mature, while the adult female does the same with her daughters. However, the parents sometimes help their offspring to carve out a territory and may even contribute a piece of their own. There is no antagonism between opposite-sex adults, and there have been reports that both sexes may mate with a member of a neighboring pair, and that pairs may "divorce" and form new pairs. These animals are highly arboreal and much communication is vocal. As in some birds, auditory communication encodes a great deal of information, including the species, sex and identity of the caller, and is involved in territorial defense; the male–female pair uses duetting to keep in touch.

Unimale Groups (Harems)

Unimale groups consist of a group of females and their young, together with a single adult male. Other adolescent and adult males are found in separate, all-male bands until they have an opportunity to replace, often aggressively, the adult male of an unimale group. Studies with langurs have shown that after a new male takes over the group, he will often succeed in killing the dependent infants of all the females in the group, thus increasing his own potential reproductive success at the expense of that of the previous male (Chapter 12, Infanticide). The sacred baboon *Papio hamadryas*, living in harsh, dry regions of Ethiopia, has a social system based on unimale groups, but many such groups gather together each evening to sleep on high cliff ledges where they are safe from predation. In the morning, they descend from the cliffs and spread out to forage over a wide terrain with little cover and sparse food; this unusual social organization appears to be an adaptation to the environment and to the high vulnerability to predation. The harem male possesses and herds several females, which he may acquire before they reach puberty. He appears to mate exclusively with his group of females, but the latter may, if unobserved by him, copulate with nearby peripheral males ("sneak copulations"). These males have no females of their own and stay near the periphery of another male's group, which gives them an opportunity to replace a harem male or to entice some of the females away. Within a unimale group, mature females tend to show menstrual synchrony, for which there is little evidence in other Old World monkeys, and, consequently, birth synchrony (Kummer, 1968). Among women living together in college dormitories, there are data pointing to menstrual synchrony.

Age-Graded Male Groups

This intermediate category between unimale and multimale systems was proposed because in some species, usually classified as multimale, the group does not contain more than one high-ranking male of a given age and there are no male alliances of the type seen in macaque and baboon societies. Instead, a dominant male tolerates younger males that in time may take the dominant rank or leave the group for another. This social system is seen most clearly in gorillas. Together with females and their young, there is usually only one socially mature "silver-backed" male in a group that controls group movements. The hair on the back changes from black to white with age. The silver-backed male is the undisputed dominant male and appears to have priority of access to sexually receptive females; mating is very rarely observed in the wild. There are typically one or two black-backed males peripheral to the group core, and also separate all-male bands. A similar kind of social system is seen in the hanuman langur (*Presbytis entellus*) in some regions of India, while in others they occur in unimale groups; it seems that low-ranking males may be tolerated within the group in more open habitats where predation poses a greater threat.

Multimale Groups

The most complex social system among nonhuman primates is that in which there is considerable tolerance between adult males, resulting in groups that contain several adult males together with adult females and their young. Among prosimians, the ring-tailed lemur (*Lemur catta*) and sifakas (*Propithecus verreauxi*) have this social system, which otherwise is largely restricted to some Old World primates, notably, macaques, most baboons, vervet monkeys, and chimpanzees. Adult males and females each have a dominance hierarchy, but alliances between individuals can make it difficult to recognize a linear ranking. Social groups are semiclosed; that is, one sex remains in the group for life, while the other sex disperses after adolescence (see below). In Old World monkeys, the group is matrilineal, that is, consists of several females and their female offspring, all of which remain in the group for life, whereas males disperse after puberty by transferring to other groups. There is a dominance hierarchy between matrilines as well as between the females of each matriline. In females, dominance rank is culturally inherited from the mother and is stable throughout life (Chapter 12). The ranks of males, on the other hand, fluctuate widely. Male rank is related to the mother's rank during the males' infancy due to her support, but during adolescence it drops to a very low level, and males then have to gain rank in their natal group by their own efforts and, subsequently, each time they emigrate into a new group. In baboons, the social group is tightly organized when progressing over open ground, a time of vulnerability to predation (Fig. 16-4). Adult males make up the front and the rear of the group; if threatened

FIGURE 16-4. Marching order of a troop of olive baboons. Dominant males accompany females with young infants in the center. Other adult males precede and follow them, while juveniles tend to be peripheral. Two estrous females (anogenital swelling shown in black) are each followed by a male. (SOURCE: Hall & DeVore, 1965)

by a predator such as a leopard, all males turn and take up the defense of the group. In contrast to monkeys, chimpanzees have a social system that is based on bonds between males, which remain in their natal community. In this case, it is the females that emigrate from their natal group and join a different one when they become sexually mature.

Mating Systems

In solitary species, females attract and mate with any male that is nearby or, as in orangutans, are mated during chance encounters. In species with male–female pairs, unimale, and age-graded male social systems, mating systems are straightforward, as described earlier for the sacred baboon. Mating systems in multimale groups vary somewhat with species. In rhesus monkeys, a male and female form temporary but exclusive pair bonds, termed *consort bonds*, staying together to feed, groom, and copulate, while distancing themselves somewhat from other group members. Consortships may last from a few hours to about 2 weeks. Either the male, the female, or both may then consort with another partner. Thus, a female may consort with one or more males during her midcycle period of maximal sexual attractiveness and receptivity but does so with only one male at a time. Although there are occasional "sneak copulations," field studies indicate that most completed copulations occur during such consortships. Baboons and vervet monkeys form similar consort bonds. In Barbary macaques (*M. sylvanus*), however, females mate with the majority of males in the group during every day of their fertile periods. The significance of this behavioral difference between two such closely related species is not clear.

Mate Competition and Mate Choice

Bateman's Principle and Trivers's Theory of Parental Investment predict that males, especially mammals, should compete for mates, while females should exercise mate choice (Chapter 13). Many studies have sought evidence for male mate competition in monkeys with multimale social systems by examining whether male dominance rank is positively correlated with mating frequency or paternity. Results have been equivocal; competition exists in some years or in some groups but not in others. However, results of a meta-analysis of data from many studies supported the existence of male mate competition in anthropoid primates (Cowlishaw & Dunbar, 1991). In laboratory studies, the existence of mate competition by both male and female macaques can be demonstrated in social settings in which the sex under investigation must compete because there is no other choice of partner. Thus, male mate competition is evident when there are two or more males but only one female (Fig. 16-5), and female mate competition is seen when there are several females but only one male (Fig. 16-6). This is what one might expect, namely, that competition implies choice for a particularly desirable partner, and that choice implies competition for a particularly desirable partner. The prediction that males compete and females choose is not absolute but a matter of statistical averages: Males are more *likely* to compete and females are more *likely* to be selective.

Parental Investment

As described in Chapter 13, the Trivers–Willard Hypothesis predicts that female mammals should adjust the sex ratio (SR) of their offspring according to prevailing conditions to produce the sex that is most likely to increase their inclusive fitness for a given amount of parental investment, generally, sons in good conditions and daughters in poor conditions. In the multimale social systems of baboons and macaques, based on

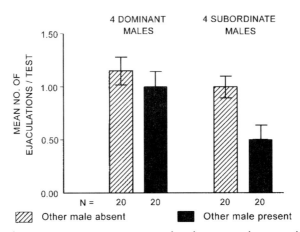

FIGURE 16-5. Male mate competition in pairs of male cynomolgus monkeys for a single female. Whereas mating frequency in dominant males was not affected by the presence of a caged subordinate male, that in subordinate males was greatly reduced in the presence of a caged dominant male. (SOURCE: Modified from Zumpe & Michael, 1990 Effects of the presence of a second male on pair-tests of captive cynomolgus monkeys (*Macaca fascicularis*): Role of dominance. Am. J. Primatol. 22, 145–148. Copyright © 1990 by John Wiley & Sons, Inc. Reprinted by permission of Wiley–Liss, Inc., a subsidiary of John Wiley & Sons, Inc.)

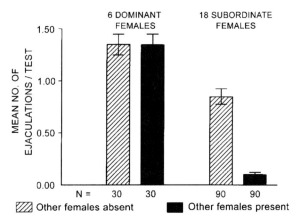

FIGURE 16-6. Female mate competition for a single male in groups of four female rhesus monkeys. Whereas mating frequency in the highest-ranking (dominant) females was not affected by the presence of three lower-ranking females, that of the lower-ranking females was suppressed in the presence of the other females including the dominant one. N = number of tests. (SOURCE: Modified from Michael & Zumpe, 1987, Relation between the dominance rank of female rhesus monkeys and their access to males. *Am. J. Primatol.* **13**, 155–169. Copyright © 1987 by John Wiley & Sons, Inc. Reprinted by permission of Wiley–Liss, Inc., a subsidiary of John Wiley & Sons, Inc.)

several matrilines of females that never leave their natal group, there is a rank order between matrilines as well as between individual females within a matriline. If a high-ranking female has a daughter, both are likely to benefit. The daughter acquires high rank for life with the help of her mother, and there is mutual aid between the mother and daughter in resource competition with other matrilines. High-ranking families are therefore likely to become larger and more powerful, resulting in increased inclusive fitness for their members. If a high-ranking mother has a son, on the other hand, his initial rank and mutual-aid advantage is lost after emigration from the natal group, as is the initial rank disadvantage of the son of a low-ranking mother. The lifetime reproductive success of both males will depend largely on their relationships with members of new groups and is unpredictable. In these social systems, then, the so-called local-resource-competition hypothesis would suggest that high-ranking mothers should raise more daughters than sons, but for species in which males remain in the natal group and females emigrate, the prediction would be the opposite, namely, that high-ranking females should have more sons than daughters. It is difficult to validate these predictions, but there is now some evidence for the predicted SR bias in two captive groups of rhesus monkeys (Nevison *et al.*, 1996) but none for yellow baboons (*Papio cynocephalus*) in the wild (Rhine *et al.*, 1992). There remains the difficult problem of identifying the proximate mechanisms by which mothers adjust SRs.

Dispersal and Inbreeding Avoidance

In many nonhuman primates, there are dispersals from the natal group around the time of puberty, and there are also subsequent dispersals. Their functions may include the reduction of infections, the reduction of competition for mates and other resources and, perhaps most importantly, the avoidance of inbreeding (Chapter 10). In species with solitary and male–female pair social systems, both sexes disperse at adolescence. In those with unimale social systems, the males disperse and groups that become too large split into smaller groups along matrilines. In age-graded male and multimale social systems, the dispersing sex varies with species. Both sexes may disperse from the natal group, perhaps primarily due to resource competition (howler monkeys); females may disperse (spider monkeys and chimpanzees), or males may disperse (baboons, macaques and vervet monkeys).

Information about possible proximate mechanisms underlying emigration comes from natal and subsequent dispersal in male macaques and olive baboons. Males are not forced out of a group but leave of their own volition. Over a period of weeks, they spend more time on the group periphery, then near the periphery of another group, and finally become integrated into the new group after mating with resident females. The importance of sexual interactions with females in the new group for acceptance of an immigrant male was demonstrated experimentally in captive male rhesus monkeys. They became integrated when introduced to a group of sexually active females during the fall mating season but not when introduced to a group of lactating females during the summer (Bernstein *et al.*, 1977).

Both anecdotal evidence and systematic studies point to a common and probably causal factor underlying male group transfers, namely, an effect of familiarity between potential partners that decreases their sexual interest in each other. Field observations on rhesus monkeys documented inhibition of mother–son mating (Sade, 1968); father–daughter relationships cannot be determined without DNA fingerprinting. Since then, other studies on macaques, baboons, and vervet monkeys have shown that (1) males transfer to new groups only at the start of the mating season in seasonally breeding species, (2) females are sexually attracted to unfamiliar males and facilitate their integration into the new group, and (3) there is a decline in the mating frequency of a male, and in the number of females with which he mates, from the first year of his residence in a group to the year he leaves again. The average length of residence in a group is about 4 years in rhesus monkeys, so that males leave a group when any daughters they may have sired during the first year become sexually mature. Laboratory studies have shown that the decline in sexual interest between long-familiar partners is independent of kinship and is a function of the time they have spent together. Males were paired with each of the same four ovariectomized, estrogen-treated females for 1 hour on 4 successive days each week for several years, during which their mating activity with these females eventually declined to low levels. This effect was immediately reversed when the females were replaced with four unfamiliar females for 4 weeks but evident again when the original females were reintroduced (Fig. 16-7). Another experiment showed that the familiarity effect does not depend on the mating interactions themselves. Males were housed individually for 1 year in full sight of each of their four future female partners but not allowed to come into tactile contact with them. When they were finally paired once for an hour with each female, sexual activity was much lower than when the same females were paired with males that had not had any visual contact with them in the preceding year (Fig. 16-8).

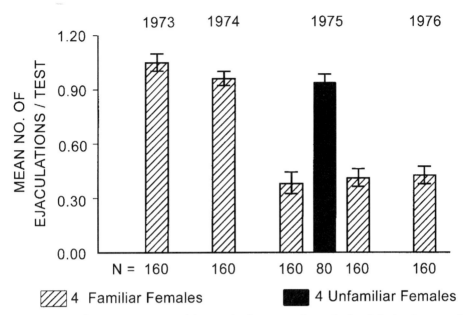

FIGURE 16-7. The mating activity of four male rhesus monkeys declined during 3 years of regular testing, a decline that was immediately but temporarily reversed when their familiar female partners were replaced by unfamiliar females. (SOURCE: Reprinted with permission from Michael & Zumpe, 1978b, Potency in male rhesus monkeys: Effects of continuously receptive females. *Science* **200**, 451–453. Copyright © 1978 by the American Association for the Advancement of Science)

Natal and subsequent dispersal by males in these species therefore appears to depend on a mechanism similar to the Westermarck effect in humans (Chapter 17). Based on the fact that females have more to lose by a single reproductive mistake (inbreeding) than do males, one might expect that the familiarity effect would be stronger in females than in males, and there is some evidence for this from a captive study on members of a group of Japanese macaques (*M. fuscata*) known for the high number of females involved in homosexual as well as in heterosexual consortships (Chapais & Mignault, 1991). The kin relationships of all individuals were known. While there were occasional heterosexual interactions between a female and a related male, there were no homosexual interactions between related females. In common chimpanzees, in which females disperse from the natal community while males remain, it has been observed that adolescent females showing their first large sexual skin swelling actively avoid the sexual advances of brothers and eventually leave to join a different community. As in macaques and baboons, sexual activity appears to serve as a passport into an unfamiliar social group of chimpanzees: Females without a sexual skin swelling may be attacked and even killed by males when they travel through an unfamiliar community.

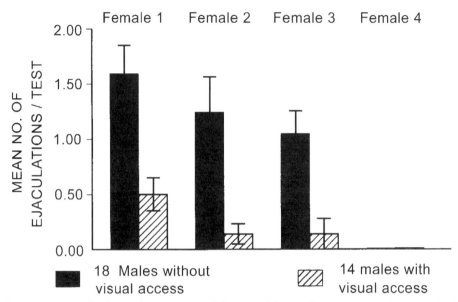

FIGURE 16-8. Males housed for 1 year in full view of their subsequent sexual partners had much lower levels of mating activity than did males without prior visual contact with the females. (SOURCE: Zumpe & Michael, 1984, Low potency of intact male rhesus monkeys after long-term visual contact with their female partners. *Am. J. Primatol.* **6**, 241–252. Copyright © 1984 by John Wiley & Sons, Inc. Reprinted by permission of Wiley–Liss, Inc., a subsidiary of John Wiley & Sons, Inc.)

Hormonal and Seasonal Influences

We confine ourselves here to the rhesus monkey, an Old World species for which a good deal is known about hormonal and seasonal influences on behavior. It is not widely recognized, even by those who should know better, that it is only the Old World primates that have a true menstrual cycle, which, in the case of the rhesus monkey, is about 28 days. Ovulation occurs near midcycle in both rhesus monkeys and humans.

Social Effects on Sexual Activity in Relation to the Menstrual Cycle

Because menstruation is difficult to detect in the field, laboratory studies were needed to establish the link between changes in the sexual activity of the pair and the stages of the female's menstrual cycle. When enough male–female pairs were observed in the absence of other conspecifics, a clear rhythmic change in copulatory activity was apparent, with a maximum occurring near the expected time of ovulation (Fig. 16-9). When radioimmunoassays became available for hormone measurements in blood, the precise relationship between behavior and hormone levels could be determined (Chapter 5). In pair tests, sexual activity does not cease or even decline to low levels at any time during the female's cycle. This supported the prevailing view that primates, unlike nonprimates, are largely "emancipated" from the effects of gonadal hormones on sexual

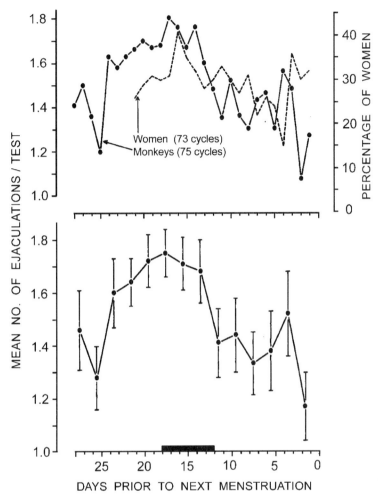

FIGURE 16-9. Top: Comparison of mating activity in rhesus monkeys and reports of sexual intercourse by women (Udry & Morris, 1968) in relation to the menstrual cycle. Bottom: Rhesus monkey data smoothed by using means of two consecutive days. Horizontal bar gives expected time of ovulation. (SOURCE: Michael & Zumpe, 1970a, with permission)

behavior. Subsequent studies on large captive groups containing many males and females clearly showed that female rhesus monkeys do, in fact, stop mating during the luteal phase of the cycle when under the influence of progesterone.

A possible explanation for the difference between pair and group observations is based on the prediction that males should try to copulate as much as possible, but females should try to be selective both with regard to the identity of their male partners and the timing of mating so as to coincide with ovulation. Thus, mating in pair tests would be a compromise between the opposing mating strategies of the male and female, with mating activity in the unfertile phases of the cycle resulting primarily from the male's attempts to maintain high levels of sexual behavior. In a group, the male can leave a luteal-phase female for a midcycle one, which results in more constant and less rhythmic mating by the male and more rhythmic mating by each female.

This prediction was examined in small groups, each consisting of one male with four ovariectomized females, two of which were given hormones to produce artificial menstrual cycles (Chapter 5). These cycles were either synchronized, or offset by 7-day intervals. During offset cycles, female mating activity was *more* rhythmic than during synchronized cycles, showing larger peaks when their own estradiol levels peaked, and lower troughs when the other females' estradiol levels peaked (Fig. 16-10). In contrast, male mating activity was *less* rhythmic in offset than in synchronized cycles (Fig. 16-11); indeed, the male mating pattern during synchronized cycles was similar to that when male–female pairs were tested alone (Fig. 16-9, bottom). These findings support the prediction and explain why female mating activity is less rhythmic in pairs than in social groups.

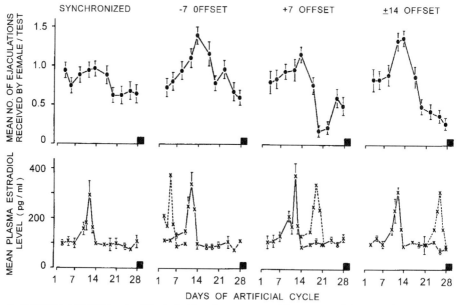

FIGURE 16-10. Differences in rhythmic patterns of ejaculations received by female rhesus monkeys ($N = 8$) during artificial menstrual cycles (top). Patterns varied in relation to changes in the timing of their plasma estradiol levels (bottom, solid lines) and the timing of the estradiol peaks of the other female in the group (bottom, broken lines). Behavioral changes were *more* rhythmic during offset than during synchronized cycles. Black squares give the onset of vaginal bleeding. (SOURCE: Michael & Zumpe, 1988, Determinants of behavioral rhythmicity during artificial menstrual cycles in rhesus monkeys (*Macaca mulatta*). Am. J. Primatol. 15, 157–170. Copyright © 1988 by John Wiley & Sons, Inc. Reprinted by permission of Wiley–Liss, Inc., a subsidiary of John Wiley & Sons, Inc.)

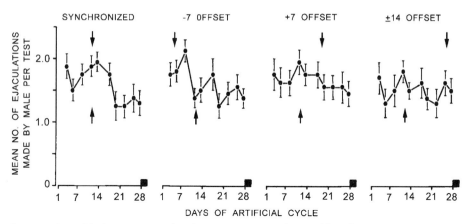

FIGURE 16-11. Mating patterns of males, in contrast to those of females, were *less* rhythmic during offset than during synchronized cycles. Arrows below indicate the day of the estradiol peak of the female whose cycle days are given on the x axis, and arrows above indicate the day of the estradiol peak of the other female. (SOURCE: Michael & Zumpe, 1988, Determinants of behavioral rhythmicity during artificial menstrual cycles in rhesus monkeys (*Macaca mulatta*). Am. J. Primatol. 15, 157–170. Copyright © 1988 by John Wiley & Sons, Inc. Reprinted by permission of Wiley–Liss, Inc., a subsidiary of John Wiley & Sons, Inc.)

Social Influences on Mating Seasonality

Like most other nonhuman primates, rhesus monkeys are seasonal breeders with a fall mating season and a spring birth season. The photoperiod is certainly involved, because monkeys maintained in the Southern Hemisphere begin to breed 6 months later than those in the Northern Hemisphere. Other exteroceptive factors such as temperature and rainfall may also contribute. Around September and October in Georgia (U.S.), and in the wild in India, the testes of males, which have been quiescent, producing practically no testosterone and no sperm since about March, suddenly begin to enlarge and produce both testosterone and sperm. This is associated with increased levels of aggression and of mounting activity toward females (Fig. 16-12). Female rhesus monkeys can breed once a year, but if impregnated late in the season, they may not breed during the next. Either way, they may undergo one or two menstrual cycles during the summer. In September and October, females begin regular menstrual cycles again and show proceptive and receptive behavior.

In addition to the photoperiod, social stimulation by sexually active animals also plays an important role in activating behavior. For example, the introduction of estrogen-treated females into a free-ranging group of rhesus monkeys during the nonmating season stimulated the sexual activity of males in the group and, within a few days, caused reddening of the scrotal skin, a sign of increased testosterone secretion. The onset of the mating season in that group was advanced by several weeks, suggesting that social cues accelerated the onset of ovulatory cycles (Vandenbergh & Drickamer, 1974). Social influences were also evident in laboratory studies. When male rhesus monkeys in Georgia were maintained in a constant photoperiod for several years and paired daily

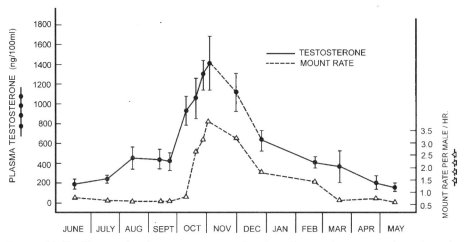

FIGURE 16-12. Changes in plasma testosterone levels and mounting rate per hour in male rhesus monkeys at the Field Station of the Yerkes Regional Primate Research Center in Georgia. (SOURCE: Gordon et al., 1976, with permission)

with ovariectomized, estrogen-treated females, their plasma testosterone levels remained high throughout the study, with peaks at approximately 13-month intervals (Chapter 7, Fig. 7-11) suggesting a free-running circannual rhythm, and mating activity continued throughout. In both studies, then, the presence of sexually receptive females appears to stimulate mating activity and plasma testosterone levels in males during the season when they would normally be low, and the end of the mating season may depend primarily on a lack of sexually receptive females due to pregnancy or lactation.

LANGUAGE IN APES

It is now generally recognized that there is a genetic, hardwired component to language acquisition and speech in humans. This is associated with morphological specializations, not possessed by apes, in the position and shape of the larynx, tongue, epiglottis, and palate, as well as the musculature around the mouth and the shape of the lips and teeth (Lenneberg, 1967). All this makes it possible for humans to *speak*. This was not recognized by early investigators, who attempted unsuccessfully to teach infant apes to speak by manipulating their mouths and lips to produce words. Later, the problem of phonation was circumvented by training chimpanzees and gorillas either to use American Sign Language or to use symbols, such as plastic chips of different shapes and colors, or geometric designs on computer keys denoting various objects, individuals, or actions. The results and interpretations of this work are still hotly debated, in part because most investigators worked with only one or two subjects. We summarize here the findings from a series of important studies by one group of investigators on several chimpanzees and bonobos (Rumbaugh & Gill, 1977; Savage-Rumbaugh, 1986).

Two undisputed characteristics of language ability are (1) stimulus equivalence, namely, the use of true symbols (words or signs) that represent objects, individuals, actions, and so forth, that permit reference to them even when they are not there, and (2) learning by exclusion, called "mapping" in children, namely, the ability to form novel sentences or phrases by combining symbols in the correct syntactical sequence so as to be readily understood by others. Apes, alone among nonhuman species, have demonstrated both of these language criteria. Evidence for the first criterion, showing that the symbols on computer keys, or "lexigrams," were not merely used as labels, was provided by two young chimpanzees, Austin and Sherman, that cooperated in solving problems by communicating with each other via computer keyboards. For example, one animal would have access to a set of tools, and the other could decide which of the six tools was needed to solve the problem. The second animal then used the keyboard to ask the first for the appropriate tool for the task and was given it; they were unable to solve the problem when the computer was shut off, showing that symbolic communication was necessary. Moreover, once Austin and Sherman had learned to equate 11 foods and 11 tools with their respective lexigrams, they could equate the lexigrams for the various foods with a lexigram for the concept of food, and the various tools with a lexigram for the concept of tool. On the first trial, they correctly categorized the various foods as "food" and the various tools as "tool." Evidence for the second language criterion includes spontaneous descriptions of novel items by Lana, another chimpanzee. When presented with an orange soda for the first time ever, she called it a "Coke which is orange," and not, for example, an "orange which is Coke," which has a different meaning.

Perhaps the most striking evidence for language ability comes from studies on bonobos, which are restricted to Zaire in Africa, where the remaining population is estimated to be less than 4,000. At Georgia State University's Language Research Center, directed by Duane Rumbaugh, which lies in a woodland area where investigators and their subjects can interact in a seminaturalistic environment, researchers attempted to train an adult, wild-caught female bonobo, Matata, to use a portable computer keyboard. Their efforts were unsuccessful; she learned just four lexigrams in 3 years. During this time, Matata was carrying an adopted infant male called Kanzi, to whom little attention was paid. It was not until Matata had to be confined indoors for a few days that the researchers realized that Kanzi had been learning language "silently," like children do, by observing humans communicate with each other with lexigrams and speech. Kanzi spontaneously began to use the keyboard to communicate with humans, and after some time, it became apparent that he could also understand human speech, a skill never learned by common chimpanzees. He acquired language capabilities comparable to those of a 2.5-year-old child. For example, Kanzi responded over 75% correctly to some 640 novel sentences giving him detailed spoken instructions even when the investigator wore a helmet to prevent unintentional cueing. When asked via headphones by someone in another room to give the researcher a photograph depicting a certain object or individual, Kanzi reliably picked the correct one from a group of photographs and gave it to the researcher, who could not know which photograph was requested. He demonstrated other remarkable cognitive abilities, such as being able to match a noise heard over headphones with a photograph of the animal or object making it (barking = dog, squealing of a deflating balloon = balloon), and using a joystick to run novel, computer-generated mazes at amazing speeds while trying to blow up a balloon.

Two aspects of these studies point to two similarities between humans and bonobos in the language-learning process. First, the fact that the adult Matata was incapable of acquiring any language skills, whereas the infant Kanzi did so without any training, suggests that there is a sensitive period for language acquisition in bonobos as there is in humans. In humans, learning the first language (and others) during infancy is a painless process and when spoken is perceived as accent-free by those from whom it is acquired. During late childhood and early adolescence, however, most people lose the ability to pick up accent-free new languages quickly: Vocabulary and grammar are learned with as much difficulty as a long poem or the Krebs cycle, and accent always betrays the foreigner. Second, in young bonobos, as in humans, language acquisition occurs silently: it does not require direct training of the type given to Matata, but it does need constant exposure to a language-rich environment. This was demonstrated by a controlled experiment with two of Matata's daughters. From infancy, one daughter, Panbanisha, was allowed to be with Kanzi and the investigators much of each day, and soon acquired language skills. In contrast, the other daughter, Tamuli, spent most of her time with Matata and was only occasionally exposed to language interactions; she could neither use lexigrams nor understand speech.

The studies on chimpanzees and bonobos suggest that these apes, while not possessing the anatomical apparatus for speech production, nevertheless have the cognitive abilities, at least in rudimentary form, for symbolic communication that is similar to our own (Geschwind & Levitsky, 1968). There is now some anatomical evidence supporting this. A recent study has shown that a region known as the planum temporale is larger on the left than on the right side of the chimpanzee brain. This region is situated deeply in the Sylvian fissure, in the posterior part of the superior temporal gyrus. The planum temporale is thought to be a homologue of Wernicke's language area in the human, which is also larger on the left than on the right side of the brain. This suggests that the anatomical substrate underlying human language may have been lateralized to the left hemisphere in the common ancestor of the chimpanzee and human some 8 million years ago (Gannon *et al.*, 1998).

In summary, there is great diversity among nonhuman primates, which range in size from the tiny mouse lemur to the huge gorilla and show similarly large differences in the degree of neocortical development, from relatively little in the insectivore-like tree shrews to very large in the great apes. Unfortunately, only a few primate species have been studied extensively both in their natural habitats and in the laboratory, and the chances of extending our knowledge to other species are slim because, although several new species have been discovered in recent years, sadly, many others, including bonobos and orangutans, are likely to become extinct in the wild before long, largely as a consequence of human activities such as habitat degradation and poaching. Nevertheless, the information summarized in this chapter indicates that even Old World nonhuman primates, with their increased cognitive abilities and behavioral flexibility resulting from increased neocortical development, exhibit many of the behavioral phenomena described earlier in connection with ethological concepts, sociobiological theories, and the hormonal regulation of sexual motivation and behavior. From this, one might suspect that the human primate is no exception, and the next chapter examines whether there is any evidence for similar behavioral predispositions in our own species.

CHAPTER **17**

Humans

HUMAN ETHOLOGY

Much of human ethology is concerned with nonverbal communication. In addition to language, humans may communicate their feelings and intentions by an estimated 100 or more postural and facial displays. Numbers are inexact because to be considered a display, that is, ritualized during evolution and genetically coded, it is necessary to show that a behavior is shared in common with all humans, independent of cultural differences. But, as described below, some behaviors occur in individuals deprived of the opportunity to learn them from others and are universal among many different and often geographically isolated populations. These observations point to genetic rather than cultural coding. Much of the credit for this painstaking work goes to Eibl-Eibesfeldt, who is regarded by many as the "father" of human ethology. He traveled and filmed human behavior throughout the world. To be certain that the behavior of people was not influenced by the knowledge that they were being filmed, he used a camera with a prism that allowed him to film people at right angles to the direction in which his head and the camera pointed. Consequently, he obtained many miles of footage on people who did not realize that they were being observed.

A few examples from several of the basic ethological concepts discussed previously are mentioned here.

FIGURE 17-1. Neonate's search for the mother's nipple is an FAP the human shares with all primates. (SOURCE: Eibl-Eibesfeldt, 1975, with permission)

Fixed Action Patterns (FAPs)

As mentioned (Chapter 2), humans and certain nonhuman primates share several behavior patterns, which are therefore likely to be FAPs. They include the side-to-side movements of the head used by the neonate to search for the mother's nipple, a behavior pattern that is common to all primates (Fig. 17-1), and smiling and laughing, which appear to be homologous with the "play-face" of the chimpanzees and bonobos (Fig. 17-2).

Evidence that human locomotion is based on FAPs comes from observations that it exists in complete form both before completion of the normal 9-month period of gestation, as shown by walking with minimal or no support in an infant born prematurely at 7 months (Fig. 17-3), and before the first attempts to walk, as shown by the patterns of kicking in full-term infants. Infants with congenital hip dysplasia, who have been maintained in casts to spread their legs for the first year of life, will walk within hours or days of release from the cast. This indicates that maturational changes, probably largely related to the proportionally high body mass of newborns, rather than the need for learning, delay the onset of walking until about the first year of life. This is analogous to the findings on flight development in chicks (Chapter 2).

The most persuasive evidence that expressions of emotion are coded genetically and require no learning comes from the study of children that are born both blind and deaf. Although deprived of the opportunity to learn behaviors such as smiling, laughing,

FIGURE 17-2. "Smiling" and "laughing" in a play context are FAPs that chimpanzees alone share with the human. (SOURCE: Eibl-Eibesfeldt, 1967. Copyright © 1967 by Piper Verlag GmbH, München)

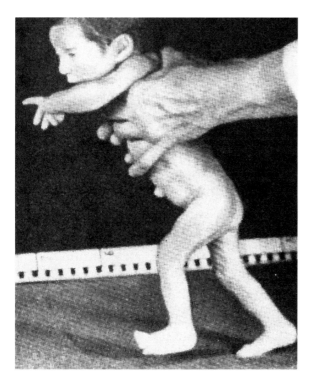

FIGURE 17-3. "Walking" by 7-month premature infant. (SOURCE: Photo by A. Peiper from Eibl-Eibesfeldt, 1975)

and crying from others, these children unmistakably show them in the appropriate context (Fig. 17-4). Another behavior that is almost certainly an FAP, because it occurs in the same form and in the same context in all cultures that have ever been studied, is the so-called "eyebrow flash." This occurs as a distance greeting, when two acquaintances make eye contact. There is a smile associated with a rapid raising and lowering of the eyebrows, held maximally high for about one-sixth of a second. We all do it quite unconsciously and respond in like manner when it is done to us, also quite unconsciously. In fact, if the person does not respond with an eyebrow flash, and it is extremely difficult not to do so, the original sender will repeat it, but more slowly—rather like raising one's voice to repeat a misunderstood statement.

Conflict Behaviors

Film analyses of people greeting their hosts at various parties and social functions have shown that greeting incorporates several behaviors that can be interpreted as intention movements and other forms of conflict behavior. As two individuals approach each other, they make eye contact and make the eyebrow flash (Fig. 17-5, panel b), but then mostly do not look at each other as they draw nearer until they are almost within touching distance (Fig. 17-6). This is reminiscent of monkeys avoiding eye contact as they approach each other. When within a few feet of each other, both individuals move an arm across their bodies (the "body-cross"), and when they are close and stationary,

FIGURE 17-4. Laughing by a 7-year-old girl born blind and deaf. (SOURCE: Eibl-Eibesfeldt, 1975, with permission)

they do not face each other directly but position themselves at an angle of about 90 degrees to each other (Fig. 17-5, panels d and e). The body-cross and the final position indicate the defensive and appeasement components contained in the greeting. This study also showed that closed-lipped smiles were directed equally at family members and at friends and acquaintances, whereas open-lipped smiles showing the teeth, a higher-intensity form of smile, were directed almost exclusively toward friends and acquaintances (Table 17-1), indicating an appeasement function. This, incidentally, strengthens the view that open-lipped smiling is homologous with the so-called fear grimace of many monkeys, including macaques.

Ritualization

There is also evidence, although not as solid as that for FAPs, that many behaviors have become ritualized into greetings, if not phylogenetically, then culturally. In some cultures, for example, Waika Indians, greeting involves a movement that is identical in form to the basic primate nipple-searching FAP of the neonate, and the nose-rubbing greeting of Eskimos involves a very similar movement. Some authorities consider that kissing may be derived from mouth-to-mouth feeding that occurs in many cultures between mother and infant, and between adults and aged, when foods need to be chewed before being given to individuals without teeth. An appeasement gesture that seems to have undergone cultural and perhaps phylogenetic ritualization is the posture that is commonly adopted by a subordinate individual toward a dominant one. This consists of

(a) p sights q

(b) p showing head toss display in distance salutation. Note the raised eyebrows and the open mouth smile.

(c) Head-dip following distance salutation. Note the eyes are now shut.

(d) p is now closer to q and is looking at her. Note that p has her head tilted to one side, an example of one of the "head-sets" characteristic of the final phase of a greeting encounter.

(e) Following the salutation p and q move out of their vis-a-vis orientation into one in which their bodies are set more nearly at 90°. Note that p has her arm across the front of her body, illustrating "body-cross".

FIGURE 17-5. Greeting sequence illustrated by traces from film records. (SOURCE: Modified from Kendon & Ferber, 1973, with permission)

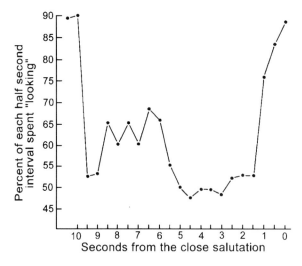

FIGURE 17-6. As two people approach each other, they first look at each other but then spend much less time doing so until they are within touching distance. (SOURCE: Kendon & Ferber, 1973, with permission)

bowing the head, kneeling, or total prostration, in order of increasing intensity. Phylogenetic ritualization is possible because some variant of this behavior is seen in the oldest depictions of human or human-like creatures, no matter what the culture. The dominant personages are generally thought to be kings or gods. So it seems to be an old and culturally very widespread behavior signaling subordination that reduces the probability of aggressive responses.

Releasers

As discussed earlier (Chapter 9), the tastes of salty, sweet, and fatty foods appear to be powerful releasers for eating, probably having evolved as a mechanism for ensuring the consumption of nutrients that were rare and difficult to find during much of human history. These tastes may well be supranormal stimuli, because we now consume salty, sweet, and fatty foods to excess. Another supranormal stimulus in the human is the "infant schema" first described by Lorenz. He drew attention to the fact that species or

Table 17-1. In Contrast to Lower-Intensity Close-Lipped Smiles, High-Intensity Open-Lipped Smiles Were Directed Mainly at Friends and Acquaintances Rather Than More Familiar Family Members

	Family	Friends/acquaintances
Closed smile	11	11
Open smile	1	20

SOURCE: Modified from Kendon & Ferber, 1973, with permission.

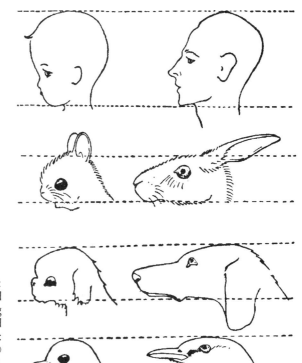

FIGURE 17-7. Top to bottom: Child, jerboa, Pekinese, and robin (left) are more appealing than man, hare, hound, and golden oriole (right). (SOURCE: Lorenz, 1965a. Copyright © 1965 by Piper Verlag, GmbH, München)

breeds of animals whose heads and especially foreheads were rounded, with large eyes set low in the face, were considered more appealing and "lovable" than species or breeds without these features (Fig. 17-7). Subsequent experiments showed that females of all ages, and postpubertal males, typically preferred less realistic, exaggerated versions of infantile heads, namely, supranormal infant schemata (Fig. 17-8). An increase in the height of the top of the cranium was found to be more important in this effect than an increase in the curvature and protrusion of the forehead. Infant schemata utilizing both variables have been used extensively in Walt Disney films (e.g., Bambi), in cartoons, in dolls (e.g., Barbie), and in advertising.

Apart from our responses to supranormal stimuli, we now have much information about our responses to the characteristics of others. Judgments are made quite unconsciously, in most cases on the basis of very simple features. We know specialized neurons in the superior temporal sulcus of the rhesus monkey brain are specialized for perceiving various aspects of the head and face (Chapter 3). People with damage in this brain region are incapable of recognizing familiar faces, including their own, a condition known as prosopagnosia. It is not surprising that facial expressions and facial morphology are foremost among the cues we use to judge both people and animals. People respond to the nose of the camel being higher than its eyes by judging it as "arrogant or aloof" (Fig. 17-9, left) because, in the human, this configuration reflects that attitude, as exemplified by the phrase "looking down one's nose" at someone. We see the bony

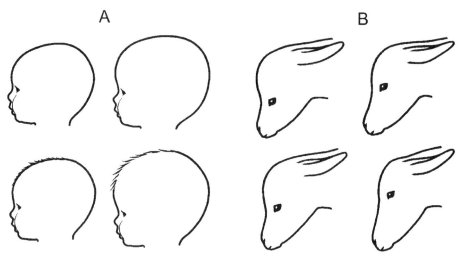

FIGURE 17-8. Although the outlines on the right of each pair in A and the ones on the left of each pair in B are exaggerations, they are typically considered "more cute" and preferred to the more realistic drawings. (SOURCE: Hückstedt, 1965. Copyright © 1965 by *Zeitschrift für Experimentelle und Angewandte Psychologie*, Hogrefe, Göttingen)

ridge above the eagle's eyes as a frown with lowered eyebrows and, together with the apparently pulled-back corners of the mouth (Fig. 17-9, right), as an expression of "proud decisiveness."

We also make character judgments on the basis of human facial characteristics. Keating and colleagues have conducted several studies on how people judge various facial characteristics and facial expressions. In one study, 5- to 7-year-old children were presented with pairs of photographs. Each pair showed the faces of two different people, for example, a man with raised eyebrows and another with lowered eyebrows, or one

FIGURE 17-9. We perceive the camel as arrogant-looking because its nose is higher than its eyes, and the eagle as proud and decisive due to the illusion of lowered eyebrows and drawn back corners of the mouth. (From Lorenz, 1965a. Copyright © 1965 by Piper Verlag, GmbH, München)

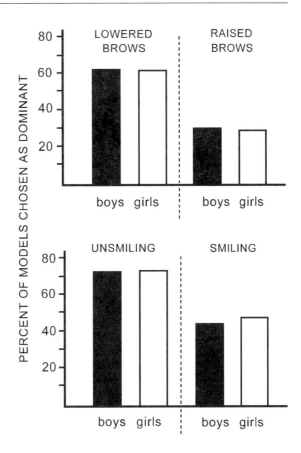

FIGURE 17-10. Young children judged photographs of models more dominant when they had lowered eyebrows or did not smile than when the same persons had raised eyebrows or smiled. (SOURCE: Figs. 5-4 and 5-5 in Keating, 1985. Copyright © 1985 by Springer-Verlag)

without a smile and the other smiling. To control for other, unrelated differences between the two people, their expressions were counterbalanced, so that a man shown smiling in one pair of photos was shown not smiling in another. The children were asked questions, phrased appropriately for their age, designed to elicit their perception of dominance versus submission. The results (Fig. 17-10) indicated that the children were more likely to perceive lowered eyebrows and an unsmiling face as reflecting social dominance. The finding that smiling was perceived as submission provides further support for the view that smiling is an appeasement gesture.

These kinds of judgments tend to become self-fulfilling prophesies, because they seem to influence how we are treated by others. It was further demonstrated that the facial changes that accompany maturation, namely, thicker eyebrows, thinner lips, and squarer jaws, are generally perceived to reflect dominance, leadership qualities, and wisdom. Four West Point cadet photos were ranked by psychology students in decreasing order of these facial characteristics. The lifetime service careers of these and other, similarly ranked cadets showed that those perceived as more dominant actually achieved higher service ranks, especially during the final 2 years at the Academy and in the later stages of their subsequent service careers (Mazur et al., 1984). Since their experiences later in life cannot have influenced their facial characteristics in youth, one is led to the view that those in authority over them must have been unconsciously influenced in their judgments and actions. This, in turn, gave a slightly positive spin to their careers in terms of better evaluations, earlier promotions, and selection for choicer commands.

Ethology in Clinical Settings

In view of these predictable behavioral interactions during normal interpersonal exchanges, ethological studies have been used to evaluate behavioral changes in psychiatric patients and to document responses to treatment. Small samples of patients diagnosed as having major depressive disorder, schizophreniform disorder, or personality disorder were observed and rated for some 100 behavior traits throughout their first 4 weeks in the hospital. There were quite distinctive differences in behavioral profiles between diagnostic categories, independent of whether or not patients improved with treatment as assessed by standard clinical criteria. For example, patients with personality disorder, unlike the other patient groups, decreased their social interactions during the study (McGuire & Polsky, 1980). There were also distinctive changes in behavior in improved but not in unimproved patients, irrespective of psychiatric diagnosis, and these were detectable as early as the second week in the hospital, before the clinical assessment of improvement. For example, improved patients directed more affiliative behavior toward, and increased their proximity to, others, and also received more assertive behavior from others (Polsky & McGuire, 1980). While ethological studies such as these might be useful in assessing early treatment efficacy, their costs in time and labor unfortunately preclude widespread clinical application.

Sensitive Periods, Imprinting

There is little solid evidence for imprinting in humans and the critical experiments would be unethical. However, language acquisition may involve imprinting-like processes similar to those in the acquisition of song by some birds, for example, male zebra finches (Chapter 3). Based on cross-cultural studies, all people appear to share in common much the same language features in terms of the basic semantics and grammar, suggesting that we are genetically preprogrammed with these universals. We then acquire the vocabulary, grammatical specifics, and pronunciation of the particular language to which we are exposed. In the absence of any exposure, a personal language may be invented by two or more individuals. For example, two Danish siblings, who were raised by a deaf-mute grandmother living in a remote rural area, were found to communicate verbally with each other in an unknown and, to others, incomprehensible language. When exposed to a language, small children rapidly acquire it and will very quickly understand a new language without benefit of translation. They rapidly acquire the vocabulary and the exact pronunciation of the speaker's dialect. Generally, this ability is lost around puberty, when it becomes increasingly difficult to understand a new language that must be painfully learned. It is extremely difficult by then to acquire the exact intonations and pronunciations, and speech will almost always be recognized by native speakers as that of a foreigner. All this points to a sensitive period for language acquisition, similar to that described in bonobos (Chapter 16, Language in Apes).

HUMAN SOCIOBIOLOGY

This is an even more challenging topic than animal sociobiology because we know for certain that most human beings no longer live in the environment that shaped their social and behavioral evolution. Consequently, in the following discussion, it is important to remember (1) that data interpretation involves much speculation for the reason given earlier; (2) that while the data may be solid for the culture from which they are derived, it should not be assumed that they would be similar in other cultures; and (3) that data are derived almost exclusively from correlational studies, which do not prove causal relationships. Moreover, humans can imitate and thereby pass on both vertically (down the generations) and horizontally (within generations) ideas, behaviors, pieces of information, instructions, beliefs, stories, and so forth. These, collectively termed "memes," may be true replicators (just like genes but totally independent of them) that are at least as important as genes in the evolution of our thoughts, attitudes, and actions (Blackmore, 1999).

Mating Systems

Extrapolating from the mating systems of nonhuman primates, it might be expected that human marriage systems would be flexible and, within certain limits, that polygyny would be the predominant form, monogamy a distant second, and polyandry vanishingly rare. In a sample of 849 human societies, 708 had polygynous marriage practices, 137 practiced monogamy, and only 4 were polyandrous (Murdock, 1967). Humans incorporate in a single species these three types of mating system in approximately similar proportions to those found in mammals in general.

Mate Competition and Mate Choice

As predicted, there is more variance in the reproductive success of men than of women, and this is illustrated by data from the polygynous Xavante Indians of Brazil (Fig. 17-11). In the Temne people of Sierra Leone, West Africa, male reproductive success increased with increasing polygyny (Fig. 17-12). Note, however, that it was less than would be expected from successive monogamous marriages with the corresponding number of wives. Female reproductive success did not increase, and even decreased somewhat. This suggests that paternal care (care by men) is important for fetal and child survival, a view supported by findings from Western societies that infant mortality before 1 year of age is higher in families without the father present than in those with the father present. So there seems to be a conflict of interests between men and women in terms of the mating system maximizing their reproductive success.

This conflict between male and female mating strategies is also evident in sex differences in premarital and extramarital coitus. In a 1972 study, married American men and women reported whether or not they had engaged in premarital and extramarital coitus: Both behaviors were reported more frequently by men than by women (Daly & Wilson, 1978) (Fig. 17-13). A change in societal norms for premarital sexual behavior by women occurred after World War II, which may be reflected here: 30–40% of older women reported premarital coitus, while 80% of the youngest women did so. Reported

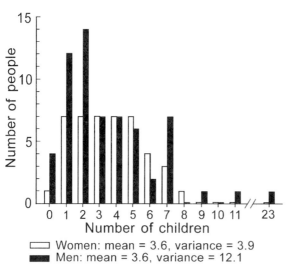

FIGURE 17-11. Among the polygynous Xavante Indians of Brazil, variance in the reproductive success of men was higher than in women. (SOURCE: Data from Salzano et al., 1967. Copyright 1967 by the University of Chicago Press)

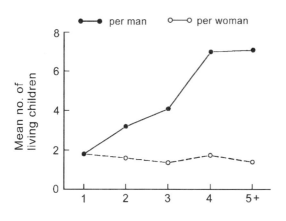

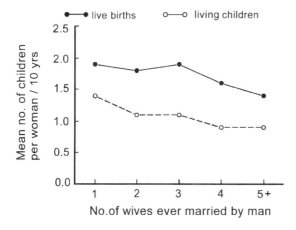

FIGURE 17-12. Polygyny and reproductive success in Temne people. Whereas the reproductive success of men increased with increasing polygyny, that of women decreased. (SOURCE: Data from Dorjahn, 1958)

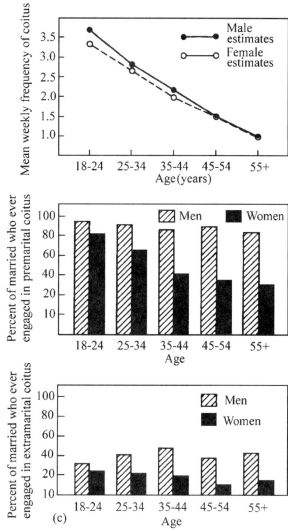

FIGURE 17-13. Gender differences in premarital and extramarital coitus reported by American adults in the 1970s. (SOURCE: Daly & Wilson, 1978, *Sex, Evolution and Behavior,* 1st ed. Copyright © 1978. Reprinted with permission of Brooks/Cole, a Division of Thomson Learning)

weekly coital frequency declined markedly with age in both sexes and was slightly greater in young men than in young women. The predicted male–female difference is more clearly apparent in customers seeking prostitutes and pornography, and in numbers of sexual partners reported by homosexual men and homosexual women. Since males can never be sure about their paternity, it is far more important for them than for women that their mates should be virgins *before* marriage and faithful *during* marriage. For the female, on the other hand, the threat of promiscuity by the mate relates primarily to his status as a provider for her and her offspring. It is devastating if his promiscuity leads him to desert her, that is, to break the pair bond, but not if he engages in "sneak copulations," as do nonhuman primates. This sex difference is apparent, for example, in the attitudes of German schoolchildren. Between ages 11 and 16 years, the rigidity of attitudes about the need for virginity in the marital partner declined in both boys and girls, but consistently more boys than girls retained this attitude at every age (Fig. 17-14, left), and from puberty onward, the genders diverged along similar lines when asked less polarized questions about the desirability of previous sexual experience in the marriage partner (Fig. 17-14, right). In all cases, boys were more likely than girls to prefer virgin spouses.

Higher rank in mammals translates into greater access to essential resources, which

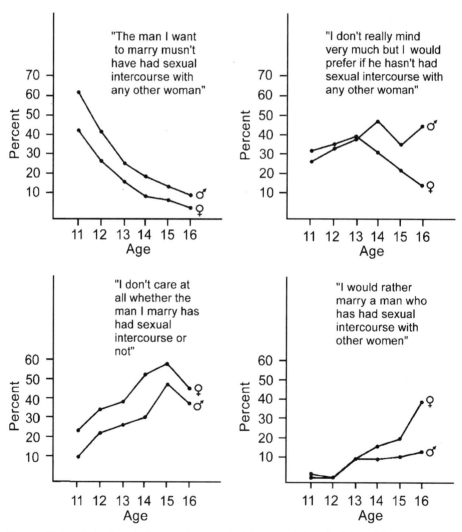

FIGURE 17-14. Attitudes of German boys and girls toward the virginity of future spouses. Questions were in a form appropriate for the sex of each child; answers given here are in the form used by girls. (SOURCE: Schoof-Tams et al., 1976, with permission)

in turn translates into higher reproductive success, especially for higher-ranking males that can secure more mates. This is also evident in the human, for example, in records of marital and extramarital reproductive success of Portuguese nobility in the Middle Ages. The same phenomenon appears in data from a polygynous society, the Yomuts of Iran, which also show greater variance in the reproductive success of men than of women (Fig. 17-15). This relationship between wealth and reproductive success no longer holds directly in modern industrial societies, where there tends to be a negative relationship between income and fertility. This is probably because of the advent of family planning, including birth control and abortion, which is more readily available for the better-off segments of the population. But among a sample of Canadian men (Fig. 17-16), the relationship can still be seen in a positive correlation between income and numbers of sexual partners, which historically would have translated into offspring.

The attitudes of German schoolchildren show that women seem attracted to more successful men, who would be able to secure more partners and be older and financially better-off, and thus likely to be better providers than younger men. Men, on the other hand, tend to be attracted to virgins, who would be younger, with a longer reproductive life ahead. This gender difference in mate choice was apparent in newspaper advertisements by men and women, in which the advertisers gave their own age and listed the age range of the desired marital partners being sought (Fig. 17-17).

Parental Investment

Another hypothesis follows from the fact that there is more variance in the reproductive success of males than that of females. The Trivers–Willard Hypothesis states that parents should invest more in daughters when conditions are poor and in sons

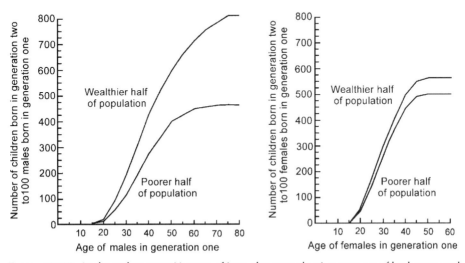

FIGURE 17-15. In the polygynous Yomuts of Iran, the reproductive success of both men and women increases with age, more so for men in the wealthier half of the population. (SOURCE: Modified from Irons, 1979, with permission)

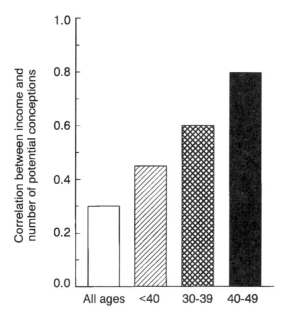

FIGURE 17-16. The positive correlation between income and number of potential conceptions during the preceding year for unmarried French Canadian men increased as they became older. (SOURCE: Modified from Pérusse, 1993, with permission from Cambridge University Press)

when conditions are good. In nonhuman species, it is thought this is expressed in the sex ratio (SR) of offspring (to understand this, see Chapter 13). In humans, the hypothesis is supported by whether sons or daughters receive the greater part of the family's inheritance. In cross-cultural comparisons, inheritance favored sons in 58% of 62 monogamous cultures, in 80% of 47 limited polygynous cultures (with less than 20% of men polygynous), and in 97% of 9 general polygynous cultures (with more than 20% of men polygynous). Thus, the greater the potential reproductive success of sons compared with that of daughters as polygyny increases, the more likely it is that sons are favored in the provision of material resources. Similar trends are evident in the relation between wealth and inheritance as documented by the wills of Canadians (Smith et al., 1987). Daughters were significantly favored over sons in wills amounting to $20,000 or less, while sons were significantly favored in wills amounting to $111,000 or more (Fig. 17-18). It is noteworthy that only the proportion allocated to sons changed consistently with the amount of the inheritance.

Because of the huge increases in divorce and remarriage in recent decades, many households contain a stepparent. Considering the costs of parental investment, it might be expected that there would be poorer parental care by stepparents than by biological parents, and there is some evidence for this, quite apart from the proverbial "wicked stepmother." There is more aggressive conflict between men and their stepchildren than between men and their biological children. Moreover, there are more reports of physical child abuse in households with a stepparent and a biological parent than in households with two biological parents (Fig. 17-19), and the perpetrators of infanticide are more likely to be stepparents than biological parents (Fig. 17-20).

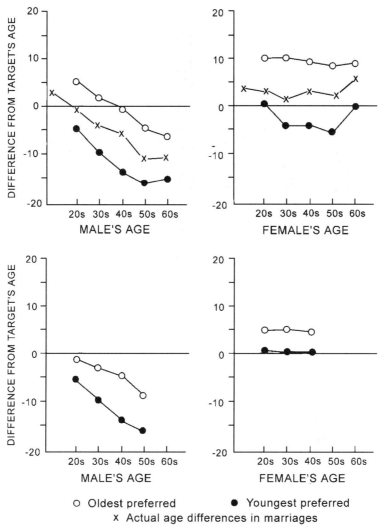

FIGURE 17-17. Gender differences in the preferred age of potential partners relative to the advertiser's age in newspaper advertisements in Arizona (top) and India (bottom). Men preferred younger partners while women preferred same-aged or older partners. (SOURCE: Modified from Kenrick & Keefe, 1992, with permission from Cambridge University Press)

Incest Avoidance

Males of many monkey species leave their natal group in adolescence and transfer into new groups about every 3–4 years, while female chimpanzees leave their natal community for another around puberty. In every case, integration into the new group appears to depend on sexual interactions with resident group members (Chapter 16). In humans, acceptance into some gangs and religious cults also depends on sexual inter-

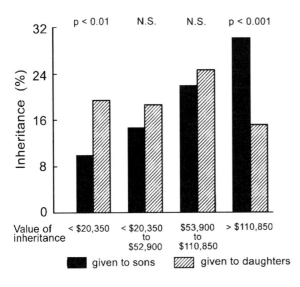

FIGURE 17-18. Relation between wealth and gender bias in inheritance decisions as reflected in wills. Daughters were significantly favored over sons in the lowest quartile of inheritance values (given in Canadian dollars) whereas sons were significantly favored over daughters in the highest quartile. (SOURCE: Data from Smith, Kish, & Crawford, 1987, Inheritance of wealth as human kin investment. *Ethol. Sociobiol.*, **8**, 171–182. Copyright 1987, with permission from Elsevier Science)

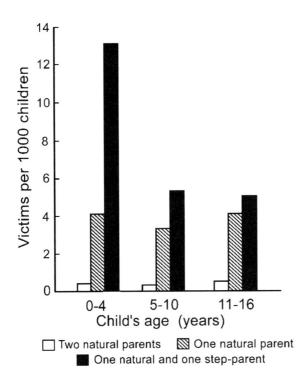

FIGURE 17-19. Abuse of children less than 5 years old in the United States is far more likely in households with a stepparent than in households with one or two biological parents. (SOURCE: Modified from Daly & Wilson, 1985, Child abuse and other risks of not living with both parents. *Ethol. Sociobiol.*, **6**, 197–210. Copyright © 1985, with permission from Elsevier Science)

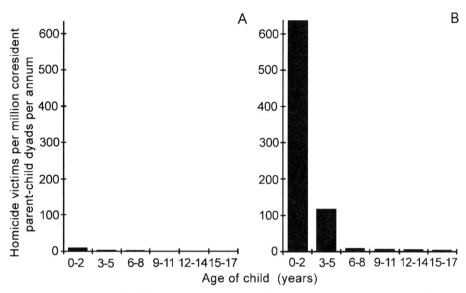

FIGURE 17-20. Rates of child homicide victimization by genetic parents (A) were vanishingly small compared with those by stepparents (B) throughout 10 years in Canada. (SOURCE: Adapted with permission from Martin Daly and Margo Wilson, 1988, *Homicide*. New York: Aldine de Gruyter. Copyright © 1988 by Aldine de Gruyter)

actions with group leaders or other members, a procedure called "sexing-in" in some teenage gangs. In nonhuman primates, natal and subsequent dispersal often seems to depend on a proximate mechanism involving a reduction in sexual interest between long-familiar partners. A study on divorces in 58 cultures between 1974 and 1981, based on the Demographic Yearbooks of the United Nations, showed that serial pair bonding is common; during the prime reproductive years, a high rate of divorce occurred at a modal interval of 4 years, especially when there were two or fewer children (Fisher, 1989). It was concluded that this reflects a hominid evolutionary adaptation to ensure infant survival until weaning. In hunter–gatherers, weaning occurred at about 4 years of age, and still does in !Kung bushmen of Africa, Yanomamo Indians of South America, and Netsilik Eskimo, among others. The high rate of divorce after 4 years is reminiscent of troop transfers by male macaques about every 4 years.

In humans, biologically determined behavioral tendencies are sometimes formalized in moral and ethical codes, taboos and laws. Incest taboos are ubiquitous, although the degree of kinship that defines incest varies between cultures. There is evidence that incest taboos are an enculturation of a biological mechanism that reduces inbreeding. Westermarck (1891) was the first to suggest that long-term familiarity between infants and young children who grow up together in the same household, usually siblings, might be responsible for the marked lack of subsequent romantic interest between them (Westermarck Hypothesis). In some cultures, for example, in China, parents sometimes arrange marriages between their children when they are still very young, called "minor" marriages, and the girl is transferred into the boy's household to grow up with the prospective bridegroom. Compared with "major" marriages between a man and woman who fall in love as adults, "minor" marriages are associated with many more problems arising from the lack of commitment between spouses, including higher rates of adultery, separation, and divorce (Wolf, 1970).

Perhaps most striking are data from Israeli kibbutzim, where children were raised in groups of four unrelated individuals, two girls and two boys, from an early age. Infantile and juvenile sexual play was not discouraged in the interests of "repression-free" sexual development, but of some 4,000 adults who were raised in this way, there was not a single marriage between members of the same childhood group, despite the fact that no incest would have been involved and the parents would often have favored such marriages (Shepher, 1971). The individuals reported that they loved their group members dearly, but exactly as they loved their siblings. They considered romantic involvement unthinkable, saying that it would have felt like "falling in love with a brother or sister." In the great tragedies involving incest (e.g., Sophocles's *Oedipus Rex*), the two individuals are described as having been separated in the infancy of one or both until they meet again as adults, and, only after falling in love, discover their blood relationship.

Hormonal and Seasonal Influences

Detecting these influences on the behavior of humans is confounded by socioeconomic, demographic, cultural, and other factors, such as work and vacation schedules, religious influences and taboos, as well as partial insulation from temperature, photoperiod, and food cycles.

Hormone-dependent variation in the sexual motivation of women has been the subject of many studies over the last 20 years, mostly on presumptively monogamous heterosexual couples, but also on lesbian couples. The expected midcycle increase in sexual activity has been slightly easier to detect in lesbian than in heterosexual women, perhaps because sexual activity in the latter reflects a compromise between differences in male and female mating strategies (Chapter 16). A midcycle increase in measures of sexual motivation, such as sexual thoughts or interest, is easier to document than a midcycle increase in sexual activity, especially when data can be aligned by the midcycle estrogen peak (Fig. 17-21).

Seasonal changes in serum testosterone levels of over 44,000 American veterans closely paralleled those in sperm concentrations determined by meta-analysis of eight studies conducted in North America and Europe (Fig. 17-22). While the amplitude of the changes is much reduced compared with seasonally breeding nonhuman primates (e.g., rhesus monkeys), the timing is similar, with higher levels occurring during fall and winter than during spring and summer. It is doubtful whether small hormonal changes have any behavioral consequences, but hormonal changes may be more robust, and perhaps of behavioral significance, in cultures less insulated from seasonal changes in exteroceptive factors such as day-length and temperature.

However this may be, the timing of seasonal changes in sexual activity, as expressed by seasonal changes in conceptions and births, varies greatly in different parts of the world. Nevertheless, even in the modern United States, changes in the frequency of some violent and sexual crimes are linked to seasonal changes in ambient temperature. In the 1840s, Adolphe Quetelet, the father of "moral statistics," formulated the so-called

HUMANS

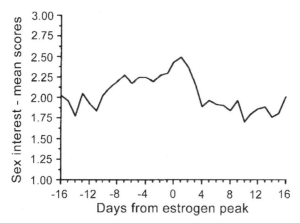

FIGURE 17-21. Mean scores of sexual interest by women (N = 160) increased during the follicular phase of the menstrual cycle, reached a clear maximum on the day after the estrogen peak (expected day of ovulation), and decreased during the luteal phase. (SOURCE: Reprinted from Dennerstein et al., 1994, The relationship between the menstrual cycle and female sexual interest in women with PMS complaints and volunteers. *Psychoneuroendocrinology* **19**, 293–304. Copyright © 1994, with permission from Elsevier Science)

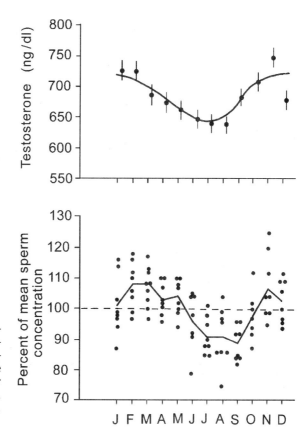

FIGURE 17-22. Serum testosterone levels (top) and sperm concentrations (bottom) of men living in the Northern Hemisphere were higher during the colder than during the warmer months. (SOURCE: Modified from Levine, 1994, with permission)

thermic law of crime, which states that crimes against the person increase, but crimes against property decrease, with seasonal and geographic increases in temperature. Thus, in preindustrial Europe, murders, assaults, rebellions, and various sexual crimes, including rape, all showed sharp peaks during summer and increased from northern to southern locations. On the other hand, robberies, burglaries, forgeries, and larceny all peaked in winter and increased from southern to northern locations. According to data obtained from recent FBI Uniform Crime Reports of sixteen different locations in the United States, annual rhythms in aggravated assaults and rapes are still closely correlated both with each other and with local annual temperature rhythms (Fig. 17-23). Robberies, in contrast, tend to be inversely related to temperature. Murder shows no significant

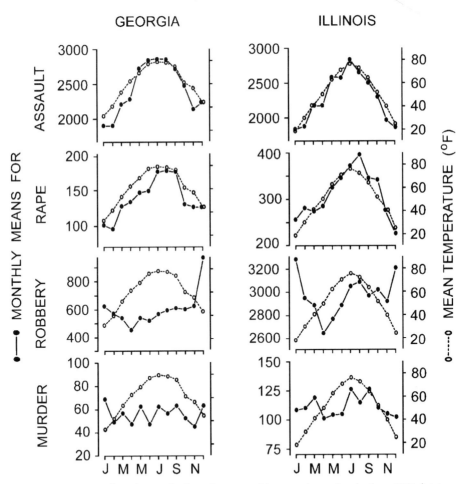

FIGURE 17-23. Assault and rape rhythms, but not robbery and murder rhythms (FBI data) are closely correlated with local temperature rhythms in Georgia (left) and Illinois (right). (SOURCE: Michael & Zumpe, 1983, Sexual violence in the United States and the role of season. *Am. J. Psychiat.* **140**, 883–886. Copyright © 1983 by the American Psychiatric Association. Reprinted with permission)

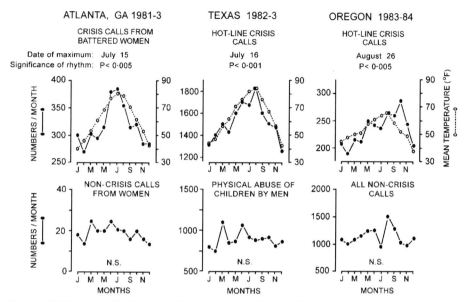

FIGURE 17-24. Annual changes in hotline crisis calls from battered women or their friends are closely correlated with local annual temperature rhythms in three states (top), but control calls to the same women's shelters showed no seasonal changes (bottom). (SOURCE: Michael & Zumpe, 1986, An annual rhythm in the battering of women. *Am. J. Psychiat.* **143**, 637–640. Copyright © 1986 by the American Psychiatric Association. Reprinted with permission)

seasonal rhythm, probably because many crimes against property, such as robbery, now often culminate in a fatal shooting. A very similar phenomenon has been documented for spouse abuse, as measured by crisis calls from battered women (Fig. 17-24). There is now no longer any evidence that these crimes occur more frequently in overall warmer than in cooler locations.

As noted at the outset, it is difficult to verify the existence of ethological and sociobiological phenomena in humans because data are generally correlational, and experimental techniques applied to animals cannot be used with humans. Most people no longer live under the conditions that presumably shaped their behavioral evolution, and early experiences in cognitively highly evolved species can result in a variety of behavioral effects and coping responses later in life that differ from the norm. Nevertheless, there is compelling evidence for the existence of phenomena such as FAPs, conflict behaviors, supranormal releasers, and imprinting. Until many more studies on different cultures and larger samples are available, caution is obviously required when interpreting human data that appear to support sociobiological predictions. Some might argue that the observed phenomena result from purely cultural and social influences that shape uniquely human attitudes and behavior. While such influences are clearly powerful, both theoretical considerations and available evidence suggest that they, too, reflect evolved psychological and behavioral predispositions. This view, although controversial, has led to the growing new academic discipline of evolutionary psychology, which is providing important insights into many aspects of human behavior, including reproductive behavior (LeCroy & Moller, 2000).

References

Aisner, R. & Terkel, J. (1992). Ontogeny of pine cone opening behaviour in the black rat, *Rattus rattus*. *Anim. Behav.* **44**, 327–336.
Alcock, J. (1979). *Animal Behavior: An Evolutionary Approach*, 2nd ed. Sinauer Associates: Sunderland, MA.
Alcock, J. (1993). *Animal Behavior: An Evolutionary Approach*, 5th ed. Sinauer Associates: Sunderland, MA.
Allen, L. S. & Gorski, R. A. (1990). Sex difference in the bed nucleus of the stria terminalis of the human brain. *J. Comp. Neurol.* **302**, 697–706.
Amoroso, E. C. & Marshall, F. H. A. (1960). External factors in sexual periodicity. In: *Marshall's Physiology of Reproduction*, Vol. 1, Part 2 (ed. A. S. Parkes), pp. 707–831. Longmans Green & Co.: London.
Anand, B. K. & Brobeck, J. R. (1951). Hypothalamic control of food intake in rats and cats. *Yale J. Biol. Med.* **24**, 123–140.
Andersson, M. (1982). Female choice selects for extreme tail length in a widowbird. *Nature* **299**, 818–820.
Andrew, R. J. (1961). The motivational organisation controlling the mobbing calls of the blackbird (*Turdus merula*): I. Effects of flight on mobbing calls. *Behaviour* **17**, 224–246.
Aschoff, J. (1978). Circadiane Rhythmen im endocrinen System. *Klin. Wochenschr.* **56**, 425–435.
Aschoff, J. (1981). A survey on biological rhythms. In: *Handbook of Behavioral Neurobiology: Vol. 4. Biological Rhythms* (ed. J. Aschoff), pp. 3–10. Plenum Press: New York.
Aschoff, J., Giedke, H., Poppel, E. & Wever, R. (1972). The influence of sleep-interruption and sleep deprivation on circadian rhythms in human performance. In: *Aspects of Human Efficiency* (ed. W. P. Colquhoun), pp. 135–150. English University Press: London.
Austad, S. N. & Sunquist, M. E. (1986). Sex-ratio manipulation in the common opossum. *Nature* **324**, 58–60.
Axelrod, R. & Dion, D. (1988). The further evolution of cooperation. *Science* **242**, 1385–1390.
Axelrod, R. & Hamilton, W. D. (1981). The evolution of cooperation. *Science* **211**, 1390–1396.
Babický, A., Ošťádalová, I., Pařízek, J., Kolář, J. & Bíbr, B. (1970). Use of radioisotope techniques for determining the weaning period in experimental animals. *Physiol. Bohemoslov.* **19**, 457–467.
Baerends, G. P., Brouwer, R. & Waterbolk, H. T. (1955). Ethological studies on *Lebistes reticulatus* (Peters): I. An analysis of the male courtship pattern. *Behaviour* **8**, 249–334.

Bailey, J. M. & Pillard, R. C. (1991). A genetic study of male sexual orientation. *Arch. Gen. Psychiatry* **48**, 1089–1096.
Bateman, A. J. (1948). Intra-sexual selection in *Drosophila*. *Heredity* **2**, 349–368.
Bateson, P. & Horn, G. (1994). Imprinting and recognition memory: A neural net model. *Anim. Behav.* **48**, 695–715.
Beach, F. A. (1948). *Hormones and Behavior*. Paul B. Hoeber, Inc.: New York.
Bentley, D. R. & Hoy, R. R. (1972). Genetic control of the neuronal network generating cricket (*Teleogryllus gryllus*) song patterns. *Anim. Behav.* **20**, 478–492.
Bernstein, I. S. (1987). The evolution of nonhuman primate social behavior. *Genetica* **73**, 99–116.
Bernstein, I. S., Rose, R. M. & Gordon, T. P. (1977). Behavioral and hormonal responses of male rhesus monkeys introduced to females in the breeding and nonbreeding season. *Anim. Behav.* **25**, 609–614.
Bierens de Haan, J. A. (1940). *Die tierischen Instinkte und ihr Umbau durch Erfahrung: Eine Einführung in die allgemeine Tierpsychologie*. E. J. Brill: Leiden.
Blackmore, S. (1999). *The Meme Machine*. Oxford University Press: Oxford, UK.
Bonsall, R. W., Rees, H. D. & Michael, R. P. (1989). Identification of radioactivity in cell nuclei from brain, pituitary gland and genital tract of male rhesus monkeys after the administration of [^{3}H]testosterone. *J. Steroid Biochem.* **32**, 599–608.
Bonsall, R. W., Zumpe, D. & Michael, R. P. (1978). Menstrual cycle influences on operant behavior of female rhesus monkeys. *J. Comp. Physiol. Psychol.* **92** 846–855.
Bovet, D., Bovet-Nitti, F. & Olivario, A. (1969). Genetic aspects of learning and memory in mice. *Science* **163**, 139–149.
Brobeck, J. R., Tepperman, J. & Long, C. N. H. (1943). Experimental hypothalamic hyperphagia in the albino rat. *Yale J. Biol. Med.* **15**, 831–853.
Brown, C. R. & Brown, M. B. (1986). Ectoparasitism as a cost of coloniality in cliff swallows (*Hirundo pyrrhonota*). *Ecology* **67**, 1206–1218.
Brown, C. R. & Brown, M. B. (1989). Behavioural dynamics of intraspecific brood parasitism in colonial cliff swallows. *Anim. Behav.* **37**, 777–796.
Bruce, H. M. (1959). An exteroceptive block to pregnancy in the mouse. *Nature* **184**, 105.
Butenandt, A. (1955). Über Wirkstoffe des Insektenreiches. II. Zur Kenntnis der Sexual-Lockstoffe. *Naturwiss. Rdschau* **12**, 457–464.
Chapais, B. & Mignault, C. (1991). Homosexual incest avoidance among females in captive Japanese macaques. *Am. J. Primatol.* **23**, 171–183.
Clancy, A. N. (1989). Neural and pheromonal regulation of gonadotropins in males. *Mol. Androl.* **1**, 275–297.
Clutton-Brock, T. H., O'Riain, M. J., Brotherton, N. M., Gaynor, D., Kansky, R., Griffin, A. S. & Manser, M. (1999). Selfish sentinels in cooperative mammals. *Science* **284**, 1640–1644.
Coplan, J. D., Andrews, M. W., Rosenblum, L. A., Owens, M. J., Friedman, S., Gorman, J. M. & Nemeroff, C. B. (1996). Persistent elevations of cerebrospinal fluid concentrations of corticotropin-releasing factor in adult nonhuman primates exposed to early-life stressors: Implications for the pathophysiology of mood and anxiety disorders. *Proc. Natl. Acad. Sci.* **93**, 1619–1623.
Cowlishaw, G. (1997a). Refuge use and predation risk in a desert baboon population. *Anim. Behav.* **54**, 241–253.
Cowlishaw, G. (1997b). Trade-offs between foraging and predation risk determine habitat use in a desert baboon population. *Anim. Behav.* **53**, 667–686.
Cowlishaw, G. & Dunbar, R. I. M. (1991). Dominance rank and mating success in male primates. *Anim. Behav.* **41**, 1045–1056.
Crews, D. (1987). Diversity and evolution of behavioral controlling mechanisms. In: *Psychobiology of Reproductive Behavior: An Evolutionary Perspective* (ed. D. Crews), pp. 88–119. Prentice-Hall: Englewood Cliffs, NJ.
Czeisler, C. A., Duffy, J. F., Shanahan, T. L., Brown, E. N., Mitchell, J. F., Rimmer, D. W., Ronda, J. M., Silva, E. J., Allan, J. S., Emens, J. S., Dijk, D.-J. & Kronauer, R. E. (1999). Stability, precision, and near-24-hour period of the human circadian pacemaker. *Science* **284**, 2177–2181.
Daly, M. & Wilson, M. (1978). *Sex, Evolution and Behavior*. Duxbury Press: North Scituate, MA.
Daly, M. & Wilson, M. (1985). Child abuse and other risks of not living with both parents. *Ethol. Sociobiol.* **6**, 197–210.
Daly, M. & Wilson, M. (1988). Evolutionary social psychology and family homicide. *Science* **242**, 519–524.

REFERENCES

Darwin, C. (1839). *The Voyage of the "Beagle"*. Reprinted 1960 by J. M. Dent & Sons: London.
Darwin, C. (1859). *On the Origin of Species*. Murray: London.
Darwin, C. (1871). *The Descent of Man and Selection in Relation to Sex*. Murray: London.
Darwin, C. (1872). *The Expression of the Emotions in Man and Animals*. Reprinted 1965 by University of Chicago Press: Chicago.
Davies, N. B. (1983). Polyandry, cloaca-pecking and sperm competition in dunnocks. *Nature* **302**, 334–336.
Dawkins, R. (1976). *The Selfish Gene*. Oxford University Press: New York.
Dawkins, R. (1980). Good strategy or evolutionarily stable strategy? In: *Sociobiology: Beyond Nature/Nurture? Reports, Definitions and Debate* (eds. G. W. Barlow & J. Silverberg), pp. 331–367. Westview Press: Boulder, CO.
Dennerstein, L., Gotts, G., Brown, J. B., Morse, C. A., Farley, T. M. M. & Pinol, A. (1994). The relationship between the menstrual cycle and female sexual interest in women with PMS complaints and volunteers. *Psychoneuroendocrinology* **19**, 293–304.
DeWitt, T. J. (1996). Gender contests in a simultaneous hermaphrodite snail: A size advantage model for behaviour. *Anim. Behav.* **51**, 345–351.
Dewsbury, D. A. (1972). Patterns of copulatory behavior in male animals. *Quart. Rev. Biol.* **47**, 1–33.
Dobson, F. S. (1982). Competition for mates and predominant juvenile male dispersal in mammals. *Anim. Behav.* **30**, 1183–1192.
Dollard, J., Miller, N. E., Doob, L. W., Mowrer, O. H. & Sears, R. R. (1939). *Frustration and Aggression*. Yale University Press: New Haven, CT.
Dorjahn, V. R. (1958). Fertility, polygyny and their interrelations in Temne society. *Am. Anthropol.* **60**, 838–860.
Downhower, J. F. & Armitage, K. B. (1971). The yellow-bellied marmot and the evolution of polygyny. *Am. Nat.* **105**, 355–370.
Eadie, J. & Lumsden, H. G. (1985). Is nest parasitism always deleterious to goldeneyes? *Am. Nat.* **126**, 859–866.
Eibl-Eibesfeldt, I. (1963). Angeborenes und Erworbenes im Verhalten einiger Säuger. *Z. Tierpsychol.* **20**, 705–754.
Eibl-Eibesfeldt, I. (1967). *Grundriss der vergleichenden Verhaltensforschung*. R. Piper & Co.: München.
Eibl-Eibesfeldt, I. (1975). *Ethology: the Biology of Behavior*, 2nd ed. Holt, Rinehart & Winston: New York.
Eisenberg, J. F., Muckenhirn, N. A. & Rudran, R. (1972). The relation between ecology and social structure in primates. *Science* **176**, 863–874.
Emlen, S. T. (1972). An experimental analysis of the parameters of bird song eliciting species recognition. *Behaviour* **41**, 130–171.
Ewert, J.-P. (1980). *Neuroethology*. Springer-Verlag: New York.
Ferkin, M. H., Sorokin, E. S., Johnston, R. E. & Lee, C. J. (1997). Attractiveness of scents varies with protein content of the diet in meadow voles. *Anim. Behav.* **53**, 133–141.
Fischer, H. (1965). Das Triumphgeschrei der Graugans (*Anser anser*). *Z. Tierpsychol.* **22**, 247–304.
Fisher, H. E. (1989). Evolution of human pairbonding. *Am. J. Phys. Anthropol.* **78**, 331–354.
Fisher, R. A. (1930). *The Genetical Theory of Natural Selection*. Oxford University Press: Oxford, UK.
Flynn, J. P. (1967). The neural basis of aggression in cats. In: *Neurophysiology and Emotion* (ed. D. C. Glass), pp. 40–60. Rockefeller University Press: New York.
Ford, C. S & Beach, F. A. (1951). *Patterns of Sexual Behavior*. Harper: New York.
Francis, D., Diorio, J., Liu, D. & Meaney, M. J. (1999). Nongenomic transmission across generations of maternal behavior and stress responses in the rat. *Science* **286**, 1155–1158.
Galef, B. G. & Henderson, P. W. (1972). Mother's milk: A determinant of feeding preferences of weanling pups. *J. Comp. Physiol. Psychol.* **78**, 213–219.
Gannon, P. J., Holloway, R. L., Broadfield, D. C. & Brown, A. R. (1998). Asymmetry of chimpanzee planum temporale: Humanlike pattern of Wernicke's brain language area homolog. *Science* **279**, 220–222.
Gebhard, P. H. (1972). Incidence of overt homosexuality in the United States and western Europe. In: *National Institute of Mental Health Task Force on Homosexuality: Final Report and Background Papers* (ed. J. M. Livingood), pp. 22–29. Department of Health, Education and Welfare: Washington, DC.
Geschwind, N. & Levitsky, W. (1968). Human brain: Left-right asymmetries in temporal speech region. *Science* **161**, 186–187.
Gordon, T. P., Rose, R. M. & Bernstein, I. S. (1976). Seasonal rhythm in plasma testosterone levels in the rhesus monkey (*Macaca mulatta*): A three year study. *Horm. Behav.* **7**, 229–243.

Gould, J. L. (1982). *Ethology: the Mechanisms and Evolution of Behavior*. W. W. Norton & Co.: New York.
Goy, R. W. (1968). Organizing effects of androgen on the behaviour of rhesus monkeys. In: *Endocrinology and Human Behaviour* (ed. R. P. Michael), pp. 12–31. Oxford University Press: London.
Graul, W. D., Derrickson, S. R. & Mock, D. W. (1977). The evolution of avian polyandry. *Am. Nat.* **111**, 812–816.
Greenberg, L. (1979). Genetic component of bee odor in kin recognition. *Science* **206**, 1095–1097.
Greenwood, P. J. (1980). Mating systems, philopatry and dispersal in mammals and birds. *Anim. Behav.* **28**, 1140–1162.
Griffin, D. R. & Taft, L. D. (1992). Temporal separation of honeybee dance sounds from waggle movements. *Anim. Behav.* **44**, 583–584.
Gubernick, D. J. & Alberts, J. R. (1983). Maternal licking of young: Resource exchange and proximate controls. *Physiol. Behav.* **31**, 593–601.
Gwinner, E. (1981). Circannuale Rhythmen bei Tieren und ihre photoperiodische Synchronisation. *Naturwissenschaften* **68**, 542–551.
Halberg, F. (1959). Physiologic 24-hour periodicity in human beings and mice: The lighting regimen and daily routine. In: *Photoperiodism and Related Phenomena in Plants and Animals* (ed. R. B. Withrow), pp. 803–878. American Association for the Advancement of Science: Washington, DC.
Hall, K. R. L. & DeVore, I. (1965). Baboon social behavior. In: *Primate Behavior* (ed. I. DeVore), pp. 53–110. Holt, Rinehart & Winston: New York.
Hamer, D. H., Hu, S., Magnuson, V. L., Hu, N. & Pattatucci, A. M. L. (1993). A linkage between DNA markers on the X chromosome and male sexual orientation. *Science* **261**, 321–327.
Hamilton, W. D. (1964). The genetical evolution of social behaviour. *J. Theoret. Biol.* **7**, 1–52.
Hamilton, W. D. (1966). The moulding of senescence by natural selection. *J. Theoret. Biol.* **12**, 12–45.
Hamilton, W. D. (1971). Geometry for the selfish herd. *J. Theoret. Biol.* **31**, 295–311.
Hamilton, W. D. & Zuk, M. (1982). Heritable true fitness and bright birds: A role for parasites? *Science* **218**, 384–387.
Harlow, H. F. & Harlow, M. K. (1962). Social deprivation in monkeys. *Sci. Am.* **207**, 136–146.
Harlow, H. F. & Zimmermann, R. R. (1958). The development of affectional responses in infant monkeys. *Proc. Am. Phil. Soc.* **102**, 501–509.
Harris, G. W. (1955). *Neural Control of the Pituitary Gland*. Edward Arnold: London.
Hasselmo, M. E., Rolls, E. T. & Baylis, G. C. (1989). The role of expression and identity in the face-selective responses of neurons in the temporal visual cortex of the monkey. *Behav. Brain Res.* **32**, 203–218.
Hausfater, G. (1975). Dominance and reproduction in baboons (*Papio cynocephalus*): A quantitative analysis. *Contrib. Primatol.* **7**, 1–150.
Hediger, H. (1934). Zur Biologie und Psychologie der Flucht bei Tieren. *Biol. Zbl.* **54**, 21–40.
Hediger, H. (1950). *Wild Animals in Captivity*. Butterworths Scientific Publications: London.
Helbig, A. J. (1991). Inheritance of migratory direction in a bird species: A cross-breeding experiment with SE- and SW-migrating blackcaps (*Sylvia atricapilla*). *Behav. Ecol. Sociobiol.* **28**, 9–12.
Hinde, R. A. (1970). *Animal Behavior: A Synthesis of Ethology and Comparative Psychology*, 2nd. ed. McGraw-Hill: London.
Hofman, M. A. & Swaab, D. F. (1991). Sexual dimorphism of the human brain: Myth and reality. *Exp. Clin. Endocrinol.* **98**, 161–170.
Holekamp, K. E. & Smale, L. (1998). Dispersal status influences hormones and behavior in the male spotted hyena. *Horm. Behav.* **33**, 205–216.
Hooker, T. & Hooker, B. I. (1969). Duetting. In: *Bird Vocalizations: Their Relations to Current Problems in Biology and Psychology* (ed. R. A. Hinde), pp. 185–205. Cambridge University Press: Cambridge, UK.
Hrdy, S. B. (1974). Male–male competition and infanticide among the langurs (*Presbytis entellus*) of Abu, Rajasthan. *Folia Primatol.* **22**, 19–58.
Hrdy, S. B. (1977). Infanticide as a primate reproductive strategy. *Amer. Sci.* **65**, 40–49.
Hückstedt, B. (1965). Experimentelle Untersuchungen zum "Kindchenschema". *Z. Exptl. Angew. Psychol.* **12**, 421–450.
Irons, W. (1979). Cultural and biological success. In: *Evolutionary Biology and Human Social Behavior: An Anthropological Perspective* (eds. N. A. Chagnon & W. Irons), pp. 257–272. Duxbury Press: North Scituate, MA.
Isbell, L. A., Cheney, D. L. & Seyfarth, R. M. (1990). Costs and benefits of home range shifts among vervet monkeys (*Cercopithecus aethiops*) in Amboseli National Park, Kenya. *Behav. Ecol. Sociobiol.* **27**, 351–358.

Jolly, A. (1966). *Lemur Behavior: A Madagascar Field Study*. University of Chicago Press: Chicago.
Jost, A. (1971). Embryonic sexual differentiation (morphology, physiology, abnormalities). In: *Hermaphroditism, Genital Anomalies and Related Endocrine Disorders* (eds. H. W. Jones & W. W. Scott), pp. 16–64. Williams & Wilkins: Baltimore.
Kaitz, M., Good, A., Rokem, A. M. & Eidelman, A. I. (1987). Mothers' recognition of their newborns by olfactory cues. *Devel. Psychobiol.* **20**, 587–591.
Karlson, P. & Lüscher, M. (1959). Pheromones: A new term for a class of biologically active substances. *Nature* **183**, 55–56.
Keating, C. F. (1985). Human dominance signals: The primate in us. In: *Power, Dominance and Nonverbal Behavior* (eds. S. L. Ellyson & J. F. Dovidio), pp. 89–108. Springer-Verlag: New York.
Keeton, W. T. (1969). Orientation by pigeons: Is the sun necessary? *Science* **165**, 922–928.
Kendon, A. & Ferber, A. (1973). A description of some human greetings. In: *Comparative Ecology and Behaviour of Primates* (ed. R. P. Michael & J. H. Crook), pp. 591–668. Academic Press: London.
Kendrick, K. M., Da Costa, A. P., Broad, K. D., Ohkura, S., Guevara, R., Levy, F. & Keverne, E. B. (1997). Neural control of maternal behaviour and olfactory recognition of offspring. *Brain Res. Bull.* **44**, 383–395.
Kenrick, D. T. & Keefe, R. C. (1992). Age preferences in mates reflect sex differences in reproductive strategies. *Behav. Brain Sci.* **15**, 75–133.
Kenward, R. E. (1978). Hawks and doves: Factors affecting success and selection in goshawk attacks on woodpigeons. *J. Anim. Ecol.* **47**, 449–460.
Kinsey, A. C., Pomeroy, W. B. & Martin, C. E. (1948). *Sexual Behavior in the Human Male*. W. B. Saunders: Philadelphia.
Klapow, L. A. (1972). Natural and artificial rephasing of a tidal rhythm. *J. Comp. Physiol.* **79**, 233–258.
Kramer, G. (1952). Experiments on bird orientation. *Ibis* **94**, 265–285.
Krebs, J. R. & Davies, N. B. (1978). *An Introduction to Behavioural Ecology*. Sinauer Associates, Inc.: Sunderland, MA.
Kummer, H. (1968). *Social Organization of Hamadryas Baboons*. University of Chicago Press: Chicago.
Kummer, H. (1971). Immediate causes of primate social structures. *Proc. Int. Congr. Primatol.* **3**, 1–11.
Kühn, A. (1919). *Die Orientierung der Tiere im Raum*. Fischer: Jena.
Lashley, K. (1950). In search of the engram. *Soc. Exp. Biol. Symp.* **IV**, 454–482.
LeCroy, D. & Moller, P., Eds. (2000). *Evolutionary Perspectives on Human Reproductive Behavior*. Ann. NY Acad. Sci. **907**, 1–233.
Lehrman, D. S. (1959). Hormonal responses to external stimuli in birds. *Ibis* **101**, 478–496.
Lehrman, D. S. (1961). Gonadal hormones and parental behavior in birds and infrahuman mammals. In: *Sex and Internal Secretions* Vol. II (ed. W. C. Young), pp. 1268–1382. Williams & Wilkins: Baltimore.
Lenneberg, E. H. (1967). *Biological Foundations of Language*. New York: John Wiley & Sons.
LeVay, S. (1991). A difference in hypothalamic structure between heterosexual and homosexual men. *Science* **253**, 1034–1037.
Levine, R. J. (1994). Male factors contributing to the seasonality of human reproduction. *Ann. NY Acad. Sci.* **709**, 29–45.
Leyhausen, P. (1973). Verhaltensstudien an Katzen. *Z. Tierpsychol.* Suppl. 2, 6–232.
Liberg, O. & von Schantz, T. (1985). Sex-biased philopatry and dispersal in birds and mammals: The Oedipus hypothesis. *Am. Nat.* **126**, 129–135.
Lorenz, K. (1963). *Das Sogenannte Böse*. Borotha-Schoeler: Vienna.
Lorenz, K. (1965a). Ganzheit und Teil in der tierischen und menschlichen Gemeinschaft: Eine methodologische Erörterung (1950). In: *Über Tierisches und Menschliches Verhalten: Aus dem Werdegang der Verhaltenslehre* Vol. 2 (ed. K. Lorenz), pp. 114–200. R. Piper & Co.: München.
Lorenz, K. (1965b). Über die Wahrheit der Abstammungslehre. In: *Darwin Hat Recht Gesehen* (ed. K. Lorenz), pp. 55–74. Günther Neske Pfullingen: Stuttgart.
Lorenz, K. (1965c). Vergleichende Bewegungsstudien an Anatinen (1941). In: *Über Tierisches und Menschliches Verhalten: Aus dem Werdegang der Verhaltenslehre* Vol. 2 (ed. K. Lorenz), pp. 13–113. R. Piper & Co.: München.
Lorenz, K. (1981). *The Foundations of Ethology*. Springer: Wien–New York.
Marler, P. (1959). Developments in the study of animal communication. In: *Darwin's Biological Work* (ed. P. R. Bell), pp. 150–206. Cambridge University Press: Cambridge, UK.

Marshall, F. H. A. (1936). The Croonian lecture: Sexual periodicity and the causes which determine it. *Phil. Trans. Roy. Soc. Lond. B* **226**, 423–456.

Masters, W. & Johnson, V. (1966). *Human Sexual Response*. Little, Brown: Boston.

Maxson, S. J. & Oring, L. W. (1980). Breeding season time and energy budgets of the polyandrous spotted sandpiper. *Behaviour* **74**, 200–263.

Maynard Smith, J. (1976). Evolution and the theory of games. *Amer. Sci.* **64**, 41–45.

Mazur, A., Mazur, J. & Keating, C. (1984). Military rank attainment of a West Point class: Effects of cadets' physical features. *Am. J. Sociol.* **90**, 125–150.

McCann, T. S. (1981). Aggression and sexual activity of male southern elephant seals, *Mirounga leonina*. *J. Zool.* **195**, 295–310.

McDougall, W. (1936). *An Outline of Psychology*, 7th ed. Methuen: London.

McGuire, M. T. & Polsky, R. H. (1980). Ethological assessment of stable and labile social behaviors during acute psychiatric disorders: Clinical applications. *Psychiat. Res.* **3**, 291–306.

Meire, P. M. & Ervynck, A. (1986). Are oystercatchers (*Haematopus ostralegus*) selecting the most profitable mussels (*Mytilus edulis*)? *Anim. Behav.* **34**, 1427–1435.

Michael, R. P. (1961) An investigation of the sensitivity of circumscribed neurological areas to hormonal stimulation by means of the application of oestrogens directly to the brain of the cat. In: *Regional Neurochemistry* (eds. S. S. Kety & J. Elkes), pp. 465–480. Pergamon Press: Oxford, UK.

Michael, R. P. & Bonsall, R. W. (1977a). A 3-year study of an annual rhythm in plasma androgen levels in male rhesus monkeys (*Macaca mulatta*) in a constant laboratory environment. *J. Reprod. Fertil.* **49**, 129–131.

Michael, R. P. & Bonsall, R. W. (1977b). Peri-ovulatory synchronisation of behaviour in male and female rhesus monkeys. *Nature* **265**, 463–465.

Michael, R. P., Bonsall, R. W. & Zumpe, D. (1976). Evidence for chemical communication in primates. *Vit. Horm.* **34**, 137–186.

Michael, R. P., Wilson, M. & Zumpe, D. (1974). The bisexual behavior of female rhesus monkeys. In: *Sex Differences in Behavior* (eds. R. C. Friedman, R. M. Richart, & R. L. Van de Wiele), pp. 399–412. John Wiley & Sons, Inc.: New York.

Michael, R. P. & Zumpe, D. (1970a). Rhythmic changes in the copulatory frequency of rhesus monkeys (*Macaca mulatta*) in relation to the menstrual cycle and a comparison with the human cycle. *J. Reprod. Fertil.* **21**, 199–201.

Michael, R. P. & Zumpe, D. (1970b). Sexual initiating behaviour by female rhesus monkeys (*Macaca mulatta*) under laboratory conditions. *Behaviour* **36**, 168–186.

Michael, R. P. & Zumpe, D. (1978a). Annual cycles of aggression and plasma testosterone in captive male rhesus monkeys. *Psychoneuroendocrinology* **3**, 217–220.

Michael, R. P. & Zumpe, D. (1978b). Potency in male rhesus monkeys: Effects of continuously receptive females. *Science* **200**, 451–453.

Michael, R. P. & Zumpe, D. (1982). Influence of olfactory signals on the reproductive behaviour of social groups of rhesus monkeys (*Macaca mulatta*). *J. Endocrinol.* **95**, 189–205.

Michael, R. P. & Zumpe, D. (1983). Sexual violence in the United States and the role of season. *Am. J. Psychiatry* **140**, 883–886.

Michael, R. P. & Zumpe, D. (1986). An annual rhythm in the battering of women. *Am. J. Psychiatry* **143**, 637–640.

Michael, R. P. & Zumpe, D. (1988). Determinants of behavioral rhythmicity during artificial menstrual cycles in rhesus monkeys (*Macaca mulatta*). *Am. J. Primatol.* **15**, 157–170.

Michael, R. P. & Zumpe, D. (1990). Behavioral changes associated with puberty in higher primates and the human. In: *Control of the Onset of Puberty* (eds. M. M. Grumbach, P. C. Sizonenko, & M. L. Aubert), pp. 574–587. Williams & Wilkins: Baltimore.

Michael, R. P., Zumpe, D., & Bonsall, R. W. (1982). Behavior of rhesus monkeys during artificial menstrual cycles. *J. Comp. Physiol. Psychol.* **96**, 875–885.

Miller, N. E. (1956). Effects of drugs on motivation: The value of using a variety of measures. *Ann. NY Acad. Sci.* **65**, 318–333.

Miller, N. E. (1959). Liberalization of basic S-R concepts: Extensions to conflict behavior, motivation, and social learning. In: *Psychology: A Study of Science* Vol. 2 (ed. S. Koch), pp. 196–292. McGraw-Hill: New York.

Moore-Ede, M. C., Sulzman, F. M. & Fuller, C. A. (1982). *The Clocks That Time Us*. Harvard University Press: Cambridge, MA.

Morgan, C. L. (1894). *An Introduction to Comparative Psychology*. Scribner: New York.
Morton, E. S. (1975). Ecological sources of selection in avian sounds. *Am. Nat.* **109**, 17–34.
Moyer, K. E. (1976). *Psychobiology of Aggression*. Harper & Row: New York.
Munn, C. A. (1986). Birds that "cry wolf." *Nature* **319**, 143–145.
Murchison, C. (1935). The experimental measurement of a social hierarchy in *Gallus domesticus*: IV. Loss of body weight under conditions of mild starvation as a function of social dominance. *J. Gen. Psychol.* **12**, 296–311.
Murdock, G. P. (1967). *Ethnographic Atlas*. University of Pittsburgh Press: Pittsburgh.
Nastase, A. J. & Sherry, D. A. (1997). Effect of brood mixing on location and survivorship of juvenile Canada geese. *Anim. Behav.* **54**, 503–507.
Nevison, C. M., Rayment, F. D. G. & Simpson, M. J. A. (1996). Birth sex ratios and maternal social rank in a captive colony of rhesus monkeys (*Macaca mulatta*). *Am. J. Primatol.* **39**, 123–138.
Packer, C., Collins, D. A., Sindimwo, A. & Goodall, J. (1995). Reproductive constraints on aggressive competition in female baboons. *Science* **373**, 60–63.
Packer, C., Herbst, L., Pusey, A. E., Bygott, J. D., Hanby, J. P., Cairns, S. J. & Borgerhoff Mulder, M. (1988). Reproductive success of lions. In: *Reproductive Success: Studies of Individual Variation in Contrasting Breeding Systems* (ed. T. H. Clutton-Brock), pp. 363–383. University of Chicago Press: Chicago.
Panksepp, J. (1974). Hypothalamic regulation of energy balance and feeding behavior. *Fed. Proc.* **33**, 1150–1165.
Pavlov, I. P. (1927). *Conditioned Reflexes* (trans. G. V. Anrep). Oxford University Press: London.
Penfield, W. & Rasmussen, T. (1950). *The Cerebral Cortex of Man: A Clinical Study of Localization of Function*. Macmillan: New York.
Perrett, D. I., Smith, P. A. J., Potter, D. D., Mistlin, A. J., Head, A. S., Milner, A. D. & Jeeves, M. A. (1985). Visual cells in the temporal cortex sensitive to face view and gaze direction. *Proc. Roy. Soc. Lond. B* **223**, 293–317.
Petrie, M. (1992). Peacocks with low mating success are more likely to suffer predation. *Anim. Behav.* **44**, 585–586.
Pérusse, D. (1993). Cultural and reproductive success in industrial societies: Testing the relationship at the proximate and ultimate levels. *Behav. Brain Sci.* **16**, 267–322.
Pillard, R. C. & Weinrich, J. D. (1986). Evidence of familial nature of male homosexuality. *Arch. Gen. Psychiatry* **43**, 808–812.
Polsky, R. H. & McGuire, M. T. (1980). Observational assessment of behavioral changes accompanying clinical improvement in hospitalized psychiatric patients. *J. Behav. Assess.* **2**, 207–223.
Provine, R. R. (1981). Development of wing-flapping and flight in normal and flap-deprived domestic chickens. *Devel. Psychobiol.* **14**, 279–291.
Pusey, A., Williams, J. & Goodall, J. (1997). The influence of dominance rank on the reproductive success of female chimpanzees. *Science* **277**, 828–831.
Pusey, A. E. & Packer, C. (1987). The evolution of sex-biased dispersal in lions. *Behaviour* **101**, 275–310.
Rasa, O. A. E. (1971). Appetence for aggression in juvenile damsel fish. *Z. Tierpsychol.* Suppl. 7, 1–67.
Rasa, O. A. E. (1976). Aggression: Appetite or aversion?—An ethologist's viewpoint. *Aggr. Behav.* **2**, 213–222.
Reeve, H. K., Westneat, D. F., Noon, W. A., Sherman, P. W. & Aquadro, C. F. (1990). DNA "fingerprinting" reveals high levels of inbreeding in colonies of the eusocial naked mole-rat. *Proc. Natl. Acad. Sci.* **87**, 2496–2500.
Renner, M. (1957). Neue Versuche über den Zeitsinn der Honigbiene. *Z. Vergl. Physiol.* **40**, 85–118.
Rhine, R. J., Norton, G. W., Rogers, J. & Wasser, S. K. (1992). Secondary sex ratio and maternal dominance rank among wild yellow baboons (*Papio cynocephalus*) of Mikumi National Park, Tanzania. *Am. J. Primatol.* **27**, 261–273.
Richter, C. (1976). *The Psychobiology of Curt Richter* (ed. E. M. Blass). York Press: Baltimore.
Roeder, K. D. (1963). *Nerve Cells and Insect Behavior*. Harvard University Press: Cambridge, MA.
Roeder, K. D. & Treat, A. E. (1961). The detection and evasion of bats by moths. *Amer. Sci.* **49**, 135–148.
Rosenblum, L. A. & Paully, G. S. (1984). The effects of varying environmental demands on maternal and infant behavior. *Child Dev.* **55**, 305–314.
Rosenfield, R. L. & Lucky, A. W. (1993). Acne, hirsutism and alopecia in adolescent girls: Clinical expressions of androgen excess. *Endocr. Metab. Clin. N. America* **22**, 507–532.

Rumbaugh, D. M. & Gill, T. V. (1977). Lana's acquisition of language skills. In: *Language Learning by a Chimpanzee: The Lana Project* (ed. D. M. Rumbaugh), pp. 165–192. Academic Press: New York.

Russell, E. S. (1938). *The Behaviour of Animals: An Introduction to Its Study*. Edward Arnold: London.

Sade, D. S. (1968). Inhibition of son–mother mating among free-ranging rhesus monkeys. *Sci. Psychoanal.* **12**, 18–35.

Sakurai, T., Amemiya, A., Ishii, M., Matsuzaki, I., Chemelli, R. M., Tanaka, H., Williams, S. C., Richardson, J. A., Kozlowski, G. P., Wilson, S., Arch, J. R. S., Buckingham, R. E., Haynes, A. C., Carr, S. A., Annan, R. S., McNulty, D. E., Liu, W.-S., Terrett, J. A., Elshourbagy, N. A., Bergsma, D. J. & Yanagisawa, M. (1998). Orexins and orexin receptors: A family of hypothalamic neuropeptides and G protein-coupled receptors that regulate feeding behavior. *Cell* **92**, 573–585.

Salzano, F. M., Neel, J. V. & Maybury-Lewis, D. (1967). Further studies on the Xavante Indians: I. Demographic data on two additional villages: Genetic structure of the tribe. *Am. J. Hum. Genet.* **19**, 463–489.

Savage-Rumbaugh, E. S. (1986). *Ape Language*. Columbia University Press: New York.

Schenkel, R. (1956). Zur Deutung der Phasianidenbalz. *Ornithol. Beobacht.* **53**, 182–201.

Schjelderup-Ebbe, T. (1922). Beiträge zur Sozialpsychologie des Haushuhns. *Z. Tierpsychol.* **88**, 225–252.

Schleidt, M. & Genzel, C. (1990). The significance of mother's perfume for infants in the first weeks of their life. *Ethol. Sociobiol.* **11**, 145–154.

Schleidt, W. M. (1961). Reaktionen von Truthühnern auf fliegende Raubvögel und Versuche zur Analyse ihres AAM's. *Z. Tierpsychol.* **18**, 534–560.

Schneirla, T. C., Rosenblatt, J. S. & Tobach, E. (1963). Maternal behavior in the cat. In: *Maternal Behavior in Mammals* (ed. H. L. Rheingold), pp. 122–168. John Wiley & Sons: New York.

Scholz, A. T., Horrall, R. M., Cooper, J. C. & Hasler, A. D. (1976). Imprinting to chemical cues: The basis for home stream selection in salmon. *Science* **192**, 1247–1249.

Schoof-Tams, K., Schlaegel, J. & Walczak, L. (1976). Differentiation of sexual morality between 11 and 16 years. *Arch. Sex. Behav.* **5**, 353–370.

Schöne, H. (1962). Optisch gesteuerte Lageänderungen (Versuche an Dytiscidenlarven zur Vertikalorientierung). *Z. Vergl. Physiol.* **45**, 590–604.

Scott, J. P. (1960). *Aggression*. University of Chicago Press: Chicago.

Scott, J. P. (1972). *Animal Behavior*, 2nd ed. University of Chicago Press: Chicago.

Selye, H. (1973). The evolution of the stress concept. *Amer. Sci.* **61**, 692–699.

Shaw, C. E. (1948). The male combat "dance" of some crotalid snakes. *Herpetologica* **4**, 137–145.

Shepher, J. (1971). Mate selection among second generation kibbutz adolescents and adults: Incest avoidance and negative imprinting. *Arch. Sex. Behav.* **1**, 293–307.

Skinner, B. F. (1953). *Science and Human Behavior*. Macmillan: New York.

Smith, M. S., Kish, B. J. & Crawford, C. B. (1987). Inheritance of wealth as human kin investment. *Ethol. Sociobiol.* **8**, 171–182.

Soumah, A. G. & Yokota, N. (1991). Female rank and feeding strategies in a free-ranging provisioned troop of Japanese macaques. *Folia Primatol.* **57**, 191–200.

Soumah, A. G. & Yokota, N. (1992). Rank-related reproductive success in female Japanese macaques. In: *Topics in Primatology: Vol. 2. Behavior, Ecology and Conservation* (eds. N. Itoigawa, Y. Sugiyama, G. P. Sackett, & K. R. Thompson), pp. 11–22. University of Tokyo Press: Tokyo.

Spalding, D. A. (1873). Instinct: With original observations on young animals. *MacMillans Magazine* **27**, 282–293 (reprinted 1954 in *Brit. J. Anim. Behav.* **2**, 1–11).

Sperry, R. W. (1958). Physiological plasticity and brain circuit theory. In: *Biological and Biochemical Bases of Behavior* (eds. H. F. Harlow & C. N. Woolsey), pp. 401–424. University of Wisconsin Press: Madison.

Stephan, H. (1963). Vergleichend-anatomische Untersuchungen am Uncus bei Insectivoren und Primaten. *Prog. Brain Res.* **3**, 111–121.

Stephens, D. W. & Krebs, J. R. (1986). *Foraging Theory*. Princeton University Press: Princeton.

Sugiyama, Y. (1965). On the social change of hanuman langurs (*Presbytis entellus*) in their natural condition. *Primates* **6**, 381–418.

ten Cate, C. (1989). Behavioral development: Toward understanding processes. *Perspect. Ethol.* **8**, 243–269.

Thorndike, E. L. (1898). Animal intelligence: An experimental study of the associative process in animals. *Psych. Monogr.* **2**(4), 1–109.

Tinbergen, N. (1951). *The Study of Instinct*. Oxford University Press: Oxford, UK.

Tinbergen, N., Broekhuysen, G. J., Feekes, F., Houghton, J. C. W., Kruuk, H. & Szulc, E. (1962). Egg shell

removal by the black-headed gull, *Larus ridibundus* L.: A behaviour component of camouflage. *Behaviour* **19**, 74–117.
Tinbergen, N. & Kuenen, D. J. (1939). Über die auslösenden und die richtunggebenden Reizsituationen der Sperrbewegung von jungen Drosseln (*Turdus m. merula* L. und *T.e. ericetorum* Turton). *Z. Tierpsychol.* **3**, 37–60.
Tolman, E. C. (1932). *Purposive Behaviour in Animals and Men*. Appleton: New York.
Trivers, R. L. (1971). The evolution of reciprocal altruism. *Quart. Rev. Biol.* **46**, 35–57.
Trivers, R. L. (1972). Parental investment and sexual selection. In: *Sexual Selection and the Descent of Man* (ed. B. Campbell), pp. 136–179. Aldine: Chicago.
Trivers, R. L. (1974). Parent–offspring conflict. *Am. Zool.* **14**, 249–264.
Trivers, R. L. & Willard, D. E. (1973). Natural selection of parental ability to vary the sex ratio of offspring. *Science* **179**, 90–92.
Trumler, E. (1959). Das "Rossigkeitsgesicht" und ähnliches Ausdrucksverhalten bei Einhufern. *Z. Tierpsychol.* **16**, 478–488.
Tryon, R. C. (1940). Genetic differences in maze-learning ability in rats. *Yrbk. Nat. Soc. Stud. Educ.* **39**, 111–119.
Udry, J. R. & Morris, N. M. (1968). Distribution of coitus in the menstrual cycle. *Nature* **220**, 593–596.
van den Höövel, H. (1973). Social subordination, renal function and behaviour in the field vole. *Naturwissenschaften* **60**, 434–435.
van der Lee, S. & Boot, L. M. (1956). Spontaneous pseudopregnancy in mice. *Acta Physiol. Pharmacol. Neer.* **5**, 213–215.
Vandenbergh, J. G. & Drickamer, L. C. (1974). Reproductive coordination among free-ranging rhesus monkeys. *Physiol. Behav.* **13**, 373–376.
von Frisch, K. (1911). Beiträge zur Physiologie der Pigmentzellen in der Fischhaut. *Pflüger's Arch. Ges. Physiol.* **138**, 319–387.
von Frisch, K. (1967). *The Dance Language and Orientation of Bees*. Harvard University Press: Cambridge, MA.
von Holst, D. (1972). Renal failure as the cause of death in *Tupaia belangeri* exposed to persistent social stress. *J. Comp. Physiol.* **78**, 236–273.
von Holst, E. (1950). Quantitative Messung von Stimmungen im Verhalten der Fische. *Symp. Soc. Exp. Biol.* **4**, 143–172.
von Holst, E. & von Saint Paul, U. (1960). Vom Wirkungsgefüge der Triebe. *Naturwissenschaften* **47**, 409–422.
von Uexküll, J. (1921). *Umwelt und Innenwelt der Tiere*, 2nd ed. J. Springer: Berlin.
Watson, J. B. (1919). *Psychology from the Standpoint of a Behaviorist*. Lippincott: Philadelphia.
Watson, J. D. & Crick, F. H. C. (1953). Genetical implications of the structure of deoxyribonucleic acid. *Nature* **171**, 964–967.
Westermarck, E. (1891). *The History of Human Marriage*. Macmillan: New York.
Wheeler, D. A., Kyriacou, C. P., Greenacre, M. L., Yu, Q., Rutila, J. E., Rosbash, M. & Hall, J. C. (1991). Molecular transfer of a species-specific behavior from *Drosophila simulans* to *Drosophila melanogaster*. *Science* **251**, 1082–1085.
Whitten, W. K. (1956). Modifications of the oestrous cycle of the mouse by external stimuli associated with the male. *J. Endocrinol.* **13**, 399–404.
Williams, G. C. (1966). *Adaptation and Natural Selection: A Critique of Current Evolutionary Thought*. Princeton University Press: Princeton, NJ.
Williams, G. C. (1975). *Sex and Evolution*. Princeton University Press: Princeton, NJ.
Wilson, E. O. (1975). *Sociobiology: The New Synthesis*. Harvard University Press: Cambridge, MA.
Wittenberger, J. F. (1979). The evolution of mating systems in birds and mammals. In: *Handbook of Behavioral Neurobiology: Vol. 3, Social Behavior and Communication* (eds. P. Marler & J. G. Vandenbergh), pp. 271–349. Plenum Press: New York.
Wolf, A. P. (1970). Childhood association and sexual attraction: A further test of the Westermarck hypothesis. *Am. Anthropol.* **72**, 503–515.
Wynne-Edwards, V. C. (1962). *Animal Dispersion in Relation to Social Behaviour*. Oliver & Boyd: Edinburgh.
Yasukawa, K. (1981). Song repertoires in the red-winged blackbird (*Agelaius phoeniceus*): A test of the Beau Geste hypothesis. *Anim. Behav.* **29**, 114–125.

Yasukawa, K. & Searcy, W. A. (1985). Song repertoires and density assessment in red-winged blackbirds: Further tests of the Beau Geste hypothesis. *Behav. Ecol. Sociobiol.* **16**, 171–175.

Yeaton, R. I. & Cody, M. L. (1974). Competitive release in island song sparrow populations. *Theoret. Popul. Biol.* **5**, 42–58.

Zach, R. (1979). Shell dropping: Decision making and optimal foraging in Northwestern crows. *Behaviour* **68**, 106–117.

Zahavi, A. (1975). Mate selection—a selection for handicap. *J. Theoret. Biol.* **53**, 205–214.

Zhang, Y., Proenca, R., Maffel, M., Barone, M., Leopold, L. & Friedman, J. M. (1994). Positional cloning of the mouse *obese* gene and its human homologue. *Nature* **372**, 425–432.

Zumpe, D. & Michael, R. P. (1970). Redirected aggression and gonadal hormones in captive rhesus monkeys (*Macaca mulatta*). *Anim. Behav.* **18**, 11–19.

Zumpe, D. & Michael, R. P. (1984). Low potency of intact male rhesus monkeys after long-term visual contact with their female partners. *Am. J. Primatol.* **6**, 241–252.

Zumpe, D. & Michael, R. P. (1987). Relation between the dominance rank of female rhesus monkeys and their access to males. *Am J. Primatol.* **13**, 155–169.

Zumpe, D. & Michael, R. P. (1990). Effects of the presence of a second male on pair-tests of captive cynomolgus monkeys (*Macaca fascicularis*): Role of dominance. *Am. J. Primatol.* **22**, 145–158.

Author Index

Aisner, R., 133, 134, 313
Alberts, J. R., 250, 251, 316
Alcock, J., 187, 228, 313
Allan, J. S., 314
Allen, L. S., 247, 313
Amemiya, A., 320
Amoroso, E. C., 78, 313
Anand, B. K., 145, 146, 313
Andersson, M., 235, 313
Andrew, R. J., 183, 313
Andrews, M. W., 314
Annan, R. S., 320
Aquadro, C. F., 319
Arch, J. R. S., 320
Armitage, K. B., 262, 315
Aschoff, J., 101, 110, 313
Austad, S. N., 227, 313
Axelrod, R., 9, 313

Babický, A., 250, 251, 313
Baerends, G. P., 43, 313
Bailey, J. M., 59, 314
Barone, M., 322
Bateman, A. J., 224–226, 258, 277, 314
Bateson, P., 48, 314
Baylis, G. C., 316

Beach, F. A., 5, 245, 314, 315
Bentley, D. R., 65, 314
Bergsma, D. J., 320
Bernstein, I. S., 12, 273, 280, 314, 315
Bíbr, B., 313
Bierens de Haan, J. A., 3, 314
Blackmore, S., 299, 314
Bonsall, R. W., 82, 84, 87, 114, 314, 318
Boot, L. M., 190, 321
Borgerhoff Mulder, M., 319
Bovet, D., 62, 314
Bovet-Nitti, F., 314
Broad, K. D., 317
Broadfield, D. C., 315
Brobeck, J. R., 145, 146, 313, 314
Broekhuysen, G. J., 320
Brotherton, N. M., 314
Brouwer, R., 313
Brown, A. R., 315
Brown, C. R., 158, 159, 314
Brown, E. N., 314
Brown, J. B., 315
Brown, M. B., 158, 159, 314
Bruce, H. M., 190, 233, 314
Buckingham, R. E., 320

Butenandt, A., 189, 314
Bygott, J. D., 319

Cairns, S. J., 319
Cannon, W. B., 90
Carr, S. A., 320
Chapais, B., 281, 314
Chemelli, R. M., 320
Cheney, D. L., 316
Clancy, A. N., 194, 314
Clutton-Brock, T. H., 150, 314
Cody, M. L., 200, 322
Collins, D. A., 319
Cooper, J. C., 320
Coplan, J. D., 143, 314
Cowlishaw, G., 139, 140, 278, 314
Crawford, C. B., 306, 320
Crews, D., 222, 240, 314
Crick, F. H. C., 55, 321
Czeisler, C. A., 105, 106, 314

Da Costa, A. P., 317
Daly, M., 299, 301, 306, 307, 314
Darwin, C., 2, 4–6, 56, 132, 200, 201, 216, 217, 224, 315
Davies, N. B., 30, 232, 315, 317
Dawkins, R., 6, 7, 10, 315

323

Dennerstein, L., 309, 315
Derrickson, S. R., 316
DeVore, I., 277, 316
DeWitt, T. J., 223, 315
Dewsbury, D. A., 243, 244, 315
Dijk, D.-J., 314
Dion, D., 9, 313
Diorio, J., 315
Dobson, F. S., 161, 162, 315
Dollard, J., 204, 315
Doob, L. W., 315
Dorjahn, V. R., 300, 315
Downhower, J. F., 262, 315
Drickamer, L. C., 239, 285, 321
Duffy, J. F., 314
Dunbar, R. I. M., 278, 314

Eadie, J., 164, 315
Eibl-Eibesfeldt, I., 21, 25, 206, 207, 209, 289–292, 315
Eidelman, A. I., 317
Eisenberg, J. F., 273, 274, 315
Elshourbagy, N. A., 320
Emens, J. S., 314
Emlen, S. T., 184, 315
Ervynck, A., 136, 318
Ewert, J.-P., 181, 315

Farley, T. M. M., 315
Feekes, F., 320
Ferber, A., 293, 294, 317
Ferkin, M. H., 234, 315
Fischer, H., 242, 315
Fisher, H. E., 307, 315
Fisher, R. A., 223, 234, 315
Flynn, J. P., 202, 315
Ford, C. S., 245, 315
Francis, D., 63, 64, 315
Freud, S., 2, 51, 204
Friedman, J. M., 322
Friedman, S., 314
Fuller, C. A., 318

Galef, B. G., 133, 315
Gannon, P. J., 288, 315
Gaynor, D., 314
Gebhard, P. H., 245, 315
Genzel, C., 190, 320
Geschwind, N., 288, 315
Giedke, H., 313
Gill, T. V., 286, 320
Good, A., 317
Goodall, J., 215, 319
Gordon, T. P., 286, 314, 315
Gorman, J. M., 314

Gorski, R. A., 247, 313
Gotts, G., 315
Gould, J. L., 10, 11, 316
Goy, R. W., 76, 316
Graul, W. D., 263, 316
Greenacre, M. L., 321
Greenberg, L., 166, 316
Greenwood, P. J., 160, 316
Griffin, A. S., 314
Griffin, D. R., 177, 316
Gubernick, D. J., 250, 251, 316
Guevara, R., 317
Gwinner, E., 113, 316

Halberg, F., 100, 316
Hall, J. C., 321
Hall, K. R. L., 277, 316
Hamer, D. H., 59, 316
Hamilton, W. D., 6, 7, 9, 153, 155, 164, 236, 252, 313, 316
Hanby, J. P., 319
Harlow, H. F., 143, 269, 271, 273, 316, 320
Harlow, M. K., 273, 316
Harris, G. W., 68, 91, 92, 316
Hasler, A. D., 320
Hasselmo, M. E., 41, 42, 316
Hausfater, G., 229, 316
Haynes, A. C., 320
Head, A. S., 319
Hediger, H., 203, 316
Helbig, A. J., 64, 316
Henderson, P. W., 133, 315
Herbst, L., 319
Hinde, R. A., 5, 316
Hofman, M. A., 247, 316
Holekamp, K. E., 163, 316
Holloway, R. L., 315
Hooker, B. I., 182, 316
Hooker, T., 182, 316
Horn, G., 48, 314
Horrall, R. M., 320
Houghton, J. C. W., 320
Hoy, R. R., 65, 314
Hrdy, S. B., 159, 215, 316
Hu, N., 316
Hu, S., 316
Hückstedt, B., 296, 316

Irons, W., 303, 316
Isbell, L. A., 160, 316
Ishii, M., 320

Jeeves, M. A., 319
Johnson, V., 247, 318

Johnston, R. E., 315
Jolly, A., 271, 317
Jost, A., 223, 317

Kaitz, M., 48, 317
Kansky, R., 314
Karlson, P., 189, 317
Keating, C. F., 296, 297, 317, 318
Keefe, R. C., 305, 317
Keeton, W. T., 125, 126, 317
Kendon, A., 293, 294, 317
Kendrick, K. M., 48, 317
Kenrick, D. T., 305, 317
Kenward, R. E., 155, 317
Keverne, E. B., 317
Kinsey, A. C., 245, 317
Kish, B. J., 306, 320
Klapow, L. A., 111, 317
Kolář, J., 313
Kozlowski, G. P., 320
Kramer, G., 124, 125, 317
Krebs, J. R., 30, 135, 317, 320
Kronauer, R. E., 314
Kruuk, H., 320
Kuenen, D. J., 40, 321
Kummer, H., 12, 273, 276, 317
Kühn, A., 117, 119, 317
Kyriacou, C. P., 321

Lack, D., 256
Lashley, K., 5, 317
LeCroy, D., 311, 317
Lee, C. J., 315
Lehrman, D. S., 5, 241, 317
Lenneberg, E. H., 286, 317
Leopold, L., 322
LeVay, S., 247, 317
Levine, R. J., 309, 317
Levitsky, W., 288, 315
Levy, F., 317
Leyhausen, P., 29, 317
Liberg, O., 162, 317
Liu, D., 315
Liu, W.-S., 320
Long, C. N. H., 314
Lorenz, K., 2, 3, 16–18, 22, 23, 30–32, 38, 41, 45, 46, 51, 52, 64, 197, 199, 204, 216, 294–296, 317
Lucky, A. W., 94, 319
Lumsden, H. G., 164, 315
Lüscher, M., 189, 317

Maffel, M., 322
Magnuson, V. L., 316

AUTHOR INDEX

Manser, M., 314
Marler, P., 184, 317, 321
Marshall, F. H. A., 52, 78, 313, 318
Martin, C. E., 317
Masters, W., 247, 318
Matsuzaki, I., 320
Maxson, S. J., 263, 318
Maybury-Lewis, D., 320
Maynard Smith, J., 9, 318
Mazur, A., 297, 318
Mazur, J., 318
McCann, T. S., 229, 318
McDougall, W., 3, 48, 318
McGuire, M. T., 298, 318, 319
McNulty, D. E., 320
Meaney, M. J., 315
Meire, P. M., 136, 318
Michael, R. P., 35, 76, 80, 81, 84–86, 114, 191, 193, 218, 246, 270, 278, 279, 281–285, 310, 311, 314, 316–318, 322
Mignault, C., 281, 314
Miller, N. E., 49, 50, 315, 318
Milner, A. D., 319
Mistlin, A. J., 319
Mitchell, J. F., 314
Mock, D. W., 316
Moller, P., 311, 317
Moore-Ede, M. C., 103, 104, 107, 318
Morgan, C. L., 4, 319
Morris, N. M., 283, 321
Morse, C. A., 315
Morton, E. S., 185, 319
Mowrer, O. H., 315
Moyer, K. E., 199, 319
Muckenhirn, N. A., 315
Munn, C. A., 173, 319
Murchison, C., 210, 319
Murdock, G. P., 299, 319

Nastase, A. J., 164, 319
Neel, J. V., 320
Nemeroff, C. B., 314
Nevison, C. M., 279, 319
Noon, W. A., 319
Norton, G. W., 319

Ohkura, S., 317
Olivario, A., 314
O'Riain, M. J., 314
Oring, L. W., 263, 318
Ošťádalová, I., 313
Owens, M. J., 314

Packer, C., 161, 211, 226, 319
Panksepp, J., 145, 319
Pařízek, J., 313
Pattatucci, A. M. L., 316
Paully, G. S., 141, 142, 319
Pavlov, I. P., 3, 4, 319
Penfield, W., 198, 319
Perrett, D. I., 41, 319
Petrie, M., 236, 319
Pérusse, D., 304, 319
Pillard, R. C., 59, 246, 314, 319
Pinol, A., 315
Polsky, R. H., 298, 318, 319
Pomeroy, W. B., 317
Poppel, E., 313
Potter, D. D., 319
Proenca, R., 322
Provine, R. R., 20, 21, 319
Pusey, A. E., 161, 210, 319

Quetelet, A., 308

Rasa, O. A. E., 205, 216, 319
Rasmussen, T., 198, 319
Rayment, F. D. G., 319
Rees, H. D., 314
Reeve, H. K., 154, 319
Renner, M., 109, 319
Rhine, R. J., 279, 319
Richardson, J. A., 320
Richter, C., 113–115, 319
Rimmer, D. W., 314
Roeder, K. D., 186, 319
Rogers, J., 319
Rokem, A. M., 317
Rolls, E. T., 316
Ronda, J. M., 314
Rosbash, M., 321
Rose, R. M., 314, 315
Rosenblatt, J. S., 320
Rosenblum, L. A., 141, 142, 314, 319
Rosenfield, R. L., 94, 319
Rudran, R., 315
Rumbaugh, D. M., 286, 287, 320
Russell, E. S., 3, 48, 320
Rutila, J. E., 321

Sade, D. S., 280, 320
Sakurai, T., 145, 320
Salzano, F. M., 300, 320
Savage-Rumbaugh, E. S., 286, 320

Schenkel, R., 33, 320
Schjelderup-Ebbe, T., 208, 320
Schlaegel, J., 320
Schleidt, M., 190, 320
Schleidt, W. M., 42, 320
Schneirla, T. C., 253, 320
Scholz, A. T., 128, 320
Schoof-Tams, K., 302, 320
Schöne, H., 121, 320
Scott, J. P., 199, 204, 320
Searcy, W. A., 174, 322
Sears, R. R., 315
Selye, H., 90, 320
Seyfarth, R. M., 316
Shanahan, T. L., 314
Shaw, C. E., 202, 320
Shepher, J., 308, 320
Sherman, P. W., 319
Sherry, D. A., 164, 319
Silva, E. J., 314
Simpson, M. J. A., 319
Sindimwo, A., 319
Skinner, B. F., 4, 320
Smale, L., 163, 316
Smith, M. S., 304, 306, 320
Smith, P. A. J., 319
Sorokin, E. S., 315
Soumah, A. G., 210–212, 320
Spalding, D. A., 2, 20, 320
Sperry, R. W., 5, 320
Stephan, H., 191, 320
Stephens, D. W., 135, 320
Sugiyama, Y., 159, 320
Sulzman, F. M., 318
Sunquist, M. E., 227, 313
Swaab, D. F., 247, 316
Szulc, E., 320

Taft, L. D., 177, 316
Tanaka, H., 320
ten Cate, C., 47, 320
Tepperman, J., 314
Terkel, J., 133, 134, 313
Terrett, J. A., 320
Thorndike, E. L., 4, 320
Tinbergen, N., 2, 9, 15, 18, 19, 25, 28, 38–40, 42, 44, 52, 119, 120, 122, 197, 208, 320, 321
Tobach, E., 320
Tolman, E. C., 3, 48, 321
Treat, A. E., 186, 319
Trivers, R. L., 163, 224–227, 249, 250, 252, 254, 258, 277, 278, 303, 321

Trumler, E., 175, 321
Tryon, R. C., 61, 321

Udry, J. R., 283, 321

van den Höövel, H., 215, 321
van der Lee, S., 190, 321
Vandenbergh, J. G., 239, 285, 321
von Frisch, K., 2, 18, 106, 197, 321
von Holst, D., 215, 321
von Holst, E., 19, 22, 24, 120, 321
von Saint Paul, U., 24, 321
von Schantz, T., 162, 317
von Uexküll, J., 37, 321

Walczak, L., 320
Wasser, S. K., 319
Waterbolk, H. T., 313
Watson, J. B., 4, 321
Watson, J. D., 55, 321
Weinrich, J. D., 246, 319
Westermarck, E., 281, 307, 321
Westneat, D. F., 319
Wever, R., 313
Wheeler, D. A., 66, 321
Whitten, W. K., 190, 321
Wilkins, M. H. F., 55
Willard, D. E., 226, 227, 278, 303, 321
Williams, G. C., 6, 222, 252, 254, 321
Williams, J., 319
Williams, S. C., 320
Wilson, E. O., 6, 135, 171, 176, 199, 255, 256, 321
Wilson, M., 299, 301, 306, 307, 314, 318
Wilson, S., 320
Wittenberger, J. F., 259, 321
Wolf, A. P., 308, 321
Wynne-Edwards, V. C., 6, 321

Yanagisawa, M., 320
Yasukawa, K., 174, 321, 322
Yeaton, R. I., 200, 322
Yokota, N., 210–212, 320
Yu, Q., 321

Zach, R., 136, 322
Zahavi, A., 30, 235, 236, 322
Zhang, Y., 145, 322
Zimmermann, R. R., 269, 271, 316
Zuk, M., 236, 316
Zumpe, D., 35, 86, 193, 218, 270, 278, 279, 281–285, 310, 311, 314, 318, 322

Subject Index

Acoustic cues/stimuli. *See also* Communication, auditory
  in orientation, 117, 127
Acquired immunodeficiency syndrome (AIDS), 222
Acrophase, 99
Action-specific potentials, 51, 52
Adaptation, 2
  to selection pressure, 12-13
  somatic, 146-148
  to stress, 90, 95-96
Adaptive functions, of behavior patterns, 16, 18
Adaptive radiation, 132, 201
Adenine, as deoxyribonucleic acid (DNA) component, 55
Adolescents, suicide by, 98
Adoption, 164
Adrenal cortex
  anatomy of, 92-94
  hormone production by, 92-94
  in stress response, 91-92
Adrenal gland
  hormone production by, 67, 70
  stress-related hypertrophy of, 96
Adrenal hormones, 67, 70, 92-94. *See also* Adrenocorticoids; Glucocorticoids; Mineralocorticoids

Adrenalin, 91
Adrenal medulla, in stress response, 91
Adrenarche, 93
Adrenocorticoids
  anti-inflammatory effects of, 95, 97
  immunostimulatory effects of, 95
  metabolism of, 93-95
  production of, 92-93
Adrenocorticotropic hormone, 67-68, 92
Affective disorders, 115-116
Affiliative behavior, 268
"After-reaction," of female cats, 81
Aggregations, 151, 152
Aggression
  androgen-induced, 86
  antipredatory, 199, 203, 213
  during courtship, 240-241
  during diestrus, 80-81
  expressed as alternation behavior, 27, 28
  expressed as ambivalent behavior, 27, 29
  genetic factors in, 204-205
  by humans, 216-219
  interspecific, 199, 200-203
  intraspecific, 199, 203-215
    categories of, 206-215
    as consummatory response, 205
    dominance-related, 199, 206-207, 208-211

Aggression (*cont.*)
  intraspecific (*cont.*)
    reduction of costs of, 205–206
    territorial, 199, 206–208
  during mating, 86, 213, 285
  parental, 199, 213
  parent-offspring, 199, 213–214
  passive, 218
  predatory, 199, 202
  property-protective, 216
  redirected, 24–25, 32, 34–35, 268
  self-defensive, 216
  sexual, 85, 199, 211–213, 218, 228–230
  between siblings, 214
  spontaneous appearance of, 205
  tactile communication during, 195
  territorial, 138, 152, 199, 207–208
Agonism
  antipredatory, 203
  definition of, 199–200
  interspecific, comparison with intraspecific agonism, 216
Agonistic behavior. *See also* Aggression; Agonism
  cognitive rehearsal of, 272–273
  during courtship, 240–241
  definition of, 199–200
  facial expressions in, 267, 268
  interspecific, 199, 200–203
Agranulocytosis, cyclic, 113, 115
AIDS (acquired immunodeficiency syndrome), 222
Alarm calls, 181–182
Alarm reaction, 90, 91
Alarm signals, chemical and olfactory, 190–191
Allele recognition, as kin recognition mechanism, 167
Alleles, 56–57
  exchange of, 65–66
  identical, of inbred strains, 61–62
  "silenced," 57
  of twins, 59
Alternation, 27, 28
Altricial species, maturation of innate behavior in, 19–20
Altruism
  definition of, 164–165
  evolution of, 6, 7
  reciprocal, 150, 163–164
Ambivalence, 27, 29, 30
Ambushing, as predatory technique, 148
American Sign Language, 286
Anabolic steroids, 86, 88. *See also* Androgens
Analogy, 16, 17
Androgen insensitivity syndrome, 73–74
Androgens. *See also* Dihydrotestosterone; Testosterone
  effect on male sexual activity, 238

Androgen insensitivity syndrome (*cont.*)
  molecular structure of, 93
  production of, 70, 92–93
  effect on sexual differentiation, 244–245
  effects on tissues, 72
Androstenol, 192
Anestrus, 79
Aneuploidy, 56
Antiandrogens, 73
Antidiuretic hormone, 51
Antiestrogens, 73
Ant lions, biological rhythms in, 112
Ants, navigation by, 122–123
Anxiety attacks, 90
Anxiety disorder, 143
Appeasement gestures, 292, 294. *See also* Submissive behavior
Appetite, 131, 145–146
Appetitive behavior, 50, 52, 205
Aquatic species
  fertilization in, 243
  orientation behavior of, 119–121, 127
Aristotle, 1
Armor, as antipredator defense, 149
Arousal, sympathetic, 91
Artificial insemination, 228
Artificial selection, 60, 204
Assaults, seasonal variation in, 310
Association hypothesis, of parental care, 254–255
Attention, perception *versus*, 37–38
Autoimmune disorders, stress response in, 97
Autosomes, 55

Baboons
  dominance hierarchy of, 209–211, 229–230
  harems of, 154, 167–168, 229, 275–276
  inbreeding avoidance by, 280, 281
  infanticide by, 159
  mate competition by, 278
  mating system of, 277
  parental investment by, 278–279
  predation response of, 139, 140
  social system of, 154, 275–277
Bacteria
  magnetic orientation by, 127
  role in digestion, 146
Bands, 168
Bateman's Principle, of variance in reproductive success, 224, 225, 277–278
Bats
  echolocation by, 117, 188
  moth's auditory detection of, 185–187
Bees. *See also* Honeybees
  kin recognition among, 166
Beetles, maternal care by, 257

SUBJECT INDEX

Behavior
  evolutionary basis of, 2, 15–17
  hierarchical organization of, 52
  study of, historical background of, 1–13
    classical ethology, 1–3
    comparative psychology, 4–5
    theory and terminology, 6–11
Behavioral differences, 57, 59–66
Behavioral similarities, 57–59
  in geographically-isolated populations, 57–58
Behaviorism, 4–5
Benefit-cost ratio, 7, 9
  of parental care, 252–253, 254
Benign prostatic hyperplasia, 73
Biological (internal) clock, 102, 105–108, 113
Biological rhythms, 99–116
  amplitude of, 99, 100
  circadian, 101–110
    desynchronization of, 105
    double-plotting of, 102, 103
    entrainment in, 101, 102–104, 106, 108, 112, 113
    free-running, 101, 102–104, 105, 109, 112
    genetic factors in, 66
    internal clock in, 102, 105–108, 113
    effect of light on, 102–103, 105–106
    in orientation, 100, 124, 125–126
    periodicity of, 101, 108
    phase shifts in, 101–102
    of plasma hormone levels, 70
    properties of, 101–102
    temperature compensation in, 102
    ubiquity of, 101
  circalunar, 101, 111–112
  circannual, 101, 113
  circatidal, 101, 111–112
  in disease, 113–116
  frequency of, 99, 100
  functions of, 99–101
  in humans, 100, 103, 105, 106, 108–110, 112, 113–116
  monthly, 101, 112
  periods in, 99, 100
  phases in, 99, 100
Bioluminescence, 180, 239–240
Birds. *See also* Bird song; Bower birds; Chickens; Darwin's finches; Ducks; Eagles; Geese; Hawks; Homing pigeons; Owls; Peacocks; Penguins; Turkeys
  adaptive radiation of, 132, 201
  adoption in, 164
  beak morphology of, 10–11
  bower decoration by, 178, 261
  brood parasitism in, 44, 159, 173
  circannual rhythms in, 113
  communal defense of offspring by, 157

Birds (*cont.*)
  courtship behavior of, 31, 32, 58
  deceptive communcation by, 173–174
  egg recognition by, 45
  facial characteristics of, 295–296
  fixed action patterns in, 19, 22–24
  foraging behavior of, 135–136
  habituation in, 41–43
  hybridization of, 64
  imprinting in, 45–48, 58, 240
  killing of conspecific's offspring by, 159
  lek polygyny of, 230, 239, 261
  male territoriality in, 230
  migration by, 124–125, 128
  monogamy (pair bonding) of, 161, 241, 242, 262–263
  multiple paternity in, 230
  navigation by, 122–125, 126, 127–128
  nestlings, gaping response of, 119
  parental behavior of, 58
  parental care by, 233
  sexual preferences of, 58
  siblicide by, 214
  swallow-bug infestations of, 157–159
  vigilance/flock size relationship in, 155
Bird song
  alarm calls, 184
  as antiphonal singing (duetting), 181, 182
  frequency of, 185
  imprinting of, 46, 47, 240
  informational content of, 183–184
  role in territorial defense, 173–174
Bisexual/homosexual behavior
  anterior hypothalamus in, 247
  genetic factors in, 59, 246
  in humans, 59, 245–247, 301, 308
  lesbianism, 308
Bitter taste, 145
Blind and deaf children, smiling and laughing by, 290–291, 292
Body temperature, circadian rhythms in, 105, 109, 110
Bombykol, 189, 193
Bonding. *See also* Pair bonding
  homosexual, 246
  through grooming behavior, 195
  through redirected behavior, 25
Bonobos, language use by, 286, 287, 288
Bower birds
  bower decoration by, 178, 261
  lek polygyny of, 261
Brain. *See also* Cerebral cortex; Corpus callosum; Hypothalamus
  inferior temporal cortex, in response to facial expression, 40–41, 42
  internal clock, 105–108, 113

Brain (cont.)
  language area, 288
    sexual dimorphism of, 247
  olfaction area, 190, 191
  preoptic area, sexual dimorphism of, 247
  sexual dimorphism of, 77, 247
Breast cancer, tamoxifen treatment of, 73
Breast feeding, 144
Breast milk, 190–191
Breeding. *See also* Mating
  seasonality of, 78
Breeding systems, 258
Breeds, of animals, 60
Brood parasitism, 44, 159, 173
Bruce Effect, 190, 232–233

Caloric intake, 134–135, 157
Camels, facial characteristics of, 295, 296
Cannibalism, 9, 249
Carbohydrates, 144
Caregiving behavior. *See also* Maternal care;
    Parental care; Paternal care
  releasers for, 272
Carnivores, 131
  primates as, 144
Castration, 85–86, 212–213
Cats
  ambivalent behavior of, 27, 29
  courtship and mating behavior of, 80–81
  predatory behavior of, 21, 132, 202
Cattle, freemartins in, 75, 244
Central nervous system, relationship with
    endocrine system, 68
Centromeres, 55
Cerebral cortex
  role in learning, 5
  sensory and motor function areas of, 197, 198
  superior temporal sulcus
    in facial recognition, 39, 41
    visual signal processing by, 179
Certainty of paternity hypothesis, of male parental
    care, 254
Cetaceans, sonar of, 188
Chasing, as predatory technique, 148–149
Cheetahs, prey chasing by, 148–149
Chemical stimuli. *See also* Communication,
    chemical
  in orientation, 117
Chemotaxis, 119
Chickens
  brain stimulation in, 22
  flight development in, 20–21, 64
  "pecking order" of, 208–209, 210
Chiefdoms, 168–169
Child abuse, by stepparents, 304, 306, 307

Children, stress response in, 97
Chimpanzees
  diet of, 144
  inbreeding avoidance by, 281
  language use by, 286, 287, 288
  smiling and laughing by, 290
  social system of, 276, 277
  xenophobia among, 214
Chipmunks, foraging behavior of, 137
Cholesterol, 93, 94
Chromatids, 56
Chromosomes. *See also* Genes
  homologous, 55–56
  recombination of, 56
  sex, 55, 222–223
Circadian rhythms, 101–110
  desynchronization of, 105
  double-plotting of, 102, 103
  entrainment in, 101, 102–104, 106, 108, 112, 113
  free-running, 101, 102–104, 105, 109, 112
  genetic factors in, 66
  internal clock in, 102, 105–108, 113
  effect of light on, 102–103, 105–106
  in orientation, 100, 124, 125–126
  periodicity of, 101, 108
  phase shifts in, 101–102
  of plasma hormone levels, 70
  temperature compensation in, 102
  ubiquity of, 101
Circalunar rhythms, 101, 111–112
Circannual rhythms, 101, 113
Circatidal rhythms, 101, 111–112
Civetone, 192
Classical conditioning, 4
Clinging response, of infants, 267, 269, 271
Coelenterates, colonies of, 152
Color, as visual signal, 148, 149, 178
Color vision, 179
Communication, 171–198
  adaptive value of, 171
  auditory, 176, 177, 180–188
    intrinsic and extrinsic mechanisms of, 183
    range and frequency of, 180–181, 182, 183, 184–185
    receiver mechanisms of, 185–188
    sender mechanisms of, 183–185
    signal properties in, 180–182, 183
  chemical (olfactory), 176, 177, 188–195
    intrinsic and extrinsic mechanisms of, 191–192
    by lemurs, 192, 270–271
    receiver mechanisms of, 193–195
    sender mechanisms of, 191–193
    signal properties of, 188–191
  communicatory signals in, 174–176
  definition of, 171

Communication (*cont.*)
  effectiveness in darkness, 176, 177
  electrical, 176, 197–198
  energetic expense of, 177
  environmental barriers to, 176, 177
  fadeout time of, 176, 177
  functions of, 171–172
  in geographically-isolated populations, 57–58
  "honesty" and "deception" in, 172–174
  indexical capacity of, 177
  nonverbal, in humans, 289
  range of, 176, 177
  representational capacity of, 177
  sender, receiver, response components of, 171
  tactile, 176, 177, 195–197
  transmission rate of, 176, 177
  visual, 175, 176, 177–180
    intrinsic and extrinsic mechanisms in, 178
    by lemurs, 270–271, 272
Compass navigation, 122–123, 124–126
Competition, 204
  between conspecifics, 157
  as constraint on foraging, 135, 138
  for mates. *See* Mate competition
  reduction of, by natal dispersal, 279–280
  reproductive, parent-offspring, 162
Compromise, 28
Computer keyboards, apes' use of, 287
Conditioned stimulus, 4
Conditioning, operant and classical, 4
Configurational stimuli, 38, 40, 42–43, 178–179
Conflict behaviors, 22, 24–30
  alternation, 27, 28
  ambivalence, 27, 29, 30
  compromise, 28
  displacement, 25–27, 30
  of humans, 291–292
  intention movements, 27
  of nonhuman primates, 268
  psychiatric manifestations of, 28, 30
  redirection, 24–25, 30
Conflict model, of parent-offspring relationship, 250–252
Congenital adrenal hyperplasia, 76–77, 94–95
Consort bonds, 231, 277
Conspecifics, isolation from, effect on behavior patterns, 58
Consummatory response, 50, 52
Cooperation (mutualism), 163
  evolution of, 8–9, 162–165
  game theory of, 8–9
Coping
  with changes in food supply, 139–143
  with stress, 90
Coprophagy, 146

Copulation
  definition of, 243
  patterns of, 243–244
    in nonhuman primates, 267–268, 269
  premarital, 299, 301–302
  relationship with menstrual cycle, 283
  "sneak," 230
  synchronous behavior leading to, 242
Copulatory lock/plug, 231, 243–244
Copulins, 192–193
Corpus callosum, sexual dimorphism of, 77, 247
Corpus luteum, 69–70
Correlational studies, 11–12
Corticosteroids
  metabolism of, 93–95
  molecular structure of, 93–94
  production of, 92–93
  stress-related increase in, 96
Corticosterone, effect on migratory behavior, 128
Corticotropin-releasing factor, 91–92, 143
Cortisol, secretion of
  circadian rhythm of, 106, 108, 110
  foraging-related decrease in, 143
  in major depressive disorder, 97–98
Cost-benefit analysis, 7
Countershading, 147
Courtship behavior and displays, 31–32, 33, 34–35, 178–179
  aggressive interactions in, 86, 213
  from conflict behavior, 24
  exteroceptive stimuli in, 238
  of females (proceptive behavior), 79–82, 83, 285
  functions of, 239–243
    aggression reduction, 240–241
    behavioral and physiological synchronization, 241–242
    gender identification, 240
    individual recognition, 241
    signaling of competitive and parental abilities, 242–243
    species and strain identification, 239–240
  hormonal factors in, 238, 241–242
  imprinting of, 46
  lekking, 230, 239, 261
  between males, 46
  morphological traits in, 58–59
  pecking behavior in, 31, 33
  phylogenetic relationships in, 58
  redirected behavior in, 24–25, 31–32
  seasonal factors in, 238
  social factors in, 239
  of stickleback fish, 27, 28, 38, 40
Courtship songs
  of crickets, 64, 65
  of fruit flies, 65–66

Crabs, courtship behavior of, 31, 58–59
Crickets, hybridization of, 64, 65
Crime
  seasonal influences on, 308, 310–311
  thermic law of, 308, 310
  violent, 218
Cross-fostering studies, 58, 62–64
Crypsis, 147
Cues
  navigational, 125, 127–128
  olfactory, 188–189
Cultural influences, on behavior, 167
  in humans, 168–169
  in nonhuman primates, 167–168
Cyproterone acetate, 73
Cystic fibrosis, 221
Cytosine, as deoxyribonucleic acid (DNA) component, 55

Darwin, Charles, 200–201
  Descent of Man and Selection in Relation to Sex, 224
  evolution theory of, 2, 4, 6
Darwin's finches. *See also* Galapagos finches
  adaptive radiation of, 132, 201
  divergent evolution of, 132, 200–201
Daughters
  inheritance patterns of, 303–304, 306
Dead-reckoning navigation, 123
"Dear enemy effect," 208
Death
  feigning of, 149
  food shortage-related, 140–141
Deception, in communication, 172–174
Deer, male breeding behavior of, 83–85
Delayed sleep phase insomnia, 106
Deletions, chromosomal, 56, 64–65
Dentition, relationship with diet, 132
Deoxyribonucleic acid (DNA), structure of, 55
Depression, in domesticated animals, 98
Descent of Man and Selection in Relation to Sex (Darwin), 224
Dexamethasone suppression test, 97–98
Diabetes insipidus, 51
Diestrus, 79, 80, 81
Diethylstilbestrol, 77
Digestion, in herbivores, 146
Dihydrotestosterone, 69, 72
Dilution effect, in predation, 156, 164
Diploidy, 153, 165
Direct effects, on reproductive behavior, 237
Directional navigation, 122, 123–126, 128
Direct selection
  in guarding behavior, 150
  in parental care, 165

Disease, biological rhythms in, 113–116
Dispersal, 159–162
  as inbreeding avoidance mechanism, 160, 279–282
  natal, 159–160
  by nonhuman primates, 276, 279–282
Displacement activity, 25–27
  displacement *versus* disinhibition hypotheses of, 26
  in humans, 26, 30
  in nonhuman primates, 268
  as psychiatric term, 27
Displays, 30–35
Distance, individual, violations during mating, 241
Divergent evolution, 16, 132, 200–201
Diversion, as antipredator defense, 149
Divorce, 304, 307, 308
Dogs
  artificial selection in, 60
  displacement behavior in, 26
Dolphins
  sonar of, 188
  songs of, 184
Domesticated animals
  artificial selection in, 60
  castration of, 86, 212–213
  depression in, 98
Dominance
  aggression associated with, 199, 206–207, 208–211
  based on facial characteristics, 297
  territoriality associated with, 207
Dominance hierarchy, 208–211
  as a characteristic of individualized groups, 208
  in feeding behavior, 138
  intergroup, 208
  intragroup, 208
  of nonhuman primates, 157, 209–211, 229–230, 276–277
  relationship with mating activity, 229–230
  as response to conspecific competition, 157
  sociality-related, 207
Dominant males, 276
Dopamine, 91
"Dorsal light reaction," 119
Double helix structure, of deoxyribonucleic acid (DNA), 55
Drinking. *See also* Thirst
  as displacement behavior, 25–26
Drive concept and model, of motivation, 48–53
Drive state, effect on orientation, 119–121
Drones, 153
Drugs, circadian rhythm-based efficacy of, 109
Ducks
  courtship behavior of, 31, 32
  fixed action patterns in, 22, 23

Duetting, 181, 182, 275
Duplications, chromosomal, 56
Dynamic selection, 7

Eagles, facial characteristics of, 295–296
Ear, anatomy of, 186–187
"Ear wiggling," as female courtship behavior, 79–80
Eating. *See also* Feeding behavior
　as displacement behavior, 25–26
　proximate and ultimate functions of, 15–16
Echolocation, 117, 127
Eclosion, circadian rhythm of, 66
Ecology, 1
　behavioral, 6
Egg incubation, 242
Eggs
　mimicry of, 178, 249–250
　parent birds' recognition of, 45
Electrical communication, 176, 197–198
Electrical cues/signals, in orientation, 117, 127
Embryo, cannibalization of, 249
Endocrine system. *See also* names of specific hormones
　relationship with central nervous system, 68
Endocrinology, behavioral
　adrenal hormones and stress effects, 89–98
　　adrenal cortex and, 91–92
　　adrenal medulla and, 91
　　corticosteroid metabolism, 93–95
　　corticosteroid production, 92–93
　　hypothalamus and, 91–92
　　in major depressive disorder, 97–98
　　psychosomatic medicine and, 96–97
　　in suicide, 98
　gonadal hormones, 67–88
　　action mechanisms of, 71–74
　　activational effects of, in adults, 77–78, 282–283, 308, 309
　　organizational effects of, during development, 74–77
　　receptors, 88
　　synthesis and sites of production of, 69–70
　　transport of, 71
Entrainment, of biological rhythms, 101, 102–104, 106, 108, 112, 113
Environmental influences
　biological rhythm-based adaptation to, 99–100
　on fixed action patterns, 18, 19–21
　interaction with genetic effects, 57
Epinephrine, 91
Eskimos, 292, 307
Estradiol, 83
　active and inactive forms of, 93–94
　effect on estrous behavior, 79

Estradiol (*cont.*)
　in fetal masculinization, 88
　effect on sexual behavior, 34, 88
　synthesis of, 69
　effects on tissues, 71–72
Estrogen antagonists, 73
Estrogens, 192
　cyclic production of, 238
　levels during ovulation, 227–228
　molecular structure of, 93
　positive feedback effects of, 69–70
　effects on tissues, 71–72
Estrous cycle, 79–82
　phases of, 79
　effect of pheromones on, 190
　proceptive and receptive behavior during, 79–82
　relationship with menstrual cycle, 70
Estrus, 79
　postpartum, 82
Ethogram, 17–18
Ethology
　classical, 1–3
　comparison with sociobiology, 6
　development of, 1
　objectives of, 17
　theoretical foundation of, 15
Eusociality, 7
　evolution of, 165
　in insects, 152–154
Evolution
　convergent, 16
　　of birds' alarm calls, 184
　　of social behavior, 154
　Darwinian theory of, 2, 4, 6
　divergent, 16, 132, 200–201
Evolutionary stable strategies (ESS)
　mixed, 9
　prediction of, 8–9
　pure, 9
　reciprocal altruism as, 164
Exploratory behavior, effect of maternal care on, 62–63
Exteroceptive sensations, 197
Exteroceptive stimuli
　in courtship behavior, 238
　in mating, 78, 238, 285
　in orientation, 117
Extinction
　hunting-related, 217
　of nonhuman primates, 266–267, 288
Extramarital sexual activity, gender differences in, 299, 301
Eye
　compound, 179
　vertebrate, 179, 180

"Eyebrow flash," 267, 291, 293
Eyespots, 149, 179

Faces, recognition of, 39, 40–41, 42, 295
Facial characteristics, 294–298
Facial expressions, of nonhuman primates, 40–41, 267, 268, 270
Familiarity, effect on sexual interest, 166, 234, 280, 281, 282
Fat, as human dietary component, 44, 144–145
Fatty acids, short-chain, 192
Fatty foods, humans' preference for, 44
FBI Uniform Crime Reports, 218, 310
Fear, effect of maternal care on, 62–63
Fear grimace, 267, 268, 292
Feces
  as communicatory signals, 176
  eating of, 146
  as pheromone source, 191, 192
Feeding behavior, 131–134
  of humans, 144–145
  physiological aspects of, 145–146
Feeding territory, 138
Fermenters, foregut or hindgut, 146
Fertility, relationship with income, 303, 304
Fertilization, 56, 222–223
  external, 243
  internal, 243–244
Fetus, sexual differentiation of, 75, 76–77, 88, 222–223, 244–245
Fight-or-flight response, 90, 91, 200
Filter-feeders, 131
Fire fly, bioluminesence of, 239–240
Fish
  adoption in, 164
  aggressive behavior of, 58–59
  courtship behavior of, 27, 28, 38, 39, 40, 43
  electrical communication by, 197–198
  external fertilization in, 243
  fixed action patterns of, 19
  food competition by, 138
  orientation and navigation by, 119–120, 127, 128
  parental care by, 249–250, 255
  redirected behavior of, 25
  spawning by, circalunar rhythm of, 112
  visual mimicry in, 178
Fisher's runaway selection model, of evolution of male traits, 234–235
Fisher's Theorem, of sex ratio, 223–224
Fishing territories, 208
Fitness, 6–7
  inclusive, 7–8, 151, 153–154, 278
  altruism-related increase in, 165
Fixed action patterns (FAPs)
  chains of, 21, 241

Fixes action patterns (FAPs) (cont.)
  as displays, 31, 32
  in humans, 290, 291, 292, 294
  in mating, 241
  in nonhuman primates, 267–268
  properties of, 18–22
    brain stimulation evoked, 22, 24
    coordination of muscles, 19
    environmental influences, 19–21
    genetic control (species specific), 22, 23, 58
    unvarying form (stereotypy), 18–19
"Flehmen," 194
Flight, from aggressors, 215–216
Flight distance, 139, 203
Flutamide, 73
Follicle-stimulating hormone, 67–68
  secretion of, 70
    circadian rhythm of, 110
    coitus-induced, 78–79
Food, taste and odor of, 144–145
Food poisoning, 145
Food resources, interspecific conflict related to, 201–202
Foraging, 134–138
  energy per unit time expended/energy per unit gained in, 134–135
  by groups, 156
  effect on maternal-offspring relationship, 64, 141–143, 251–252
  optimal, 135–139
    constraints on, 135, 138–139
    search technique in, 137–138
Form, perception of, 197
Fostering studies, of species-specific fixed action patterns, 58
Founder effect, 56, 60
Freemartins, 75, 244
Freud, Sigmund, 2, 51, 204
Frogs, visual detection of prey by, 179
Fruit flies, courtship songs of, 65–66
Function, 12, 18
  relationship with genes, 12
  proximate, 15–16
  ultimate, 15–16

Galapagos finches, beak morphology of, 10–11
Gamete order hypothesis, of parental care, 254
Gametes, 56
  parental investment in, 249
Game theory, 8–9
  of reciprocal altruism, 164
Gang members, "sexing-in" of, 305, 307
Gang warfare, 218
Geese
  adoption in, 164

SUBJECT INDEX 335

Geese (*cont.*)
 fixed action patterns in, 22, 23
 pair bonding by, 241, 242, 262
 triumph ceremony of, 241, 242
Gender identification, based on courtship displays, 240
Gene flow, 56
General adaptation syndrome, 90, 95–96
Genes
 deletion of, 64–65
 epistatic, 57
 functions of, 56–57
 mutations of, 12, 56
 pleiotropic, 57, 64
 polymorphism of, 55–56
 relationship with function, 12
 "selfish," 6, 7
Genetic drift, 56
Genetic factors, in behavior, 5–6
 in aggressive behavior, 204–205
 assessment of, 55–66
  direct methods, 59–66
  indirect methods, 57–59
 in circadian rhythms, 66, 108
 in feeding behavior, 132
 in fixed action patterns, 19, 22, 23
 in geographically-isolated populations, 57–58
 in homosexuality, 59, 246
 in obesity, 145
 in sexual differentiation, 74
Genetics, Mendelian, 55
Genocide, 215, 217–218
Genome, 234
Genotype, 56
Geophysical factors, relationship with biological rhythms, 100, 108
Geotaxis, 119
Gibbons, social systems of, 274, 275
Glucocorticoids, 67, 92, 93
 molecular structure of, 93
Goats, diet of, 132
Gonadal hormones. *See also* Androgens; Estrogens; Progestins
 action mechanisms of, 71–74
 activational effects of, in adults, 77–88, 282–283, 308, 309
  in females, 78–83
  in males, 83–88
 organizational effects of, during development, 74–77
 receptors for, 88
 synthesis and sites of production of, 69–70
 transport of, 71
Gonadotropin-releasing hormone, 70
Gonadotropins, 67 68

Gonorrhea, 222
Goodall, Jane, 215
Gorillas, social system of, 276
Gravitational fields, effect on circatidal rhythms, 111
"Green beard effect," 167
Greeting behavior, 291–292, 293
Grooming behavior, 25–26, 195, 268
Group defense, against predators, 149–150, 155–156
Groups
 anonymous, 151
 individualized, 152
Group selection, 6
Grouse, lekking behavior of, 239, 261
Growth hormone, circadian rhythm of secretion of, 110
Guanine, as deoxyribonucleic acid (DNA) component, 55
Guarding behavior, 150, 230–231

Habituation, to sign stimulus or releaser, 41–43
Hamilton-Zuk model, of sexually-selected traits, 236
Hamsters
 circadian rhythms in, 102–104
 olfactory and chemoreceptive systems of, 194, 195
Handedness, 247
Handicap model, of evolution of male traits, 235–236
Hand-reach gesture, as sexual invitation, 269, 270
Haplodiploidy, 153–154
Haploidy, 153
Harassment, as sexual interference mechanism, 231–232
Harems (unimale groups), 154, 167–168, 228–229, 262, 274, 275–276
Hawks, group hunting by, 156
Head-bob gesture, as sexual invitation, 269, 270
Head-duck gesture, as sexual invitation, 269, 270
Health, of males, influence on female mate choice, 233, 234, 236, 242
Hemispheric lateralization, gender differences in, 247
Herbivores, 131
 digestive system of, 146
 primates as, 144
Herd effect, selfish, 155, 164
Hermaphroditism, 75
 sequential, 223
 true, 223
Heterogeneous summation, law of, 38
Heterozygosity, 56
 inbreeding-related decrease in, 61
Hibernation, 140
Hierarchical organization, of behavior, 52

Hoarding, 140
Hodgkin's disease, 115
Homing, 123, 124, 127
Homing pigeons, 123, 124, 125, 126, 127
Homogametic sex, female or male as, 223
Homology, 16
  of behavior patterns, 58
Homosexual consortships, 281
Homosexuality. See Bisexual/homosexual behavior
Homozygosity, 56
  inbreeding-related increase in, 60–61
Honeybees, 2–3
  circadian rhythms of, 108, 109
  eusociality in, 152–153, 165
  round and waggle dances of, 195–197
  sex pheromones of, 189
Hormonal factors. See also Endocrinology, behavioral; names of specific hormones
  in mating, 238, 240, 241–242
Hormone receptors, 71
Hormone replacement therapy, 71
Humans, 289–311
  aggression by, 216–219
  biological rhythms in, 100, 104–105, 108–110, 112, 113–116
  displacement behavior, 26
  ethology of, 289–298
    in clinical settings, 298
    conflict behaviors, 291–292
    fixed action patterns (FAPs), 290–291
    releasers, 295–298
    ritualization, 292–294
    supranormal stimuli, 294–295
  feeding behavior of, 142–145
  redirected behavior, 25
  sociobiology of, 299–311
    hormonal and seasonal influences, 308–311
    incest avoidance, 305, 306, 307–308
    mate competition and mate choice, 299–303, 305
    mating systems, 299
    parental investment, 303–304, 306, 307
Hunger, 131, 139, 145
Hunter, John, 239
Hunting
  cooperation in, 163
  by groups, 156
  by humans, 217
  as parental care, 257–258
Hybridization, 64
Hydrarthrosis, intermittent, 113, 114
Hydraulic model, of drive, 51, 52
Hyenas, male sexual activity of, 162, 163
Hymen, 243
Hyperphagia, 145, 146
Hypogonadism, 70

Hypophagia, 145
Hypothalamic-pituitary-adrenal axis (HPA)
  in post-traumatic stress disorder, 97
  in stress response, 91–92, 96
Hypothalamus
  anterior
    in circadian rhythms, 106
    in relation to sexual orientation, 247
  arcuate nucleus, in testosterone secretion, 108–109
  lateral area, in feeding, 145, 146
  median eminence, 91, 92
  paraventricular nuclei
    in stress, 91
    in circadian rhythms, 106
    in predatory and intraspecific aggression, 202
  suprachiasmatic nuclei
    anatomy of, 106, 107
    in seasonal breeding, 78
  supraoptic nuclei, osmoreceptors and sodium receptors of, 50–51
  ventromedial nucleus of, in feeding, 145, 146

Immobilization reflex, of females during estrus, 81
Imprinting
  of bird song, 46, 47, 240
  characteristics of, 45–46
  of courtship behavior, 46
  filial, 45, 46
  genomic, 57
  maternal, olfactory stimuli in, 48
  parental, 45, 46–47
  sensitive periods in, 45–46, 47
  sensitive periods in humans, 298
  sexual, 45, 46, 58
Inbreeding, 60–64
  effect on aggressiveness, 204
  avoidance of, 234
    by dispersal, 160, 279–281
  in eusocial species, 153–154, 165
Incest avoidance, 305, 307–308
Incest taboos, 234
Inclusive fitness, 7–8, 151, 153–154, 278
  altruism-related increase in, 165
Income, relationship with fertility, 303, 304
Indirect selection, for altruistic behavior, 165
Individual distance, in intraspecific agonism, 203–204
Individual recognition, during courtship, 241
Individual (direct) selection, in reciprocal altruism, 163
Infanticide, 215
  by conspecifics, 159
  by langurs, 159, 275
  by males, 232

Infanticide (*cont.*)
  by stepparents, 304, 307
Infants
  nipple-searching behavior of, 267, 290
  smiling and laughing by, 290–291, 292
  walking by, 290, 291
"Infant schema," 43, 294–295, 296
Infection
  avoidance of, by natal dispersal, 160, 279–280
  sociality as risk factor for, 157–159
  transmission during mating, 222
Inheritance, gender bias in, 303–304, 306
Innate releasing mechanism (IRM), 38, 43–44
Insectivores, 131
  primates as, 144
Insects. *See also* Ant lions; Ants; Bees; Beetles; Crickets; Fruit flies; Honeybees; Moths
  antiaphrodisiac substances of, 231
  aquatic, orientation behavior of, 119–120, 121, 127
  eusociality in, 152–154
  mimicry by, 172–173
  sex pheromones of, 188, 189, 193
Instinct, 3, 48
Instinct-learning intercalation, 47
Insulin, 145
Intention movements, 27, 268
Internal (biological) clock, 102, 105–108, 113
Intersex, 56, 94–95
Inuit, 168. *See also* Eskimos
Invertebrates, social behavior of, 152–154
Isolation, in determining hereditary influences on behavior, 58
Iteroparity, 256–257

Jackals, communal defense of offspring by, 157
Jacobson's organ, 194, 195

Kant, Emmanuel, 2
Kibbutzim, 308
Kinesis, 117, 118
Kin recognition, mechanisms of, 165–167
Kin selection, 165
Kissing, 292
Klinokinesis, 118
Klinotaxis, 118
Königsberg University, 2
K-selected species, 10–11
  with unpredictable food supply, 139–143
!Kung bushmen, 168, 307

Landmarks, orientation to, 119, 122
Language
  acquisition of
    in humans, 286, 298
    in nonhuman primates, 287–288

Language (*cont.*)
  learning by exclusion ("mapping") in, 286–287
  of nonhuman primates, 286–288
  stimulus equivalence in, 286–287
Langurs
  infanticide by, 159, 275
  social systems of, 275, 276
Laughing
  by blind and deaf children, 290–291, 292
  by chimpanzees, 290
Law of heterogeneous summation, 38
Learning
  associative, 4, 41
  by exclusion, 286–287
  by nonhuman primates, 272–273
  physiological psychology theories of, 5
  programmed, 44–45. *See also* Imprinting
  socially-facilitated, in feeding behavior, 132–134
  trial-and-error, 4
Lee-Boot effect, 190
Lekking species, 239
Lek mating systems, 230, 261
Leks, 261
Lemurs
  facial expressions of, 270
  olfactory communication by, 270–271
  social system of, 276
  "stink fights" of, 192, 271
  visual communication by, 271, 272
Leptin, 145
Lesbians, sexual activity of, 308
Lexigrams, 287
Leydig cells, 70
Lifestyle strategies, K-selected and r-selected, 10, 11
Light. *See also* Photoperiodicity; Sun
  orientation response to, 117, 118, 119, 120, 121
  polarized, as navigational cue, 127, 128
Light intensity, as visual signal, 178
Light-receptive organs, 179
Limbic system, in agonistic behavior, 204
Lions
  food competition by, 138
  infanticide by, 159, 215
  interspecific aggression by, 201–202
  territoriality of, 156–157
Lizards
  mating behavior of, 239, 240
  parthenogenesis in, 222
  sexual reproduction in, 222
Local-resource-competition hypothesis, of parental investment in daughters *versus* sons, 279
Locomotor activity
  circadian rhythm of, 66
  fixed action patterns of, 19–21, 290, 291

Locomotor activity (*cont.*)
    maturational processes of, 20
Lordosis response, of females, 80, 81
Lorenz, Konrad, 2, 3, 18, 22, 30, 38, 41, 45, 46, 51, 197, 199, 204, 216, 294
Luteinizing hormone, 67–68
    secretion of, 70
        circadian rhythm of, 108, 110
        coitus-induced, 78–79
    effect on testosterone production, 70, 85

Macaques (bonnet, cynomolgus, and rhesus monkeys; Japanese macaques). *See also* Nonhuman primates
    bisexual behavior of, 245, 246
    copulatory patterns of, 267–268
    cultural influences on behavior of, 167–168
    dispersal and inbreeding avoidance in, 160, 276, 279–281, 282
    dominance hierarchy of, 209–211, 229–230
    dominant/submissive postures of, 178
    facial expressions of, 267
        as conflict behaviors, 268
        as fixed action patterns, 267
        recognition mechanism of, 39–41
    food competition by, 138
    homosexual consortships of, 281
    mate competition and mate choice in, 192–193, 277–278, 279
    maternal foraging by, effects of, 64, 141–143, 251–252
    mating system of, 277
    menstrual cycles in, 82–83, 84
    mother surrogates for, 269, 271
    pair bonding (consort bonds) in, 230–231, 277
    parental investment by, 278–279
    releaser, sexual skin swelling as, 271–272
    ritualized behavior of, 268–269, 270
    seasonal aggressive and sexual behavior of, 85–86
    sensitive periods of, 273
    sex pheromones (copulins) in, 192–193
    social effects
        on behavior during the menstrual cycle, 282–285
        on mating seasonality, 285–286
    social systems of, 230–231, 276–277
    "threatening-away behavior" of, 32, 34, 35, 268–269
    visual communication by, 178
Magnetic fields, as navigational cue, 127
Major depressive disorder, 97–98, 143, 298
Major histocompatibility complex alleles, 167
Males
    parental ability of, 259, 260

Males (*cont.*)
    paternal care by, 249, 252–255, 259–263, 275
        in humans, 299
        certainty of paternity hypothesis for, 254
        in polyandrous species, 263
    potential for, influence on female mate choice, 233
    traits of, influence on female mate choice, 233–236, 242–243
"Mapping," 286–287
Marmosets, social systems of, 274, 275
Marmots, reproductive success among, 262
Marriage, "minor" and "major", 307–308
Masculinization, of females, 75, 76–77, 94–95
Mate choice, 224, 225, 226, 259, 260, 263
    criteria for, 233–236, 242–243
    by females, 233–236, 259, 260, 263
    by humans, 299, 301–303
    by males, 192–193, 236
    by nonhuman primates, 192–193, 277–278
    effect of pheromones on, 192–193
Mate competition, 224, 225, 226, 259, 261
    aggression related to, 86, 199, 211–213, 218, 228–230, 259
    by females, 233, 263, 279
    by humans, 299, 301–303
    by males, 83–84, 86, 199, 211–213, 227–233, 259, 261
    by nonhuman primates, 277–278, 279–280
Maternal care, 249–252, 253, 254, 255
    during foraging, 141–143, 251–252
    nongenomic transmission of, 62–64
    effect on offspring's feeding behavior, 133–134
    releasers for, 272
Mating
    assortative, 234, 236
    cognitive rehearsal of, 272–273
    evolution of, 237
    exteroceptive stimuli in, 78, 238, 285
    hazards of, 222
    hormonal stimulation in, 238, 241–242
    infertile, 243
    relationship with menstrual cycle, 282–285
    seasonal factors in, 238
        social influence on, 285–286
    social stimulation in, 239, 285–286
    synchronization between male and female, 241–242
Mating interference, in social species, 159
Mating systems, 258–263
    human, 299, 300
    lek, 230, 239, 261
    monogamy, 161, 162, 258, 259, 260, 262–263
        in humans, 299
        male mate choice in, 236

SUBJECT INDEX

Mating systems (*cont.*)
  monogamy (*cont.*)
    resource-defense, 161
  polyandry, 259, 260, 263
    in humans, 299
    male mate choice in, 236
    parental investment in, 225
    resource-defense, 263
  polygyny, 161–162, 259–262
    in humans, 299, 300–302, 303, 304
    male mate choice in, 236
    mate-defense, 161
    parental investment, 261–262
    pure dominance, 259, 260, 261
    reproductive success in, 262, 299, 300, 303, 304
    resource-defense, 261–262
  relationship with social organization, 154
Matrilineality, 168, 276, 278–279
Maturation, role in predatory behavior, 132
"Mechanists," 3
Meerkats, sentinel behavior of, 150
Meiosis, 56
Melatonin, 78, 105–106
Menotaxis, 119
Menstrual cycle, 82–83, 112. *See also* Estrous cycle
  artificial, 83, 284
  hormonal control of, 70
  proceptive and receptive behavior during, 83, 285
  relationship with sexual activity, 282–285
  synchronous, 276
Merkwelt, 38
Mesor, 99
Metacommunication, 175
Metestrus, 79
Mice
  allele recognition in, 167
  inbred, 62
  "knock-out," 64–65
Midges, circalunar rhythms in, 111–112
Migration, 117, 128–129, 140
  circadian rhythms in, 100
  seasonal photoperiod and temperature changes in, 128–129
Migratory direction, effect of hybridization on, 64
Mimicry, 44, 147–148, 172–173
  Batesian, 148
  of eggs, 178, 249–250
Mineralocorticoids, 67, 92, 93
  molecular structure of, 93
Mnemotaxis, 119–120, 122
Mobbing, 203
Monkeys. *See* Macaques; Nonhuman primates

Monogamy, 161, 162, 258, 259, 260, 262–263
  in humans, 299
  male mate choice in, 236
  resource-defense, 161
Moon, as navigational cue, 127
Moro reflex, 267
Morphological structures/traits
  behavioral traits associated with, 58–59
  in communication, 176
  in courtship behavior, 58–59
  in female mate choice, 234–236
  in food consumption, 131–132
  in male mate competition, 86, 88
  as releasers for aggression, 204
  in ritualized displays, 31, 33
Morphology, evolution of, 2
Mosquitoes, as r-selected species, 11
Mother-offspring relationship, effect of foraging on, 141–143, 251–252, 273
Mothers, recognition of offspring by, 165, 190
Moths
  sex pheromones of, 188, 189, 193
  sound detection by, 185–187
Motivation
  drive concept of, 48–53
  in ritualization, 32, 34–35
  sexual, 51
    female, 268–269, 270, 308, 309
    hormone-induced changes in, 268–269
Mounting activity, 245, 246, 285, 286
Movement, as visual communication mechanism, 178–179
Müllerian duct system, 74, 75
Müllerian-inhibiting factors, 74
Murder, 310–311
Musk, 192
Mutations, 12, 56
Mutual benefit (symbiosis) model, of parent-offspring relationship, 250, 251
Mutualism. *See* Cooperation

Naked mole-rat, eusociality in, 153–154
Nationalism, 218
Natural selection, 2, 6, 12
  dynamic, 7
  phenotypic, 56
  stabilizing, 7
Nature-nurture controversy, 5–6
Navigation, 121–128. *See also* Orientation
  definition of, 117
  methods of
    compass or directional, 122–123, 124–125, 127
    dead reckoning, 123
    homing, 123, 124, 127

Navigation (*cont.*)
  methods of (*cont.*)
    navigational cues, 125, 127–128
    piloting, 122, 127
Negative reinforcement, 44, 45
Negative reinforcing stimuli, 4
Neocortex, of anthropoid primates, 265–266
Nest (brood) parasitism, 44, 159, 173
Nest building, 242
Nesting territories, 208
Neural plasticity, 5
Neurohormones, 68
Neurons, "all-or-none" firing of, 51
Neuropeptides, appetite-enhancing, 145–146
Neurotransmitters, 68
  in stress response, 91
Nipple-searching behavior, of infants, 267, 290
Nobel Prize, 2, 4, 55, 197
Nonhuman primates, 265–288. *See also* Baboons; Bonobos; Chimpanzees; Gibbons; Gorillas; Langurs; Lemurs; Macaques; Prosimian species; Orangutans; Siamang; Sifakas; Tamarins
  alarm calls by, 181–182
  bisexual behavior of, 245, 246
  classification of, 266
  diet of, 144
  cultural influences on behavior of, 167–168
  ethology of, 265–273
    conflict behaviors, 268
    fixed action patterns (FAPs), 267–268
    releasers, 269–272
    ritualization, 268–269, 270
    sensitive periods, imprinting, 272–273
  facial expressions of, 267, 268, 270
  infanticide among, 159, 232, 275
  language in apes, 286–288
  sociobiology of, 273–286
    dispersal and inbreeding avoidance, 279–281, 282
    hormonal and seasonal influences in, 282–286
    mate competition and mate choice, 192–193, 277–278, 279
    mating systems, 277
    parental investment, 278–279
    social systems, 154, 230–231, 273–277
Noradrenalin, 91
Noxious substances, as defense against predators, 149

obese gene, 145
Obesity, 131
  physiological mechanisms of, 145–146
Offspring
  communal defense of, 157

Offspring (*cont.*)
  infanticide of, by conspecifics, 159
  mothers' recognition of, 165
  post-weaning dependence of, 213
  sex ratio of, 226–227, 278–279, 303–304
  size of, relationship with parental care, 255–256, 257
Olfactory bulb
  accessory, 194
  main, 194, 195
Olfactory cues/stimuli. *See also* Communication, chemical
  in maternal imprinting, 48
  in navigation, 127, 128
Olfactory epithelium, 193, 194
Omnivores, 131
  humans as, 144
Ontogeny, of behavioral traits, 15, 18
Operant conditioning, 4
Opossum, sex ratio in, 227
Optic tectum, 180, 181
Optimality theory, 7–8, 12–13, 135, 136, 137
Orangutans, solitary social system of, 273, 274, 275
Orexins, 145–146
Orientation, 117–121. *See also* Navigation
  of body, in visual communication, 178
  circadian rhythms in, 100, 124, 126
  orienting responses, classification of, 117–119, 120
    kinesis, 118
    taxis, 118–119, 120
Orthokinesis, 118
Osmoreceptors, 50–51
Outbreeding, excessive, 234
Ovariectomized females, artificial menstrual cycles in, 83
Ovaries
  gonadal hormone production by, 69–70
  of honeybees, 153
  suppression of, 233
Ovulation, 282, 283
  hormonal factors in, 69, 78–79
  reflex, 241, 244
Owls, reproductive failure in, 141
Oxford University, 2
Oxidative reactions, in corticosteroid metabolism, 94
Oxodecenoic acid, 153

Pain perception, 197
Pair bonding, 262–263
  consortships in nonhuman primates, 231, 277
  by geese, 241, 242
  serial, 307

SUBJECT INDEX

Panda, diet of, 132
Pap (Papanicolaou) smear, 79
Parental care, 249–258
  benefit-cost ratio of, 252–253, 254
  evolution of, 165, 252–258
    effect of environmental factors on, 251–252, 254, 255–257
  by females. *See* Maternal care
  by males. *See* Paternal care
  sex differences in, 252–254
Parental exploitation, by conspecifics, 159
Parental investment, 249, 250, 254. *See also* Parental care
  by humans, 303–304, 306, 307
  by nonhuman primates, 278–279
  total, 224–226
  Trivers' Theory of, 224–226, 277–278
  Trivers-Willard Hypothesis of sex ratio manipulation, 226–227, 278–279, 303–304
Parental provision model, of parent-offspring relationship, 250
Parent-offspring relationship
  aggressive behavior in, 213–214
  models of, 250–252
  reproductive competition in, 162
Parents, aggressive defense of offspring by, 213
Parthenogenesis, 221–222, 239, 240
Paternal care, 249, 253, 254, 255
  by humans, 299
  in monogamous species, 260, 262–263
  by nonhuman primates, 275
  paternity, effect on, 254
  in polyandrous species, 260, 263
  in polygynous species, 259, 260, 261–262
  potential for, influence on female mate choice, 233
Paternity
  effect on male parental care, 254
  mixed, 258
  multiple, 230
Patrilineality, 168
Patriotism, 218
Pavlov, Ivan, 3, 4
Peacocks
  male secondary sex characteristics of, 235–236
  tail feather displays by, 31, 33, 58–59
"Pecking order," 208–209, 210
Penguins, predator response of, 155–156
Peptidergic neurons, 68
Peptide hormones, 67–68
Perception
  attention *versus*, 37–38
  Gestalt, 17–18
Personality disorder, 298
Phenotype
  definition of, 56

Phenotype (*cont.*)
  sexual reproduction-related variability of, 222
Phenotype matching, as kin recognition mechanism, 166
Phenylketonuria, 221
Pheromones, 189–191
  of honeybees, 153
  primer, 190
  sexual
    of insects, 188, 189, 193
    of mammals, 85, 189–195
  signaling, 190–191
Philopatry (*versus* dispersal), 159–162, 255, 256
Philosophy, as basis for psychology, 1
Phonotaxis, 119
Photoperiod, effects of
  on estrus, 79
  on mating, 238, 285–286
  on migration, 128–129
Photoreceptors, 179, 180
Phototaxis, 119
Phylogenetic inertia, 13
Phylogenetic trees, morphological/behavioral trait-based, 23, 58
Phylogeny, of behavioral traits and patterns, 15, 18, 58
Physiological changes, as communicatory signals, 176
Physiological regulation, of behavior traits, 18
Physiological traits, food consumption-related, 131
Piloerection, 30
Piloting, 122, 127
Pineal gland, 78, 105, 106
Pituitary gland
  anatomy of, 91
  pars nervosa of, antidiuretic hormone secretion by, 51
Pituitary portal system, 91, 92
Planum temporale, role in language acquisition, 288
Plato, 1
Play, as cognitive rehearsal, 272–273
Pleiotropism, genetic, 57
Polyandry, 259, 260, 263
  in humans, 299
  male mate choice in, 236
  parental investment in, 225
  resource-defense, 263
Polygyny, 161–162, 259–262
  female defense, 258
  in humans, 299, 300–302, 303, 304
    effect on inheritance patterns, 304, 306
    reproductive success in, 299, 300, 302–303, 304
  lek, 261
  male mate choice in, 236

Polygyny (cont.)
  mate-defense, 161
  parental investment, 261–262
  pure dominance, 259, 260, 261
  reproductive success of, in nonhumans, 262
  resource-defense, 261–262
Polyps, coelenterate, 152
Population density, as territoriality cause, 207
Population growth curves, of K-selected and r-selected species, 10
Portuguese man-of-war, 152
Positive reinforcement, 44, 45, 205
Positive reinforcing stimuli, 4
Post-traumatic stress disorder, 97, 143
Postures, submissive. See Submissive behavior
Precocial species, maturation of innate behavior in, 19–20
Predation. See also Hunting; Prey species
  effect on foraging, 139
  effect on parental care, 257
  effect on polyandrous behavior, 263
Predators, aggressive behavior toward prey, 199, 202
Predatory behavior
  learning in, 132
  maturation in, 132
  techniques of, 146–149
Pregnancy, pheromone-induced disruption of, 190, 232–233
Premarin, 71
Preputial glands, 84–85
Prey species
  antipredator agonism by, 203
  predator avoidance by, 139, 140, 146–148, 149–150, 164, 199, 213
  predators' detection of, 156, 179–180
  size of, in antipredator defense, 156
  sociality in, as defense against predation, 155–156
Primates, nonhuman. See Nonhuman primates
Proceptive behavior, 79–82, 83, 285
Proestrus, 79, 80–81
Progesterone
  metabolism of, 94
  molecular structure of, 93
  secretion of, 69–70, 238
  effect on sexual activity, 34–35, 79
  effects on tissues, 71–72
Progestins, molecular structure of, 93
Prolactin, 67–68
  circadian rhythm in secretion of, 108, 110
  effect on migratory behavior, 128
Promiscuity, 259, 301
Proprioceptive sensations, 197
Prosimian species, 265, 266. See also Lemurs

Prosopagnosia, 41, 295
Prostate cancer, antiandrogen treatment of, 73
Protein, as human dietary component, 144
Proximate processes, 15–16
Pseudohermaphrodites, 75
Psychiatric disorders, aggression associated with, 219
Psychiatric patients, ethological studies of, 298
Psychology
  comparative, 4–5
  development of, 1
  physiological, 5
Psychosomatic medicine, 96–97
Puberty
  in human females, 72
  intitiation of, 93
  language acquisition ability around, 298

Quetelet, Adolphe, 308, 310

Racism, biological basis of, 214
Radiation, background, possible role in circadian rhythms, 108
Rainfall, influence on mating seasonality, 285
Random processes, in trait selection, 13
Rank order, 208–211. See also Dominance hierarchy
  relationship with reproductive success, 302–303
Rape, 230, 241
  seasonal variation in, 310
Rats
  artificial selection in, 60, 61
  maternal care transmission in, 62–64
  socially-facilitated feeding behavior of, 133–134
  thalamic-lesioned, biological rhythms in, 112
Receptive behavior, 79–82, 83, 285
Reciprocity. See Altruism, reciprocal
Recognition, of kin, 165–167
Recombinations, chromosomal, 56
Redirection, 24–25, 30
5α-Reductase, 72
5α-Reductase deficiency, 73
Reductive reactions, in corticosteroid metabolism, 94
Reflexes, 4
Reinforcement
  negative, 44, 45
  positive, 44, 45, 205
Releasers
  for caregiving behavior, 272
  definition of, 38
  intraspecific 43
  for maternal behavior, 272
  in nonhuman primate behavior, 269–272
  properties of, 38–41
  supranormal, 43

Repression, 28, 30
Reproduction. *See also* Mating; Pregnancy
  asexual, 221–222
  sexual, 221, 222
Reproductive behavior, 237. *See also* Courtship behavior and displays; Mating; Sexual behavior
  phylogenetic relationships of, in birds, 23, 58
Reproductive organs, 243
Reproductive success
  Bateman's Principle of variance in, 224, 225, 277–278
  of daughters *versus* sons, 303–304
  effect of dominance hierarchy on, 157
  effect of food shortage on, 140–141
  of humans, 299, 300, 302–303, 304
  of infanticidal males, 159
  of low-ranking females, 210, 211, 212
  male behavior and, 259, 260
  male strategies for optimization of, 227–233
  in matrilineal systems, 278–279
Reptiles, sex determination in, 223
Resources, defense of, by social group, 156–157
Rest, 135
Retinohypothalamic pathway, monosynaptic, 106
Rheotaxis, 119
Ritualization, 30–35. *See also* Courtship behavior and displays
  association with conspicuous morphological features, 31, 33
  in human behavior, 292, 294
  motivational change in, 32, 34–35
  phylogenetic, 268–269
  stereotypy in, 31, 32
  threshold changes in, 32
  typical intensity of, 31, 33
Robbery, seasonal variation in, 310
"Royal jelly," 153
r-selected species, 10, 11
  in unpredictable environments, 139
Rumbaugh, Duane, 287
Runaway selection model, of evolution of male traits, 234–235
Russell, Bertrand, 2

Salamanders, maternal care by, 257
Saliva, pheromone content of, 191, 192
Salt
  as herbivores' dietary component, 137
  as human dietary component, 44, 144–145
  taste for, 44, 144–145
Satiety mechanism, 145–146
Scapegoating, 25, 214, 218, 268
Scent glands, 191–192, 270–271
Scent marks, 177–178, 191–192
Schizophreniform disorder, 298
Schizophrenia, 98
Schreckstoff, 191
Search technique, in foraging, 137–138
Seasonality, of sexual activities, 308–311
  courtship behavior, 238
  mating, 238
  social influences on, 285–286
Secondary sex characteristics. *See also* Sexual dimorphism
  of males, 235–236
Self-grooming, 26–27
Selfish Gene, The (Dawkins), 6
"Selfish genes," 6, 7
Selfish herd theory, 155, 164
Semelparity, 256–257
Sensitive periods
  in imprinting, 45–46, 47
  in language acquisition, 288
  in social bonding, 273
Sentinel behavior, 149–150, 155, 163
Sex chromosomes, 55, 222–223
Sex hormone-binding globulin, 71
"Sexing-in," 305, 307
Sex ratio, 223–224
  maternal manipulation of, 226–227, 278–279, 303–304
Sex-role reversal, 236, 263
Sex steroids. *See also* Gonadal hormones
  secretion by adrenal cortex, 92–93
Sexual activity. *See also* Courtship behavior and displays; Mating
  relationship with menstrual cycle, social effects on, 282–285
  seasonal influences on, 308–311
Sexual behavior. *See also* Courtship behavior and displays; Mating
  fixed action patterns in, 267–268
  hormonal effects on, 75–77, 78
  ritualized displays in, 32, 34–35
Sexual differentiation, 88, 222–223, 244–245
  behavioral modification by, 75–77
  somatic modification by, 74–75
Sexual dimorphism
  of the brain, 77
  in humans, 247
  influence on female mate choice, 235–236
  morphological, 259, 260, 261, 262, 263
Sexual intercourse. *See* Copulation
Sexual interference, with male competitors, 231–232
Sexual motivation, 51
  of female, 268–269, 270
  hormone-induced changes in, 268–269
  of male, 268–269
  of women, 308, 309

Sexual mutilation, of women, 243
Sexual partners
 of homosexuals, 301
 preferred age of, 303, 305
Sexual presentation posture, of nonhuman primates, 269
Sexual selection, 221–236
 in courtship behavior, 59
 epigamic, 227
 intersexual (epigamic), 227, 233–236
 intrasexual, 227–233
 theoretical considerations in, 224–227
Sexual skin swelling, 271–272, 281
Shiftworkers, circadian rhythms in, 104–105
Siamang, social systems of, 274, 275
Siblicide, 214
Siblings
 aggression between, 214
 inbred, 61
 language of, 298
 shared traits of, 59
Sickle-cell anemia, 221
Sifakas, social system of, 274, 276
Signaling pheromones, 190–191
Signals, receivers and senders of, 37
Sign stimulus
 configurational stimulus as, 38, 41–43
 definition of, 38
 properties of, 38–41
Sleep, 135
Sleep-wake cycle, 105
Small intestine, digestion in, 146
Smiling, 290
 as appeasement behavior, 297
 by blind and deaf children, 290–291, 292
 by chimpanzees, 290
 closed-lipped and open-lipped, 292, 293, 294
Social behavior, 151–169
 benefits of sociality, 155–157
 costs of sociality, 157–159
 environmental and cultural influences on, 167–169
 evolution of cooperative behavior, 162–165
 mechanisms of kin recognition, 165–167
 philopatry and dispersal, 159–162
 social systems, 151–154
Social bonding, sensitive period for, 273
Social disruption, foraging-related, 141
Social groups
 aggregations, 151
 anonymous (open and closed), 151
 individualized groups, 151
Social organization. *See also* Dominance hierarchy; Mating systems; Social groups; Social systems; Territoriality
 environmental and cultural influences on, 167–169

Social stimulation, effect on sexual activity, 239, 285–286
Social systems, of nonhuman primates, 273–277
Sociobiology, 6
 of humans, 299–311
  hormonal and seasonal influences, 308–311
  incest avoidance, 305, 306, 307–308
  mate competition and mate choice, 299–303, 305
  parental investment, 303–304
 of nonhuman primates, 273–286
  dispersal and inbreeding avoidance, 279–282
  hormonal and seasonal influences, 282–286
  language in apes, 273, 286–288
  mate competition and mate choice, 277–278
  mating systems, 277
  parental investment, 278–279
  social systems, 273–277
Sociobiology: The New Synthesis (Wilson), 6
Sodium receptors, 50–51
Somatization, 97
Somatotype, female, 74
Songs. *See also* Bird song
 courtship, 64, 65–66
 of whales, 184, 185
Sons, inheritance patterns of, 303–304, 306
Sound vibrations
 detection of, 197
 physical properties of, 185
Sour taste, 145
Spalding, Douglas, 2
Spatial ability, gender differences in, 247
Species specificity, of courtship behavior, 239–240
Speech, 286. *See also* Language
 chimpanzees' ability to understand, 287
Sperm
 apyrene/eupyrene, 231
 in internal and external fertilization, 243
 sufficiency of, influence on female mate choice, 233
Sperm competition, 230–231, 232
Sperm count, seasonal variation in, 113, 285, 308, 309
Spider webs, 148
Spots, as protective camouflage, 147
Spouse, virginity of, 301–302, 303
Spouse abuse, 311
Squirrels, fixed action patterns in, 21
Stars, as navigational cues, 127, 128
Starvation, 131, 140–141
States (political), 168–169
Stepparents, 215, 304, 306, 307
Stereognosis, 197
Stereotyped behavior, 18–19, 21, 30, 31
Steroid abuse, 219
Steroid antagonists, 73

SUBJECT INDEX   345

Steroid hormones, 67. *See also* Gonadal hormones
"Steroid rages," 88
Stickleback fish
   aggressive behavior by, 58–59
   courtship behavior of, 27, 28, 38, 39, 40
Stimulus-response relationships, in drive (motivation) concept, 48–53
"Stink-fights," 192, 271
Stomach, multi-chambered, 146
Stress. *See also* Endocrinology, behavioral
   adaptation/habituation to, 90, 95–96
   definition of, 89
   functions of, 96
   social rank-related, 211
   types of, 89–90
Stressors
   external and internal, 90
   virulence of, 96
Stress response
   adrenal cortex in, 91–92
   adrenal medulla in, 91
   corticosteroid metabolism in, 93–95
   corticosteroid production in, 92–93
   definition of, 89–90
   foraging-related, 142
   hypothalamic-pituitary-adrenal axis in, 91–92
   major depressive disorder and, 97–98
   maternal care, effects on, 63–64
   pychosomatic medicine and, 96–97
   suicide and, 98
Stripes, as protective camouflage, 147
Strokes, 247
Structure-function relationship, of behavioral traits, 12
Submissive behavior, 206, 215–216
   postures of, 178, 217, 292, 294
Sugar
   as human dietary component, 44, 144–145
Suicide, 98
Sun, in orientation and navigation, 119, 124–125, 126, 127–128
Supranormal stimuli, 43, 44
   females' response to, 261
   in humans, 294–295, 296
   sugar, salt and fat as, 145
Swans, pair bonding by, 262–263
Sweet foods, humans' preference for, 44
Symbols, use by gorillas and chimpanzees, 286
Symmetry, as genome indicator, 234
Sympathetic nervous system, in arousal, 91
Syphilis, 222

Tabula rasa, 2
Tactile signals. *See also* Communication, tactile
   in orientation, 117

Tail beating by fish, 177
Tamarins, social systems of, 274, 275
Tamoxifen, 73
Taxis, 117, 118–119
Tay-Sachs disease, 221
Telotaxis, 119
Temne people, 299, 300
Temperature
   effect on mating seasonality, 285
   effect on migratory behavior, 128–129
   effect on sex determination, 223
Territoriality, 161
   aggression associated with, 138, 152, 199, 206–208
   dominance associated with, 207
Testicular feminizing syndrome, 73–74
Testis, androgen production by, 70
Testis-determining factor, 74
Testosterone
   action mechanisms of, 87, 88
   adrenal, 70
   constant levels of, 227–228
   effect on male breeding behavior, 83–85
   metabolism of, 69, 88
   ovarian, 70
   secretion of
      circadian rhythms in, 108–109, 110
      in response to sexually-receptive females, 285–286
      seasonality in, 86, 113, 114, 285, 286, 308, 309
   stress-related decrease in, 96
Thirst, as drive (motivation) concept, 49–51
Threat, open-mouth, 267, 268
"Threatening-away behavior", 32, 34, 35, 268–269
Thymine, as deoxyribonucleic acid (DNA) component, 55
Thyroid-stimulating hormone, 67
Tides, effect on biological rhythms, 101, 111–112
Time-energy budgets, 134, 135
Tinbergen, Nikolaas, 2, 9, 15, 18–19, 38, 52, 119, 197, 208
Toads, visual detection of prey by, 179–180, 181
Traits, 2
   adaptive, 12–13
   behavioral
      association with morphological and physiological traits, 58–59
      evolutionary basis of, 15–16
      structure-function relationship of, 12
   concomitant or consequential, 13
   dominant, 56
   of males, influence on female mate choice, 233–236, 242–243
   morphological and physiological, food consumption-related, 131–132

Traits (cont.)
  phenotypic, 6–7
  polygenic, 57
  recessive, 56, 60–61
Transsexualism, 246–247
Transvestitism, 246
Trapping, as predatory technique, 148
Trial and error learning, 4
Tribes, 168
Triumph ceremony, of geese, 241, 242
Trivers' Theory, of parental investment, 224–226, 277–278
Trivers-Willard hypothesis, of sex ratio manipulation, 226–227, 278, 303–304
Tropism, 118
Tropotaxis, 118
Turkeys
  displacement behavior of, 26
  escape from predator model, 41–43
  maternal responses of, 43
Turtles, migration by, 128–129
Twin births, in nonhuman primates, 275
Twins
  dizygotic, 59
  freemartin, in cattle, 75
  monozygotic, 56, 59
Twin studies, of homosexuality, 59, 246, 247
Tympanic membrane, 185–186

Ulceration, gastric, stress-related, 96
Ultimate processes, 15–16
Umwelt, 38
Unconditioned stimulus, 4
Urinary constituents, circadian rhythm of excretion of, 105
Urine
  as communicatory signal, 176, 191, 192
  pheromone content of, 190
Uterus, hormonal-induced growth of, 71–72

Vacuum activity, 30, 32, 34, 51
Vaginal secretions, pheromone content of, 191, 192, 194–195
Vasoconstriction, as communicatory signal, 176
Vasodilation, as communicatory signal, 176
Vegetables, as human dietary component, 144
Verbal ability. *See also* Language
  gender differences in, 247
Vibration
  in communication, 183
  detection of, 197
  physical properties of, 185

Vigilance, as group defense technique, 149–150
Violence, as stress cause, 90
Violent crime, seasonal influences on, 308, 310, 311
Virginity, 243, 301–302, 303
Visual contact, effect on sexual interest, 280, 282
Visual cues/signals. *See also* Communication, visual
  in orientation and navigation, 117, 127
  in predation, 132
  processing of, 179–180
"Vitalists," 3, 48
Vitamin C, 145
Vocal cords, 183
Vomeronasal organ (organ of Jacobson), 194, 195
von Frisch, Karl, 2–3, 18, 106, 197
von Uexküll, 37–38
Vultures, foraging behavior of, 138

Waika Indians, 292
Walking, by infants, 290, 291
War, 215, 218
Wasps
  mnemotaxis by, 119, 120
  navigational mechanism of, 122
Water currents, as orientation stimulus, 119
Wernicke's language area, 288
Westermarck effect/hypothesis, 281, 307
Whales
  baleen, as filter feeders, 131
  food trapping by, 148
  migration by, 128
  songs of, 184, 185
Whitten effect, 190
Wirkwelt, 38
Wolffian duct system, 74, 75

Xavante Indians, 299, 300
Xenophobia, 214, 218

Yanomami Indians, 307
Yomuts, 303

Zahavi's handicap model, of evolution of male traits, 235–236
Zebras, visual communication by, 175, 179
Zeitgeber, 101–102, 105, 106, 109, 112
Zona fascicularis, 92
Zona glomerulosa, 92
Zona reticularis, 92–93
Zooids, coelenterate, 152
Zoology, as basis for ethology, 1
Zygote, 56